Conservation in the Soviet Union

Conservation in the Soviet Union

PHILIP R. PRYDE

Associate Professor of Geography
California State University
San Diego

CAMBRIDGE
At the University Press 1972

CAMBRIDGE UNIVERSITY PRESS
Cambridge, New York, Melbourne, Madrid, Cape Town, Singapore, São Paulo, Delhi

Cambridge University Press
The Edinburgh Building, Cambridge CB2 8RU, UK

Published in the United States of America by Cambridge University Press, New York

www.cambridge.org
Information on this title: www.cambridge.org/9780521103510

First published 1972
This digitally printed version 2009

A catalogue record for this publication is available from the British Library

Library of Congress Catalogue Card Number: 72–182025

ISBN 978-0-521-08432-1 hardback
ISBN 978-0-521-10351-0 paperback

Contents

Contents

Illustrations

Appendices

Appendices

System of transliteration

The transliteration system used in this study is patterned after that proposed by the Board of Geographical Names of the United States Department of the Interior, and is as follows:

Cyrillic symbol	*Transliteration*
А, а	a
Б, б	b
В, в	v
Г, г	g
Д, д	d
Е, е	e, ye*
Ж, ж	zh
З, з	z
И, и	i
Й, й	y
К, к	k
Л, л	l
М, м	m
Н, н	n
О, о	o
П, п	p
Р, р	r
С, с	s
Т, т	t
У, у	u
Ф, ф	f
Х, х	kh
Ц, ц	ts
Ч, ч	ch
Ш, ш	sh
Щ, щ	shch
ь	'
ы	y

ъ	′
Э, э	e
Ю, ю	yu
Я, я	ya

* *Ye* is used at the beginning of words, after vowels, and following the soft or hard signs (′).

The English form of certain common place names has been retained, e.g. *Georgia*, *Armenia*, etc.

Preface

Natural resource management and conservation is a subject which continues to attract increasing amounts of world-wide attention, and nowhere more so than in the Soviet Union. Many studies and monographs presently exist concerning the natural resource base of the U.S S.R., which represents an appreciable fraction of the total resources of the world. However, very little has as yet been written regarding Soviet attitudes and accomplishments concerning the conservation of these natural resources. It is towards questions of this nature that the present study is directed.

Experience in the utilization of natural resources has shown that not only do all resources have inherent ecological interrelationships with other natural resources, but the uses to which they may be put by man have decisive functional relationships as well. That is, virtually any manner of developing a given natural resource will have some bearing on the human use of other resources, either as they exist in their natural state, or with regard to the manner in which they may ultimately be employed. This study will examine various instances of how the Soviet Union has approached these interdependencies in resource management, in order to attempt to summarize for the U.S.S.R. as a whole present attitudes and levels of sophistication regarding natural resource use and conservation.

The study will in no sense attempt to quantify, analyze, or compare with other countries the natural resource base of the U.S.S.R. itself; numerous such studies already exist in the literature. Nor will it have as an end in itself an exhaustive investigation into how each individual natural resource group of the Soviet Union has been developed, exploited, or conserved. Rather, it will deal mainly with the current state of Soviet efforts on behalf of natural resource conservation and, to such extent as is possible, environmental quality. Its basic objectives are two-fold.

First, the study will serve as a review of the more important or unique features of natural resource management and conservation in the Soviet Union at the present time. It will briefly summarize the administrative organization governing the use of each broad category of natural resource in the U.S.S.R., and will isolate and discuss some of the major problem areas that have arisen out of the particular Soviet approach to the management of these resources.

Its second purpose is an analytical one; to examine these Soviet approaches towards the use of their natural resources, and towards the problems which have consequently arisen, with the goal of acquiring insight into the contemporary Soviet perception of what constitutes proper natural resource conservation, and the causal attitudes and assumptions which shape these resource use patterns. In this way, the study will endeavor to determine the present understanding of the term 'conservation' in the Soviet Union, and the extent to which this understanding is broadening and deepening. It will attempt to suggest strengths and weaknesses in the Soviet approach to the employment and conservation of natural resources, as well as some of the probable trends in Soviet conservation and the quest for environmental quality in the future.

Studies such as the present ones, which would attempt to analyze Soviet actions and attitudes in a particular area of endeavor, are always subject to limitations in the availability of information, the representativeness and accuracy of that which is available, and an inability to conduct the desired intensity of field research. Nevertheless, in view of the increasing amount of descriptive, analytical, and technical material which is becoming available from the U.S.S.R. concerning the use and conservation of its natural resources, a study of this type seems timely, and should provide a useful complement to existing studies on the quantitative aspects of that endowment. In addition, growing pressures on known world resource stocks and the increasing threat of world environmental pollution lend to such an effort still further significance.

Inasmuch as this will represent one of the first broadly inclusive qualitative studies of Soviet conservation practices to be developed outside the U.S.S.R., a selected, representative bibliography is presented at the end of the book. Obviously, only a small percentage of the total number of existing references have been included, generally selected only from among those which the author was able personally to examine. An effort has been made to include as many significant sources as possible which have been translated into English, so that these listings might also be of service to students of conservation who are not familiar with the Russian language.

This study was carried out under the sponsorship of the Foreign Area Fellowship Program whose generous assistance, including partial funding of an extensive trip through the Soviet Union in the summer of 1967, greatly facilitated its completion. Appreciation is extended to Professors W. A. D. Jackson and Richard A. Cooley for their knowledgeable suggestions and constructive criticisms during the preparation of the manuscript, and to Professors Marion E. Marts, Richard L. Morrill, Denis A. Flagg, and Donald E. Bevan for their critiques of individual chapters.

Preface

The cartographic assistance of Ann Sires and the many tedious hours of typing and proofreading willingly donated by my wife Lucy are also most gratefully acknowledged. Any errors of either compilation or inference remain, of course, solely the responsibility of the author.

PHILIP R. PRYDE

1 Introduction: towards an understanding of the Soviet concept of conservation

[Conservation is] the rational use, renewal, increase, and protection of natural resources...

One of the most honored concepts in the world today is that of natural resource conservation, and its success can be measured by the increasing amount of political charisma it displays. Its magical and somewhat mystical nature was early noted by President Taft, who observed that most everyone seemed to be in favor of conservation, no matter what it meant. Although the somewhat enigmatic term originated in the United States, this concept has quickly gained favor in all parts of the world, and certainly not least in the Soviet Union.

The U.S.S.R. comprises one-sixth of the earth's land surface and probably at least an equal share of its biotic and mineral resources, and, inevitably, has become deeply concerned with the problems of judiciously managing its bountiful, but by no means inexhaustible, natural resources. The nature of this contemporary Soviet concern for natural resource conservation will be the central theme of this study. By examining a wide variety of natural resources and some of the major problems associated with their utilization in the U.S.S.R., the effectiveness of the Soviet approach to the management and conservation of these resources will hopefully be drawn more clearly into view.

Today, of course, conservation efforts cannot always be restricted to the internal territories of individual countries. Soviet and American specialists alike acknowledge that many aspects of natural resource conservation and environmental quality are not just national problems, but are of vital international concern as well. Everyone in the world must breathe the same air, extract the resources of the same world ocean, share in the use of the total land resources, and suffer the common consequences of numerous activities that may adversely affect the climatic, mineral, aesthetic, and biotic resources of the entire world – with the term 'biotic resources' being understood to include man himself. We are all ultimately dependent upon a single world environment, not scores of little ones neatly defined by invisible national boundaries.

As this concern for global environmental quality, sometimes referred to as the 'spaceship earth' concept, becomes more widely understood and accepted, the attitudes and actions of other governments in the field of conservation take on an increasing importance for us in the United States. Such activities in the Soviet Union, as the largest country in the world and a close neighbor, are perhaps of the greatest significance.

Conservation activities in the Soviet Union, as would be expected, are carried out in a considerably different political context than they are in the United States. Two factors in particular are strikingly different. The first is that the government owns all of the land and water, and all of the associated natural resources found within them. The second is that all aspects of the national economy, and hence all uses of natural resources, are centrally planned by the U.S.S.R. government. Taken together, these two factors mean that the Soviet government alone is solely responsible for the conservation of its natural riches; only the state can exploit them, only the state can conserve them. At the same time, the authoritarian nature of the Soviet political system and the paternalistic nature of its economic system have combined to discourage public participation in natural resource use decisions. As a consequence, there is less awareness on the part of the average Soviet citizen as to what constitutes good conservation practices, and to the importance of their implementation, than is the case with his American counterpart.

Before attempting to evaluate the Soviet understanding of the term 'conservation', it might be well to consider how its meaning has evolved in the United States up to the present time. When Gifford Pinchot and his compatriots gave America the word 'conservation', they unfortunately neglected to give us an adequate definition to go along with it. Most of the early definitions ('wise use', 'the greatest good for the greatest number for the greatest length of time', etc.) say at the same time everything and absolutely nothing at all. How do we determine the wisest use, the greatest good? The difficulty is that we are attempting to take a fantastic array of natural resource problems, which are often closely interrelated in both tangible and intangible ways, and to lump all possible approaches to untangling and solving these problems into a single term – 'conservation'. Lacking a precise definition, Americans have generally viewed conservation as meaning a way of acting ('wise use'), or as a set of commandments saying 'thou shalt' or 'thou shalt not' when one is in the woods, using water, farming, mining, etc. But this no longer represents the heart of the matter.

Rather, the concept of conservation today centers around the method of *approach* to resolving natural resource use conflicts; in essence, one of careful long-range planning. Stated more formally, it might be said that conservation implies an objective and comprehensive evaluation of all the various ways in which a given natural resource complex might eventually be utilized, so that short-sighted or even irreversible decisions will not preclude additional or alternate uses of the same resource in the future, as the composition of demand for natural resources changes with time.

This view of conservation obviously calls for caution in the single-purpose development of any resource in an irreversible manner; that is, in any way which would preclude some other or additional use of the resource at a later date. It also suggests that it may be pointless to base long-term planning decisions on the assumption that we have 'x numbers of years of reserves of a given resource at the present rate of use', when changes in demand or technological substitutes may greatly lessen (or increase) our need for the resource within a relatively short period of time (e.g. anthracite coal).

This concern with changes in demand and technologies reflects yet another contemporary point of view. Today, natural resources are being seen less as simply an input into production or as a way of raising the gross national product, and more as adaptable, versatile instruments for increasing the total quality of people's lives. A forest, for example, is no longer simply a source of lumber, but is also recognized as a watershed, a soil protector, an oxygen producer, a wildlife habitat, and a diverse recreational complex. If the economic use of the timber decreases any of these environmental benefits of the forest, should not this lowering of the net quality of the environment be reflected in a corresponding reduction in the gross national product? At least one Soviet economist supports this idea, and advocates lowering the annual rate of industrial growth in the U.S.S.R. to achieve a balance of nature.[1] In this manner, the understanding of the term 'conservation' is shifting from a concern with the quantity of natural resources to a concern with the quality of the total environment.

In the Soviet Union at present, this view is becoming more frequently espoused, particularly by members of the academic and scientific communities. Until rather recently, natural resource conservation was not a widely propagated concept in the U.S.S.R. The traditional love of the Russian peasant for his native land, so well reflected in Tolstoy and Sholokhov, of course could not guarantee that he would utilize that land with the trained insight of a modern conservationist. Land use practices in tsarist times, unfortunately, left much to be desired.

Even today, however, Soviet conservationists must necessarily still concern themselves to a large extent with problems of the wasteful and inefficient use of their country's natural resource endowment, and Supreme Soviet resolutions embodying 'urgent measures' for preventing erosion or pollution appear frequently in *Pravda* or *Izvestiya*. Perhaps a major reason for this is that in the past Soviet writers generally attempted to convey the idea, partly for political reasons, that their country's resources were virtually inexhaustible. Today, this view is openly challenged by those who see the end of the limitless frontier approaching for the Soviet Union, even as it did for the United States. In the words of a noted Soviet geographer, D. L. Armand:

> The theory that now, at an exacting moment in the construction of communism, we can take a 'loan' from nature, that our children will live better, and then they will return the debt to nature...the forests to barren wood-cutting areas, the fish to poisoned rivers, and so forth, must be given a resolute rebuff. This is neither a wise nor a courageous theory.[2]

Although the views of Soviet conservationists have been given increased consideration in recent years, economic planners nevertheless are still frequently cited for placing an overriding emphasis on plan fulfillment, and on obtaining the maximum possible short-term benefits from the use of Russia's natural resources. As the editors of a leading work on the U.S.S.R.'s natural resources have charged:

> Many workers in planning and economic organizations do not take into account the need for a cautious and thoughtful approach to natural resources in their practical work. They often think of nature as an inexhaustible reservoir, from which the useful product may be extracted forever, concerning themselves only with how much faster and how much more can be taken out and utilized.[3]

That many Soviet planners and administrators still look at various types of natural resources as inexhaustible should come as no surprise, for the usual system of categorizing these resources in the U.S.S.R. employs this exact same term. For example, the following classification is frequently found in Soviet works:

All natural resources:
- A. Exhaustible (*Ischerpayemyye*)
    1. Non-renewable: most mineral resources
    2. Relatively renewable: soils, mature forests
    3. Renewable: vegetation, wildlife
- B. Inexhaustible (*Neischerpayemyye*)
    1. Cosmic: solar radiation, tidal power
    2. Climatic: air and air masses
    3. Water[4]

In fact, the authors of the preceding quotation themselves use the term 'inexhaustible' in describing air and water resources.[5] It is only in very recent years that the commonplace use of this term has been called into question. It is hardly surprising, therefore, that economic workers tend to retain this concept, and some of the undesirable consequences which can arise out of such a belief are cited throughout this study. These unfortunate consequences are intensified by considerable laxity in the enforcement of the Soviet Union's existing conservation regulations. The need for economic planners and administrators to be more familiar with the basic precepts of natural resource conservation and environmental ecology is clearly a phenomenon common to all nations and economic systems of the world.

As is the case in the United States, part of the Soviet problem may arise from a lack of understanding on the part of economic planners, and others, as to what the Russian term for 'conservation' actually implies. Unlike the English word 'conservation', which suffers from a difficulty in defining it at all, the Russian term has the disadvantage of being comprised of two words whose specific individual meanings may tend to give the whole phrase too restricted a connotation.

The term for 'conservation' in Russian is *okhrana prirody. Okhrana* means 'protection' or 'safeguarding', and *prirody* means 'of nature'. Taken at face value, then, the whole phrase is somewhat limiting, as evidenced by the fact that in the early part of this century the phrase was most often used only in connection with the establishment of nature preserves.

Today, the literal meaning is clearly too narrow. To be sure, the 'preservation of nature' implies ecological studies, sustained yield logging and fishing, pollution prevention, and other examples of the beneficial, enlightened use of natural resources. But does the current popular Soviet understanding of this term include the study of conflicts between two or more beneficial uses of the same resource complex? Does it imply the consideration of all the tangible and intangible aspects of deciding, for example, whether a dam for flood control or power should submerge an agriculturally or recreationally useful valley?

For the Russian term, too, a more rigorous definition is needed, and suggestions and elaborations which embrace a broader, somewhat more sophisticated approach to the problem are frequently being put forth by biologists, geographers, ecologists, and others. For example, the internationally respected Soviet zoologist, A. G. Bannikov, writes that

> it is necessary to insure an adequate supply of natural resources through the principle of a maximum sustained yield [lit.: 'large-scale reproduction']. In this lies the main problem of conservation today.
>
> Thus, the conservation of nature from an economic point of view is managing natural resources so as to secure from them a maximum sustained yield.[6]

He appears to be thinking here primarily of biotic resources. However, a little later he adds:

> Besides its important economic significance, the problem of the conservation of nature also has tremendous aesthetic and public health significance...The aesthetic value of nature is impossible to appraise.

We preserve nature for people, for those of the present day and of future generations. We protect that environment in which people live and in which they will always have to live.[7]

Another interpretation is offered by G. P. Motovilov:

The concept of 'Protection of Nature' in...the U.S.S.R. consists not in a passive preservation of the natural wealth, but in the carrying out of a system of diverse active measures directed towards preserving, restoring, and expanding the natural wealth and regulating its exploitation in the interests of the present and future generations of mankind.[8]

A more formal definition of the term *okhrana prirody* has been put forth in a recent work on conservation in the Soviet Union: '*Okhrana prirody* – this is the complex state, international, and social measures which direct the rational use, renewal, increase, and protection of natural resources for the welfare of human society.'[9] These definitions illustrate that the literal meaning of the Russian phrase *okhrana prirody* is now broadening considerably, and that its current interpretation is becoming more similar in scope to our own understanding of the term 'conservation'.

The foregoing Soviet definitions of conservation contain several phrases which are close counterparts of traditional American conservation terms. As examples, our 'multiple use' is related to the Russian phrase 'complex use', 'wise use' becomes 'rational use', and 'sustained yield' is embraced in a Russian phrase translatable as 'large-scale reproduction' (although the latter phrase also implies expanding, if possible, the physical stock of the resource). A more thorough discussion of these Soviet terms is contained in Appendix 1.

The Soviet view of the development and conservation of natural resources is associated with an interesting conceptual consideration, involving the manner in which Soviet economic planners look at the man–environment relationship. They tend to speak in terms of the 'planned transformation of nature' (meaning extensive reclamation projects and other similar measures) as a means for improving the agricultural and industrial productivity of the country and, in turn, the well-being of the people. For Soviet planners, natural resource conservation is being increasingly viewed and defined in the light of this desired 'transformation of nature'. A leading Soviet geographer, in fact, has stated that 'under contemporary conditions the conservation of nature is becoming inseparable from its transformation, from the rational exploitation of its resources.'[10] By this he means that the extent of man's utilization, adaptation, and re-structuring of the natural environment (the 'transformation of nature') is today occurring on such a widespread and interrelated scale that in essence virtually all problems of the conservation of natural resources and of the natural environment are in reality problems involving the quality and comprehensiveness of the planning by which such transformations are carried out.

Whereas it would not be difficult for most American conservationists to agree with this analysis, they might well disagree with the extent to which Soviet planners tend to laud such transformations as the very basis for contemporary social progress and welfare. Some Soviet economic planners advocate future developments of unbelievable scope, which would literally transform the basic hydrologic cycle and biological relationships of entire continents (damning the Congo River, the Oresund, the Bering Straits, etc.). Rarely do they dwell on the difficulty of forecasting all of the possible ecological consequences of these changes, only the supposed economic benefits.

Gerasimov, however, is a far more responsible academician than this, and he does go on in his article to make a plea for a more thorough procedure for ascertaining in advance all the likely consequences of such transformations. The need for this type of approach has been increasingly stressed in recent years, and appears to reflect a growing sophistication on the part of Soviet natural resource specialists. Putting such sophistication into actual practice by local natural resource planners and users, however, remains as much of a problem in the Soviet Union as it does in the United States. Neither country's planning system seems capable, at present, of adequately appraising the full range of social costs which are involved in major natural resource development projects, and in 'large scale transformations of nature'. Hence, a somewhat greater degree of caution in initiating such projects would seem advisable in both countries.

It should be pointed out that most Soviet works on conservation, particularly those designed primarily for educational purposes within the U.S.S.R., are careful to present their material on the use of natural resources in a proper Marxist context. Among other things, this precludes any suggestion that the physical environment is a controlling determinant of societal development ('environmental determinism'). It also involves the ideological assumptions that natural resources will invariably be wastefully exploited in a capitalistic economy, but that under a state form of planned economy they will necessarily be utilized in the wisest possible manner. The possibility, under capitalism, of either private enlightenment or effective governmental supervision in the field of natural resource use seems to be ideologically unacceptable to Soviet theorists. More importantly, they are also reluctant to admit that a central planner, having the primary goal of rapidly developing the economy, might easily make the same inefficient use of natural resources as would a selfish private businessman who was trying to maximize his output, and for the same reasons. The point being made here, one which will be repeatedly illustrated in the ensuing chapters, is that political ideologies are less important for natural resource conservation than are implicit national developmental priorities, and that enlightened conservation practices are not second nature to any form of economic system.

Perhaps in partial recognition of this, it has become common in recent years for knowledgeable Soviet citizens to call attention to short-sighted or wasteful natural resource use practices, and public education regarding proper natural resource usage is being more widely encouraged. This developing trend towards broader public participation in conservation affairs has great potential importance for the Soviet Union, for in a centrally planned economy the people represent the only independent check on poorly planned natural resource usage. The need for a much greater public understanding of the concepts of natural resource conservation represents an account which has been overdue in the Soviet Union for years, even as it was, and to an extent still is, in the United States.

Present Soviet attitudes towards the use, transformation, and conservation of the U.S.S.R.'s natural resources will be considered in more detail in the concluding chapter of this study, following an examination of current Soviet approaches and problems in each of the major areas of natural resource management.

2 The historical and institutional framework of conservation in the Soviet Union

...Stalin's genius will...make it possible to master the forces of nature in the U.S.S.R.

The history of the state management of certain natural resources in Russia has an early beginning. Under the tsars, however, this management was rather sporadic, and reflected to a large extent the personality of the particular tsar in power. As a result, the natural resources of the country were generally not too well cared for, and a spontaneous conservation movement began to develop around the turn of the present century. This movement, however, was destined to be cut short by the 1917 revolution. Today, conservation is a state responsibility, and in the past decade has received a greatly increased amount of attention. The brief review which follows of natural resource conservation in the tsarist period will help to put current Soviet practices into a clearer historical perspective.

The pre-revolutionary conservation movement in tsarist Russia

In the early years of Russian tsarism, conservation was a very pragmatic affair, and quite limited in scope. Since at that time hunting was both the main form of recreational activity (for the nobility at least) and an important economic activity, decrees regulating the killing of wildlife constituted the majority of the earliest laws concerning the management of Russia's natural resources.

Peter the Great (1682–1725) represents a notable exception to this generally subjective approach to natural resource management as practiced by the Russian tsars. Peter was the first and perhaps the only tsar to appreciate the need for conserving natural resources in general as a desirable long-range national policy, rather than as isolated responses to immediate needs or crises.

He was particularly concerned with preserving the forests of Russia and, in a very progressive step for its time, ordered timber companies to separate their holdings into twenty-three to thirty sections, only one of which could be cut a year. He also divided all of Russia's forests into restricted and non-restricted categories, and during his reign the first afforestation efforts were carried out in portions of the southern steppe regions.[1] In another act, Peter prohibited the cutting of forests in a belt 50 versts (53 kilometers or 33 miles) wide along major rivers and 20 versts (21 km or 13 miles) along smaller ones. It is interesting to note that this is a wider restricted zone than exists today, which is at the most 20 km wide (Table 6.1). He issued decrees for the protection of sable, elk, and beaver, for outlawing rapacious fishing practices, and even for controlling pollution in St Petersburg's waterways. All his conservation measures were enforced by extremely harsh penalties.

Table 2.1 *Decreases in forested area in tsarist Russia*

Region	Forested area (per cent)			
	1696	1796	1888	1914
North (Arkhangel'sk, Vologda and Olonets guberniyas)	72.9	72.3	69.5	66.9
Central (Vladimir, Kaluga, Moskva, Ryazan, Smolensk, Tver, and Tula guberniyas)	53.2	41.6	30.7	22.2
Baltic (Estland, Lifland, and Kurland guberniyas)	52.0	43.7	25.6	23.2
Forest–steppe (Voronezh, Kursk, Poltava, and Khar'kov guberniyas)	18.4	12.1	8.1	6.8
Southeast (Astrakhan' and Stavropol guberniyas and Don oblast)	2.4	1.9	0.98	0.97

From M. A. Tsvetkov, *Izmeneniye lesistosti Yevropeyskoy Rossii* (Moskva: Izdatel'stvo Akademii nauk S.S.S.R., 1957) pp. 126–31.

The tsars of the latter part of the eighteenth century had little of Peter's insight into the need for conserving natural resources. The forests began to be cut again, particularly after 1782 when Catherine the Great in effect rescinded Peter's forest management regulations. A somewhat greater concern for the preservation of wildlife was maintained after Peter's death, and in 1763 the concept of a hunting season began in Russia when Catherine issued a decree banning hunting from March 1 to June 29 to allow animals to bear their young.[2]

During the nineteenth century, as the tempo of natural resource utilization increased greatly with the beginnings of industrialization, the incidence of shortsighted and wasteful exploitation of these resources increased as well. The stocks of elk began to be depleted again, and the forests continued to be heavily cut. The areas that suffered the most were the lightly forested regions of central and southern European Russia (Table 2.1), and their depletion led to the Forest Protection Law of 1888 to curtail the further cutting of forests in these areas. The loss of these forests had unfortunate consequences for other local resources, particularly for the quality of soils and streams in these regions. This deforestation, combined with poor agricultural practices on the part of the newly emancipated serfs, led to a significant increase in the incidence of gully erosion in European Russia during the latter part of the nineteenth century.[3]

The hunting laws of Catherine the Great existed basically unchanged until 1892, but the law that updated them in that year placed virtually no restrictions on hunting on a landowner's own property. A noticeable decline in the number of freshwater fish in certain areas led to laws regulating fishing on the Ural River (1803), regulating methods of fishing (1828), and protecting spawning areas (1835).

On the whole, however, the nineteenth century witnessed a significant decline in both the quantity and quality of Russia's biotic resources. Further, the rapid development of the Baku oil deposits and the Donbass coal mines foreshadowed the emergence of an entirely new set of environmental problems. The urgency of the situation began to impress itself upon the more farsighted of Russia's citizenry.

For the most part, natural resource utilization during the tsarist era had proceeded on the assumption, common in many countries at that time, that man is the owner and self-benevolent dispatcher of all the wealth of his environment. In neither Russia nor the United States had the understanding yet developed that man himself is an inseparable and interdependent component of that environment. Prior to about 1890, little concern was evidenced at the rate at which the forests of European Russia were being used up, at the increasing extent of soil erosion in the drier areas of the country, at the loss of wildlife habitats, or at other similar environmental problems.

Almost at the same time as in the United States, however, there arose in Russia a growing concern over the manner in which the country's natural resources were being exploited with, in general, insufficient regulation to ensure their continued availability. This concern initially developed among the scientific members of Russia's academic community. Unlike the situation in the United States, members of the government seemed to take little interest in this crucial problem.

Associations composed of scientists interested in natural history had begun to appear during the nineteenth century. The first in Russia was the Moscow Society of Naturalists, established in 1805. After a lapse of half a century, similar societies began to appear: a Committee on Acclimatization of Animals and Plants in 1857 (which published *Akklimatizatsiya*, the first Russian periodical dealing with conservation problems); a Russian Entomological Society in 1859; and natural history societies in Moscow in 1863 and in St Petersburg and Kazan' in 1868.[4] In 1881, a Russian Society for Fish Breeding and Fishing was organized. As can be seen from their titles, most of these were essentially scientific associations rather than conservation organizations *per se*. The first local society whose scope encompassed the entire gamut of conservation problems was established in 1910 in Khortitsa, a town across the Dnepr from Zaporozh'ye.

Two of the more active men in these early movements, the St Petersburg botanist I. P. Borodin and the Moscow zoologist G. A. Kozhevnikov, represented Russia at the first international conference on conservation problems, held in Bern in 1913. Borodin was also instrumental in establishing in 1912 a Permanent Nature Conservation Commission within the Russian Geographical Society.

From around 1895 until World War I, the Russian Society for Fish Breeding and Fishing carried out significant pilot programs in artificial fish propagation, particularly with regard to sturgeon in the Ural River. A well equipped fish hatchery was in operation on the Volga near Saratov by 1910, and fisheries research centers were operating in Astrakhan', Baku, Krasnoyarsk, and Khabarovsk prior to 1917. A congress on water quality control in river basins was held in 1896. Efforts to conserve the North Pacific fur seal, which had been badly over-harvested for a century, were begun in 1897. In that year, Russia participated in an international conference on the fur seal, which eventually led to a four power treaty signed in 1911 with Japan, Canada, and the United States.[5]

An air pollution control bill was introduced into the Duma as early as 1913, but was not enacted.

In agriculture, soil research under V. V. Dokuchayev was quite advanced, and the agricultural reforms of Interior Minister P. A. Stolypin, although administrative in nature, would probably have had certain beneficial consequences for soil conservation had they been carried out to completion. By 1913, about four million hectares of land had been provided with irrigation networks, and almost three million hectares had been drained (a metric conversion table is presented in Appendix 2). Extensive efforts at reforesting the steppe lands of European Russia were also carried out during the nineteenth century, but with little lasting success.

One of the most noteworthy of the pre-revolutionary writers on the natural environment was the climatologist A. I. Voyeykov (1842–1916). Following in the footsteps of George Perkins Marsh, Voyeykov was one of the first in Russia to think in terms of man and nature as an interacting whole. He was particularly interested in man's capacity for altering his natural environment, both for better and for worse, especially in the less humid regions. One of his major works is a lengthy, two-part article entitled 'The influence of man upon nature', published in *Zemleveden'ye* in 1894. This work, together with others, has been collected into a book with the same title.[6] Voyeykov's opposition to the environmental determinists of his day, together with his support for beneficial nature transformation projects, has kept him in good stead with contemporary Soviet geographers and historians.

Among the goals foremost in the minds of the members of the early societies was the establishment of protected natural areas, for the preservation and study of native vegetation and wildlife. One of the earliest calls for a state system of such natural areas was made in 1908 by G. A. Kozhevnikov.[7] I. P. Borodin, V. P. Semenov-Tyan-Shanskiy, and others made similar appeals, and in 1909 and 1915 proposals enumerating specific areas to be set aside were put forth.

Numerous *de facto* preserves were then in existence, in such forms as large private hunting estates and extensive church-owned lands. Efforts were made to persuade the owners of a few of these areas to turn them into fully protected preserves, and in some cases, such as the Lagodekhi preserve in Georgia, they were successful. The pre-revolutionary establishment of these preserves is examined in more detail in Chapter 4; in all about eight or ten existed in one form or another by 1917.

Prior to the revolution, then, a conservation movement of sorts was beginning to evolve in Russia. Existing essentially only among the academic community, and born like its American counterpart out of the threat of imminent and irreversible natural losses, it nevertheless was gaining momentum in the last decade of tsarist rule. It is possible only to speculate upon the degree of success it might have attained had the existing political–social order not been so fundamentally and irrevocably altered.

Conservation measures taken by the Soviet government

After 1917, the management of Russia's natural resources was carried out within a totally new context. No longer were they available for development by private individuals or concerns; they were now the exclusive property of the Soviet state.

The awareness by many in the pre-revolutionary intelligentsia of the worsening state of Russia's natural resource endowment was shared by many of the revolutionaries themselves, most notably Lenin. As practicing Marxists, however, they held the assumption that the irresponsible use and resulting deteriorated condition of Russia's resources was a consequence of the capitalist economic system which had existed in the country, and that under socialist management these resources would necessarily be properly utilized in the future. Today, conservationists better understand that there are numerous complexities, aside from simply waste or mismanagement, which are inherent in the farsighted development of natural resources and in the preservation of healthy biomes, and that many of these complexities will not be significantly lessened by simply altering the type of economic system involved. However, this fact was poorly understood by the Bolsheviks of that day, and would have been ideologically unacceptable to them in any event.

During the first years of Soviet power, numerous decrees were put forth to remedy what were felt to be the major deficiencies in current natural resource management policies. At that time, conditions were quickly becoming considerably worse as a result of the very destructive civil war and famine which immediately followed the revolution.

The breakdown in transportation in the years after 1917 led to a massive assault on the forests of European Russia as a source of fuel. Concern for the state of Russia's forest resources led to a decree in 1918 charging local authorities not only with the responsibility of managing forest activities, but also with seeing to their reforestation where needed, and with protecting outstanding natural features found in them. City parks were also being destroyed in the search for fuel, and this prompted a further resolution in 1920 aimed at the preservation of urban parks and green belts.

A decree was signed in 1919 to promote the conservation of wildlife. It banned the hunting of elk, established seasons for hunting, and contained procedures for creating wildlife preserves. This decree was elaborated upon by another signed the following year, and the list of protected wildlife species was expanded by a subsequent act in 1924.

A Central Directorate for Fishing and the Russian Fishing Industry was organized in 1918, and a decree regulating fishing activities was put forth in 1920. It delimited restricted spawning areas, prescribed fishing seasons, and banned certain fishing methods and equipment. The following year, a decree to safeguard fish and wildlife habitats in the Arctic was enacted. The first Soviet fisheries research institute was created on March 12, 1922.

A comprehensive resolution for the advancement of agriculture, the prevention of soil erosion, and the development of irrigation was signed in April of 1921. However, the advent of the New Economic Policy delayed the centralized management of agriculture for several years.

A number of new natural preserves were established in the first decade after the revolution. They included a wildlife refuge near Astrakhan' in 1919, the Il'men mineralogical preserve near Chelyabinsk in 1920, and about eight other new preserves in the period from 1921 to 1930. In addition, a significant letter of instruction from the Central Forestry Department of the Agricultural Commissariat was handed down to all regional forestry subdepartments on November 4, 1919. It stated that lumbering would be inadmissible in those areas of forests which were suitable for future national parks and natural monuments, such as areas with rare or unusual vegetation, geologically interesting sites, etc. Further, the local offices were to submit suggestions for areas in their districts suitable for such classification. Decrees authorizing and defining the objectives of natural preserves were passed in 1921 and 1924 (Appendix 10).

By 1921, basic laws regulating the use of land, forest, wildlife, and fisheries resources of the country had been enacted by the Soviet government. However, the host of internal difficulties confronting the new Soviet state could not help but greatly restrict the enforcement of these early conservation measures, and the pace of conservation achievements during the 1920s was not great.

Efforts to coordinate work on the conservation of Russia's biotic resources during the 1920s was the responsibility of an 'Interdepartmental State Committee on Nature Preservation'. It consisted of representatives from the Commissariats of Internal Affairs, Workers and Peasants Inspectorate, Gosplan, Finance, Agriculture, Forestry, Education, and others, as well as representatives from scientific and public organizations. Similar interdepartmental organizations were established at the province level.

In 1933 the Interdepartmental Committee was abolished and was replaced by a 'Committee on Natural Preserves' under the Presidium of the All-Union Central Executive Committee (VTsIK). This new committee had responsibility not only over preserves, but also for formulating and carrying out all measures necessary for the preservation, optimal use, and restoration of any of the country's valuable or endangered biotic resources.

During 1938–9, this work was transferred from a central agency at the national level to corresponding bodies in each of the Union republics. In the Russian Republic, a 'Central Directorate for Natural Preserves' was established under the Council of People's Commissars of the R.S.F.S.R., and similar departments were formed in each of the other republics.

In 1929, the 'First All-Russian Congress on the Conservation of Nature' was held in Moscow. Although called an 'All-Russian' congress, there were representatives present from many parts of the Soviet Union outside the Russian Republic. The Congress discussed the present state of natural resource conservation in the country, the need for popularizing this concept, questions relating to the establishment and functions of natural preserves, and desirable future courses of action.[8] The first conference to be designated as an 'All-Union Congress on the Conservation of Nature' was convened in 1933.

Historical and institutional framework

In the years of Stalin's rule, the overriding goal was rapid industrialization, and policies concerning the utilization of natural resources were generally formulated from this point of view. This is not to say that conservation practices were ignored, but nevertheless it is certain that much inefficient or even wasteful use of natural resources took place in these years as a consequence of the emphasis on the most rapid possible pace of economic development.

Whatever improvements in the condition of Russia's natural resources the new regulatory measures may have brought about were greatly reduced by the events of World War II. Forest resources were particularly hard hit; the approaches to Leningrad were almost completely denuded of vegetation during the seige of that city. Animal resources in European Russia were decimated again, and re-establishing agricultural production became one of the most crucial post-war tasks.

In the twenties, Stalin had assumed that the needed increases in agricultural productivity would be realized by the carrying out of collectivization. By the late forties, however, it became clear that collectivization alone was not going to be a long-range solution to all of the Soviet Union's agricultural problems. It was then decided that something must be done to remedy the climatic limitations on Russia's agricultural potential, and, as a result, the 'Great Plan for the Transformation of Nature' was born.

The 'Great Plan' was, in essence, a grandiose scheme to make a large portion of the semi-arid belt in the European part of the Soviet Union agriculturally productive. It involved a vast network of shelter belts, plus extensive river control and irrigation systems.[9] Most of these plans were technically feasible, but would have involved astronomical expenditures of capital and human resources. Also, their long-range effect on regional climatic patterns, soils, and other related resources was poorly studied. Nevertheless, in the rhetoric of the day it was modestly asserted that

> the party...and the Soviet government are doing everything possible to transform nature, to do away with the deserts, to attain a further big rise in agricultural productivity...The grand projects outlined by Stalin's genius will...make it possible to master the forces of nature in the U.S.S.R.[10]

Conservationists today would be inclined to ask whether the ultimate desirability (in terms of costs, environmental changes, agricultural alternatives, and ecological losses) of 'doing away with the deserts' was ever fully debated. It is rather doubtful that it was, at least at the high Party level.

Although certain individual portions of this plan (the Crimean canal and some areas of shelter belts) have been constructed or begun, most of it has been set aside indefinitely either as too expensive or because serious questions have arisen as to possible adverse environmental effects which might result from its full implementation.

The 'Great Plan' clearly reflected a view of man as the master and perfector of his natural environment, rather than as an integral and interdependent component of it. This view was at times presented so forcefully as to declare that man must wage war against his natural environment.[11] The 'Great Plan' may be considered as indicative of Stalin's basically domineering attitude towards natural resource exploitation and conservation, an attitude that is not in official favor in the Soviet Union today. For example, the Chairman of the Soviet Academy of Sciences' Commission for the Study of Productive Forces and Natural Resources stated in 1967 that 'we have had many troubles because people who attempted to do great deeds in "transforming" nature approached this job very shortsightedly'.[12] The probable adverse effects of these projects, it should be noted, were well known even in the 1930s.[13]

Following Stalin's death, the concept of man as the absolute master of nature generally gave way to more pragmatic approaches. The opening of the Virgin Lands in Kazakhstan to cultivation only a year after Stalin's death is perhaps the most graphic example of this. In place of the grandiose scheme to transform nature to man's needs, Soviet planners were now committing themselves instead to the gamble of developing a decidedly marginal new agricultural area.

After Stalin's death, a much wider degree of debate and independent opinion was permitted in the resolving of questions relating to natural resources. The utilization and conservation of these resources began to be approached in a more comprehensive and sophisticated manner, and with more attention to the indirect consequences to society of their use, such as air and water pollution, recreational opportunities, and changes in the physical environment. As was noted in the preceding chapter, the phrase 'the transformation of nature' is still widely used even at present, but it is today used in a very broad sense to describe any alteration of the natural environment by man, even on a relatively small or local scale. When such alterations involve large-scale plans, it is now usually acknowledged that such plans should be dependent upon detailed studies of their effects on all the interdependent natural complexes involved.[14]

In the post-war years, many new laws governing the use and preservation of specific natural resources, particularly biotic ones, were put into effect.[15] Between 1957 and 1964 all fifteen of the Union republics of the U.S.S.R. wrote comprehensive laws governing the manner in which the natural resources of those republics were to be managed and conserved. The first to appear was that of Estonia on June 7, 1957. The law for the Russian Republic, having force over three-quarters of all the land and water resources of the country, was passed on October 27, 1960. As it represents the single most important statute in the Soviet Union for the management of that country's natural resources, sections of it are included in the appendices. The preamble and certain of its general and implementing articles appear as Appendix 3. The extent to which its various provisions are complied with will be noted in succeeding chapters.

Various sections of the criminal code of the Russian Republic, which went into effect January 1, 1961, stipulate the penalties for violations of the R.S.F.S.R.'s conservation law. Among the areas covered are the setting of forest fires (sections 98 and 99), poaching (sections 163, 164, and 166), illegal timber cutting (section 169), and air and water pollution (section 223). The fact that these penalties exist, however, does not necessarily mean that they are rigidly enforced. Examples of leniency are cited frequently throughout this study, and this 'tolerance policy' represents one of the major shortcomings of the Soviet conservation effort.

A few generalized observations concerning the need for natural resource conservation appeared in the Program of the 22nd Congress of the Communist Party of the Soviet Union, held in 1961. As a result of increased efforts by conservation agencies and organizations, broader though still generalized statements appeared in the Program of the 23rd Party Congress, held in 1966. The Directives of this Congress for the 1966–70 Five-Year Plan, under the heading 'Main Tasks of Economic Development', called for the '...elaboration and implementation of measures for the better protection of nature, with a view to the more effective utilization of land, forests, reservoirs, rivers, game animals, fish, and other natural riches of the country...'.[16] Other sections of this document make passing reference to specific problems requiring attention, such as air and water pollution, afforestation, soil erosion, and urban beautification. In the Russian Republic, however, a detailed five-year program of conservation activities is to be drawn up and approved simultaneously with the drafting of the 1970–5 R.S.F.S.R. Five-Year Plan.[17]

The current institutional framework of Soviet conservation

The current increased concern for conservation matters finds institutional expression within the Soviet government and throughout Soviet society in a number of ways. At the government level, day-to-day attention to conservation problems is usually entrusted to those administrative agencies which have the responsibility for managing the use of a given resource. For example, the U.S.S.R. and republic Ministries of Fisheries are charged with ensuring that sustained yield practices are utilized by fishing enterprises. For the extraction of forest resources, this responsibility falls to the State Forestry Committee of the U.S.S.R. Council of Ministers and to the Forestry Ministries of the various Union republics. Sport and commercial hunting are managed by central directorates for hunting and preserves in each of the Union republics.

In a like manner, the development and use of water resources is supervised by the U.S.S.R. Ministry of Reclamation and Water Resources Management, and its counterparts at the republic level. The prevention of soil erosion and related problems is the responsibility of the Ministry of Agriculture and of the State Inspectorate for the Supervision of Land Use and Soil Conservation. Activities directed towards the prevention of both water and air pollution are the responsibility of agencies within the U.S.S.R. and republic Ministries of Public Health, who maintain a nationwide network of sanitation inspectors to report pollution law violations.

The task of maintaining an overall view of the state of conservation in the U.S.S.R. and, to a degree, of trying to coordinate the activities of the above agencies is entrusted to a central commission under the U.S.S.R. Council of Ministers. First established on March 11, 1955, it was initially called the Commission for the Conservation of Natural Resources and at first existed under the Biological Department of the U.S.S.R. Academy of Sciences (later under Gosplan). Its first chairman was G. P. Dement'yev. Since 1965 this work has been carried out by the Central Directorate for the Conservation of Nature, Natural Preserves, and the Hunting Economy of the U.S.S.R., which operates under the U.S.S.R. Ministry of Agriculture. The Central Directorate is headed by B. N. Bogdanov. A separate Council on the Conservation of Nature still exists under the Academy of Sciences. The idea of a powerful All-Union State Committee on Conservation which would effectively coordinate the conservation work of all the various governmental ministries has been advanced, but no action has been taken on this suggestion as yet. The need for such a coordinating body has been frequently noted by Soviet conservationists. In the words of Bogdanov:

> The branch inspection services, because of their narrowly departmental approach, have not provided in full the necessary control over conservation, especially since many of them have acquired economic and production functions that distract them from their principal duties. The system of departmental inspection services, working in dissociation, without proper coordination, is unable to take into account and foresee the possible pernicious consequences of violations in the utilization of individual resources for nature as a whole.[18]

Within the Central Directorate exists a Central Laboratory for Nature Conservation, headed by L. K. Shaposhnikov. One of its primary functions is carrying out applied research on the conservation of biotic resources. Much of this research is conducted within the state network of natural preserves, and twelve of them considered to be of national significance are administered by the Central Directorate.

There is also a Central Board for Nature Conservation, made up of about a hundred scientists who act as consultants or advisors for the Ministry of Agriculture. It has a ten-member Presidium, comprised of leading scientists in the biological sciences. In addition, all fifteen of the Union republics have established state committees on conservation under their own Councils of Ministers.

A Scientific Council for the Study of Natural Resource Problems was established in July 1966, by a joint resolution of the State Committee for Science and Technology (under the U.S.S.R. Council of Ministers) and the Presidium of the U.S.S.R. Academy of Sciences. Named as its head was the prominent Soviet geographer I. P. Gerasimov who, as noted in the preceding chapter, is a leading proponent of the carefully planned 'transformation of nature' in the Soviet Union. The Council is to be concerned with organizing and coordinating scientific and theoretical research work on major problems of natural resource use and conservation, such as averting floods and erosion, predicting environmental consequences of resource exploitation, and planning the optimum long-term use of natural resources for the development of the national economy.

On May 27, 1967, a 'Commission for the Study of Productive Forces and Natural Resources' was established within the U.S.S.R. Academy of Sciences, with N. V. Mel'nikov as its chairman. One of its main tasks will be to ensure the U.S.S.R. an adequate supply of natural resources through the year 2000, and to study the problems of the integrated use of the country's mineral, water, timber, and other natural resources. The manner in which the duties of this Commission differ from those of the Council described in the preceding paragraph has not been made clear, unless it is that the Commission will be primarily concerned with quantitative questions and the Council primarily with qualitative ones. Other research institutes working on problems of conservation and the environment exist, such as the recently established (1970) Institute of Animal and Plant Ecology, which publishes the journal *Ekologiya*.

Public conservation organizations also exist in the Soviet Union, usually in the form of conservation societies within the Union republics. The most important of these is the All-Russian Society for the Conservation of Nature, founded in 1924. At first more of a scientific or scholarly society, it now embraces about 11,000,000 members, most of whom are school children in its youth division. Branches of the Society exist in all oblasts, krays, and autonomous republics of the R.S.F.S.R. The 1960 conservation law of the Russian Republic permits the Society to conduct public conservation inspections, in which it checks on the local status of wildlife and vegetation conservation, air and water pollution, and soil preservation (Appendix 3, Article 16). A council of technically qualified advisors within the Society expresses the Society's position on questions of the most desirable use for given natural resources or natural resource complexes. The R.S.F.S.R. Council of Ministers is required to at least formally review the statements and proposals presented to it by the All-Russian Society.[19]

Other societies concerned with various aspects of conservation exist throughout the U.S.S.R., as for example the Moscow Society of Naturalists (*ispytateley prirody*). Founded in 1805, it is one of the oldest conservation societies still existing in the Soviet Union today.

It goes without saying that public conservation organizations in the Soviet Union, such as the All-Russian Society, do not enjoy the political freedom or leverage that an organization such as the Sierra Club is able to exercise in the United States. Still, Soviet societies can be effective at the local level, if they so choose, in checking on compliance with existing conservation laws, advancing conservation education, and bringing to the attention of government organs areas where improvements would be possible or desirable.

Organizations and knowledgeable individuals often draw up resolutions, write articles, and send letters to newspapers and periodicals on both general and specific conservation issues, such as the need to combat air and water pollution or to safeguard endangered species of wildlife. One of the most notable recent examples of this has been the effort by leading scientists and writers to ensure that Lake Baykal will not be polluted by effluents from new industrial enterprises (Chapter 8). Nor are such communications necessarily polite or inhibited in the statement of their case. A 1965 article by the prominent writer and conservationist Leonid Leonov, published in *Literaturnaya gazeta*, is an excellent example of the pointed attacks put forth by those who are in the vanguard of the current conservation movement in the U.S.S.R. In it, he uses discreetly tempered sarcasm to attack 'the waste – which is becoming habitual – of priceless living raw materials, waste that can exhaust even the richest treasury'. Major excerpts from it have been included as Appendix 20.

In general, it is possible today in the Soviet Union for knowledgeable groups or individuals to speak out on conservation problems as long as they do not imply that these problems exist because of faulty decisions on the part of the Party. Objective criticism of apparent cases of mismanagement by the administrative and planning bureaucracies is permitted, but policy matters traceable to past Communist Party decisions are above public reproach. Rarely are major policy decisions which have since been carried into reality openly criticized by anyone except, occasionally, the Party leadership. Presumably, though, past misjudgements are reviewed, and, if possible, corrected in the course of subsequent planning.

Conferences for the discussion of natural resource utilization and conservation problems are held frequently. Since 1958, an 'All-Union Conference on Nature Conservation' has been held approximately once a year, each time in a different Union republic capital. The 1958 conference, held in Tbilisi, was designated as the 'First All-Union Conference...'; apparently it was decided to number this new series of annual conferences starting with number one, the 1933 All-Union conference referred to earlier notwithstanding. In addition, republic, inter-republic, and regional conferences have frequently been held since the mid-fifties, as well as conferences in other fields which have dealt heavily with conservation matters. For example, the Geographical Society of the U.S.S.R., at its Third Congress (1960), resolved that problems of the proper management and conservation of natural resources should be one of its most important future tasks.[20] In June of 1969, a national conference on the legal aspects of Soviet conservation was held in Moscow.

Considerable efforts are made to teach Soviet school children some of the basic concepts of conservation. At the grade school level (that is, in all ten of the pre-university years of schooling), there are no special courses dealing with conservation. This topic, however, is actively developed in the curriculum by integrating conservation concepts into the textbooks and courses which are used for the natural (physical) sciences, biology, and geography in all grades from five through nine. For example, in the geography courses of the fifth through the eighth grades, such topics are presented as 'Conservation of Nature', 'Nature Reserves', 'Protection of Inland Waters', and 'Measures for Forest Protection'.[21]

A broad range of extracurricular activities is also available to Soviet grade school children. The Ministry of Education provides a program organized around what are known as 'young naturalist' (*yunnat*) centers; these are often run by biologists or other such specialists. A second medium for conservation education is the Pioneer organization, which is similar in some ways to the Boy Scout and Girl Scout movements but which also has a heavy emphasis on political indoctrination. Much of the contact of Pioneer members with conservation occurs during the vacation period at summer Pioneer camps. Another outlet which has already been mentioned is the youth division of the All-Russian Society for the Conservation of Nature. A fourth activity is a recent movement called the 'Green Patrol' (*Zelyonyy patrul'*), which engages school children in public conservation inspections and in simple conservation projects. It is not clear how the work of this group, which apparently arose somewhat spontaneously, differs from that performed by the previous three organizations.

Much of the activity of the school children who participate in these organizations is simple applied work, such as building birdhouses, planting trees and flowers in cities and shelter belts in the countryside, collecting seeds, removing diseased and downed timber, and, in certain advanced situations, taking part in fish and soil conservation projects and even assisting in fighting brush or forest fires. To emphasize the importance of this kind of work, many areas of the country have the school children observe special days or weeks, such as 'Bird Day', 'Garden Week', 'Forest Month', etc., in which the conservation of these particular features of nature is emphasized, and in many areas, actual conservation projects are carried out.[22]

In the higher educational institutions of the Soviet Union (universities and technical or research institutes), courses devoted entirely or partially to conservation are frequently encountered. The first in the U.S.S.R. was taught in the biology–soils department of Moscow State University. Today, specific conservation courses are encountered in the biology and geography departments of many institutions of higher learning, and some law schools offer courses in conservation law. The recommended basic format for university level courses in conservation throughout the Soviet Union is presented in Appendix 4. Many other courses in a wide variety of departments also deal in part with problems of natural resource management and conservation. Despite all this, concern is still being expressed that conservation is not being adequately taught in Soviet higher education, and suggestions for improvement occasionally appear in the press.[23]

As in the United States, many theses and research papers are devoted to questions of the wisest manner of utilizing natural resources, including numerous ones from the engineering sciences. In another analogy to American universities, many of the larger Soviet institutions, such as those at Moscow and Perm′, have associated research stations located in the countryside where environmental studies can be carried out. In addition, a special University on Nature Conservation, with a primary function of improving the conservation background of secondary school biology and geography teachers, has reportedly been established in Armenia.[24]

Outside the universities, opportunities exist for professors to act as consultants in the areas of their proficiency, for professors and students to participate in public discussions and educational programs, and for students to act as leaders in the various grade school conservation organizations outlined above.

The objective of all this is to instill an awareness of the need for conservation into the broad masses of the Soviet people. This is something that generally has been lacking in the past, and to a large degree still is even at present. A second major purpose is to broaden the base upon which major decisions regarding natural resources are made. These goals are clearly reflected in the much wider range of public conservation activities which have developed in the years since Stalin's death in 1953. If the great number of conservation activities outlined above are achieving their intended effect, and if they can find expression in governmental planning and administration as well, then the state of natural resource conservation in the Soviet Union could conceivably show marked improvement in the not too distant future.

3 The conservation of Soviet land and soil resources

Almost annually...dust storms blow away soil, damage and destroy crops...

The conservation of agricultural resources universally takes a high priority, for their preservation and improvement is fundamental to the well-being of any country's growing population. Such conservation is even more necessary in the Soviet Union, where significant physical limitations on agriculture exist and where very little potentially arable land still remains undeveloped. Although the Soviet populace is adequately fed, the labor productivity of Soviet agriculture remains low, and many problems of the conservation of agricultural resources remain unresolved. Some of the reasons for the U.S.S.R.'s low agricultural productivity are essentially of an economic or even political nature, and fall outside the scope of this study. Others, however, relate more directly to the manner of development of agricultural land, and to problems of soil and soil moisture conservation. The importance of the agricultural question to the Soviet economy warrants an examination of the causes of these latter problems, and the approaches being taken by the Soviet government towards their solution.

Problems in the conservation of agricultural resources

All land in the Soviet Union is the property of the state. Of the total land area of the U.S.S.R. of 2227 million hectares, in 1965 474.1 million hectares (21.3 per cent) was in collective farms (*kolkhozy*), and 584.4 million hectares (26.2 per cent) was in state farms (*sovkhozy*). Most of the rest (except for urbanized areas, etc.) comprises what is referred to as the 'State Land Reserve' (*Goszemzapas*). Of the area in state and collective farms, almost four-fifths is pasture, forests, or non-productive land. The remaining 223 million hectares are classified as arable land.[1]

In December of 1968, a major piece of legislation concerning the use and management of the land resources of the Soviet Union was adopted. It serves as the basic legal enactment in this field, the first major revision of Soviet land law since 1928. Entitled 'Principles of Land Legislation of the U.S.S.R.', it defines categories of land within the Soviet Union, conditions of its use, responsibility for its proper management and conservation, and procedures for settling claims and conflicts (Appendix 5). It was first issued in draft form for purposes of soliciting public commentary, and received in response some 3000 perfecting suggestions. Among these were many relating to soil conservation, the possibility of charging for the use of land, and compensation for losses of agricultural land withdrawn for urban and industrial uses. As a result, numerous changes relating to land conservation appeared in the final form of this law as officially adopted.

A 'Model Charter for Collective Farms' was adopted by the Third All-Union Congress of Collective Farmers in the fall of 1969. Among its provisions is a section entitled 'The Land and its Use', which states in Article 9 that

> the collective farm is obligated to make the fullest and most correct utilization of the land allotted to it, constantly to improve this land, to increase its fertility; to bring unused land into agricultural production; to carry out measures for land irrigation and drainage, for combatting soil erosion and for creating field shelter-belt plantings; to care for collective farm land and strictly protect it from wasteful use; to observe established regulations for the protection of nature and for the use of forests, sources of water and useful minerals (sand, clay, stone, peat, etc.).[2]

The proper management and upgrading of the nation's agricultural resources has always been a basic problem confronting the Soviet government.[3] Both the latitude and relief of the Soviet Union tend to work to the disadvantage of agriculture. Between the arid steppes on the south and the arctic expanses to the north lies only a limited amount of land having relatively favorable soil and climatic conditions for agriculture (Figure 3.1). And even here the continentality of the climatic regime and the openness of the Russian landscape makes this belt highly vulnerable to occurrences of both severe drought and late spring and early autumn frosts. Under such conditions, it is not difficult to see why the 'transformation of nature' has been a recurrent theme in planning the development of the national economy.

In addition to the natural limitations, the Soviet government inherited a tradition of backward and inefficient agricultural practices on the part of the Russian peasant. In pre-revolutionary Russia, farming equipment was usually primitive, prevalent farming methods made inefficient if not wasteful use of the resources at hand, and the average peasant was almost totally uneducated in scientific farming methods and conservation practices. An improvement in this situation was under-way at the turn of the century, but was cut short by the revolution.

At present, the management of the Soviet Union's agriculture is the responsibility of the Ministry of Agriculture of the U.S.S.R., and of similar ministries within the governmental structure of each of the fifteen Union

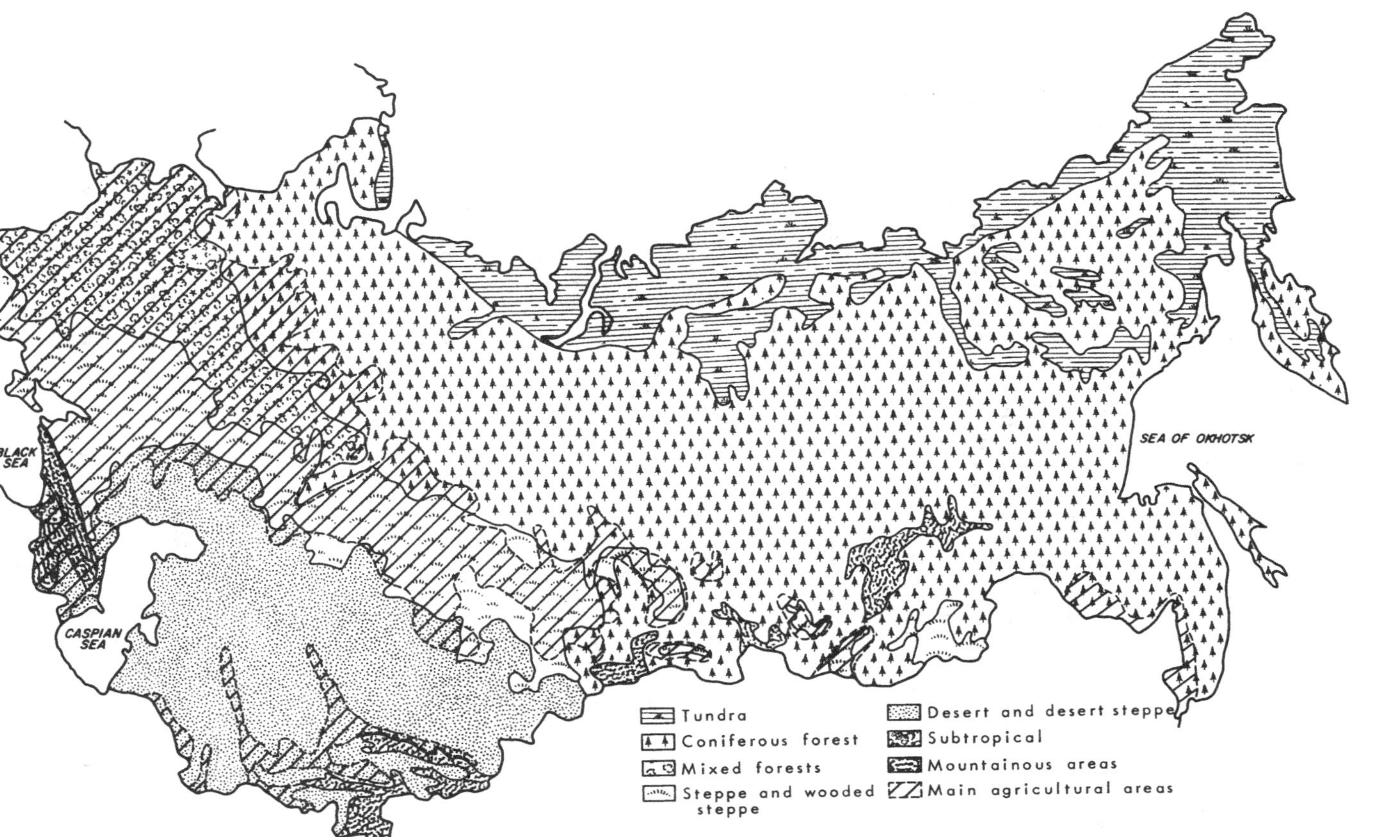

Figure 3.1. Natural zones and main agricultural areas of the Soviet Union. From *Atlas S.S.S.R.*, pp. 78 and 114–19 (modified).

republics. Although these various agencies have endeavored, with a certain degree of success, to raise the technical capacity and productivity of the U.S.S.R.'s agriculture and of its farm workers, nevertheless much of the increase in agricultural output during the Soviet period has come not through productivity increases, but through efforts to increase the total expanse of cultivated land.

In part, this preference for expanding the sown area as the simplest means of increasing agricultural output may have arisen from the normal tendency of a young nation to overestimate the 'inexhaustible resources' of its untamed frontier, even as we did in the American West. In larger part, however, it may have resulted from the ideological assumption, noted in the last chapter, that it was capitalistic limitations that made the peasant unable to improve his farming practices or to open up vast new areas of

Table 3.1 *Soviet agricultural lands*

Category	Million hectares of land		
	1953	1958	1965
Total U.S.S.R. land area	2227	2227	2227
Total area classified as agricultural land	481.6	501.0	542.8
Total area classified as arable land (*pashnya*)	188.6	217.9	223.4
Total sown area in U.S.S.R.	157.2	195.6	209.1
Sown area in Western Siberia	12.4	18.8	18.9
Sown area in Kazakhstan	9.7	28.6	30.4

From *Sel'skoye khozyaystvo S.S.S.R., statisticheskiy sbornik* (1960) pp. 126, 127, and 144; *Narodnoye khozyaystvo S.S.S.R. v 1965 g.*, pp. 278 and 294.

virgin agricultural land. Now that the country's agriculture had been socialized, it was thought, the major obstacle to opening up these lands had been eliminated, and extensive new areas could be turned into productive cropland. All of these beliefs and assumptions have been questioned by Western agricultural specialists, who view the limitations of the given climatic and soil conditions in a more critical light.[4] And whereas Soviet agronomists have not abandoned the above ideological assumptions, they do at least now view the physical constraints on expanding the tilled acreage in a more realistic manner.

The most extensive area which has been opened to intensive cultivation for the first time under the Soviet government is the vast steppe region of northern Kazakhstan and West Siberia. Between 1953 and 1958, 25.3 million hectares of new cropland were planted here (mainly to wheat).

This represented 65.9 per cent of all the new land brought into cultivation in this period (Table 3.1).

Most of this virgin steppe territory lies on the margin of the arable area in the Soviet Union, as do the western portions of the Great Plains in the United States. Consequently, when these lands were first being intensively developed, Khrushchev stated that he would be satisfied with two good crop harvests from these new areas every five years. This, in fact, has been about what the average has been, but, in addition, certain other effects have been realized which Khrushchev apparently did not anticipate. The exposure and repeated desiccation of the soil has created conditions favorable for serious dust storms; these have occurred frequently and will be discussed more fully in the following section.

Much additional land has also been brought into production through reclamation projects – irrigation and the draining of marshy lands. Large parts of European Russia and much of Western Siberia is of very low relief and has a cool, moist climate, and as a result over 200 million hectares (nearly 10 per cent of the U.S.S.R.) is made up of marshy or waterlogged lands. Within this area, large expanses of potentially valuable agricultural land have very poor surface drainage and are too marshy for crop cultivation. In addition, such swampy areas often breed mosquitoes and other pests, and as such represent a potential health hazard. As a result, the Soviet Union has devoted considerable effort toward reclaiming usable marshland. From about 3 million hectares in 1913, the total drained area in the country had grown to 10.6 million hectares in 1965. In all, about 23 to 25 million hectares of agricultural land either have or require drainage, which includes 9.5 to 12.5 million hectares of cultivated land and about 13 million hectares of hayfields and pastures.[5]

Of the 1965 total drained area of 10.6 million hectares, 7.15 million hectares was agricultural land: about 3.5 million were cultivated and 3.7 were in hayfields, pastures, and fallow.[6] The usual calculation of 'drained area' in the Soviet Union includes all land lying within 100 meters (sometimes more) of a drainage canal. The distribution of land recovered by drainage is illustrated in Table 3.2.

The lands which have been reclaimed in the Georgian Republic lie in the Colchis lowland, the only major subtropical agricultural area in the U.S.S.R. They are thus of particular importance to the Soviet economy. However, in 1970 it was reported that the arable acreage in Georgia had decreased by 38,000 hectares during the preceding decade due to erosion and other poor management practices.[7]

Despite the accomplishments to date, millions of hectares of potential agricultural land still require drainage, and drainage reclamation projects continue to have high priority. For example, the reclaimed area in Belorussia alone is scheduled to be more than doubled. However, problems associated with this work have been noted. In some drained areas, forest fires have occurred in the accumulations of newly dried organic matter. Cost cutting has sometimes resulted in the necessary supplemental structures (roads, bridges, etc.) not being built, and proper coordination is often lacking. As one report put it:

Questions of land reclamation are worked out mainly from engineering and technical positions, without integrated validation by biologists, marshland specialists, agronomists and economists...

Reclamation workers bear no responsibility for the quality of work on land turned over for use. They have only one concern – to hand over hectares to the collective and state farms; what will grow on these hectares or whether anything will grow at all seems to have nothing to do with them.[8]

Table 3.2 *Areas reclaimed by drainage as of 1965*

Republic	Total agricultural land having drainage networks (thousand hectares)	Total drained area cultivated to crops (thousand hectares)
R.S.F.S.R.	2177.5	864.9
Latvia	1434.6	863.6
Ukraine	1079.4	420.9
Belorussia	1024.6	449.5
Lithuania	915.0	610.3
Estonia	394.5	154.3
Georgia	75.2	55.8
Moldavia	49.3	43.6
Total U.S.S.R.	7150.1	3462.9

From *Narodnoye khozyaystvo S.S.S.R. v 1965 g.*, p. 366.

Such deficiencies in inter-agency coordination, unfortunately, are all too familiar to conservationists the world over, and seem often to reach a maximum in water resource projects.

In addition to the drainage of water-logged lands, Soviet specialists have done considerable work on the reclamation of saline soils, both natural (which are widespread in the drier areas of the country), and those induced by irrigation. The latter, in 1962, took in about a million hectares of land, or more than a tenth of the total irrigated area.[9] Numerous other problems surround irrigation developments in the Soviet Union, and warrant a detailed discussion. Inasmuch as irrigation projects usually involve river water diversions, their associated conservation problems are discussed in Chapter 7 under the heading of water resources.

Today, it is becoming more widely understood that, except for relatively modest increases in reclaimed land, there is very little possibility of greatly expanding the arable area of the U.S.S.R., that the old idea of 'unlimited agricultural resources' is a misleading one, and that future increases in agricultural output will have to come almost exclusively from increases in input productivity.

An example in support of this might be cited. In 1960, when 28 million hectares were being plowed in Kazakhstan (20 million of them new 'virgin lands'), it was reported that 6–7 million more virgin hectares would be sown in the immediate future. However, in 1965 the sown area of Kazakhstan stood at only 30.4 million hectares (Table 3.1). In a related vein, it is interesting to note that the 1960 Russian Republic conservation law forbids the development of agriculturally marginal lands (Appendix 6).

Measures taken to increase the productivity of agricultural land are, in essence, economic responses to economic problems. However, to the extent that some of these measures might have adverse effects on the usability of other natural resources, or fail to protect adequately the soil resources which they are intended to improve, they frequently become conservation issues as well.

The decision to fertilize, for example, or to use chemical pesticides or herbicides to increase the productivity of a cultivated area is basically an economic one. But when the use of these agents poses a threat to other resources, such as wildlife or water supplies, serious conservation problems can arise. Since the most common victims of the injudicious use of pesticides are fish and wildlife, the Soviet response to the potential dangers represented by these compounds is discussed in Chapter 5, which deals with the conservation of fauna resources.

With regard to herbicides, their use appears to be increasing rapidly, as serious weed infestations have made their appearance in the virgin lands and other agricultural areas. In 1969, herbicides were applied to 28.5 million hectares in the Soviet Union, whereas in 1970 the area treated was expected to rise to 34–5 million hectares.[10] The latter figure is two and a half times as great as the amount used five years previously. In 1970, 65 per cent of all herbicide applications were carried out by means of aerial spraying.[11] No discussion of the possible mutagenic effects of certain kinds of herbicides, a subject of much concern in the United States, has yet been seen in the Soviet Press.

Other measures employed by Soviet agronomists for increasing farmland productivity include the planting of shelter belts to improve the conservation of soil moisture, crop rotation, and fallowing.

The planting of shelter belts in the drier agricultural zones of the Soviet Union has been emphasized for many years, and such work was begun long before the revolution. These broad rows of trees, stretching for dozens of kilometers across the steppes, are seen as abetting agriculture in two primary ways. First, they act as snow sheds in the winter, as year-round moisture retainers, and as modifiers of the local microclimate, thereby improving subsurface hydrologic conditions. They also help to maintain a more even protective snow cover over winter grains. Secondly, shelter belts act as a windbreak, and by reducing the velocity of the wind they serve both to reduce evaporation and to prevent dust storms.

The importance of conserving snow accumulations as a source of soil moisture is best illustrated by certain steppe areas, where snow melt-water accounts for as much as 50 per cent of the annual precipitation. If this melt-water could be effectively absorbed and retained in the soil, the moisture supply of the fields in these areas could be increased by 30–50 per cent. The common American practices of mulching and of retaining and sowing on stubble are also being urged as useful moisture-holding practices in the dry steppes and virgin lands.[12]

However, the Soviet shelter belt planting program has had serious shortcomings. Although three-quarters of a million hectares of land have been planted to windbreaks and shelter belts in the European part of the U.S.S.R., and 324,000 more hectares are scheduled to be planted in 1968–70, a leading Soviet geographer estimates that a coverage in excess of 3 million hectares is needed.[13] In addition, as a result of either improper planning or subsequent neglect, about two-thirds of the total area planted to shelter belts prior to 1956 has been lost. Even today, many of these shelter belts are not being maintained satisfactorily, while others were planted in regions too dry to permit adequate growth. So poor were the results of the early efforts that the planting of shelter belts, effective when done properly, was greatly curtailed for several years afterward. Currently, a new effort is being made to increase the tempo of this work and to see that it is done correctly, but the pace in general still remains slow.

The soil-conserving practices of contour plowing and beneficial crop rotation are ones that have largely had to be introduced into Russian agriculture during the Soviet period. In the years since the revolution, both crop rotation and the practice of fallowing have had a hard fight against opposing points of view. Fallowing in particular, especially in the newly plowed virgin lands, has often been overlooked, undoubtedly to the long-term detriment of soil productivity in these areas.[14] Although in the late 1960s, at the urging of knowledgeable agronomists, the fallow areas in northern Kazakhstan rose to 10–15 per cent of the total arable land, this is still less than half the amount left fallow in reasonably analogous areas of Saskatchewan. But the most pressing problem in both the old and the new farming regions is that of erosion.

The prevention of soil erosion and deflation

Although the foregoing problems concerning the conservation of agricultural resources are all of considerable importance within the U.S.S.R., by far the most serious single problem has been and continues to be that of erosion. About two-thirds of the cultivated land in the Soviet Union lies in areas affected by erosion. It has been estimated that 5 to 6 million hectares of sown land are damaged by wind erosion alone every year, and that on a rough average 0.5 to 1.5 million hectares of cropland annually are totally ruined.[15] In addition, these losses occur on the best agricultural soils of the country: chernozems, chestnut browns, and grey-brown podzols. A review of the erosion problem at the highest level in 1967 stated that 'the Central Committee of the C.P.S.U. and the U.S.S.R. Council of Ministers regard the struggle against wind and water erosion as one of the most important State tasks...for further developing agricultural production in the country'.[16] The prevention of erosion is a primary concern of the U.S.S.R. and republic Ministries of Agriculture, which operate a State Inspectorate for Soil Conservation. In addition, the U.S.S.R. and republic Ministries of Reclamation and Water Resources Management, the State Forestry Committee of the U.S.S.R. Council of Ministers, and, of course, the local governmental and agricultural entities themselves, are all directly involved in this work. Their efforts are supported by numerous research organizations. The systematic investigation of soil erosion in Russia was begun before the revolution by V. V. Dokuchayev, and is carried out today by a number of special experimental stations in various parts of the country. This work is also conducted at several agricultural institutes, such as those at Khar'kov, Poltava, Gor'kiy, and Cheboksary, and at such soil institutes as the ones in Khar'kov, Tbilisi, and Tashkent. The leading journal of soils research in the U.S.S.R. is *Pochvovedeniye*, published monthly by the V. V. Dokuchayev Institute of Soil Sciences of the U.S.S.R. Academy of Sciences. It is translated into English under the title *Soviet Soil Science*, and many issues contain articles on erosion control, reclamation, and other subjects of a conservation nature.

Although the results of wind and water erosion are often found in the same region, the causes of these two types of erosion are sufficiently independent of one another so that they may be considered as separate phenomena.

The effects of water erosion – networks of gullies and ravines – are widespread over much of the steppe, forested steppe, and mixed forest zones of European Russia, as well as in certain areas of Siberia (Figure 3.2). Early estimates of the total area affected by water erosion in the Soviet Union varied between 50 and 70 million hectares, and more recent

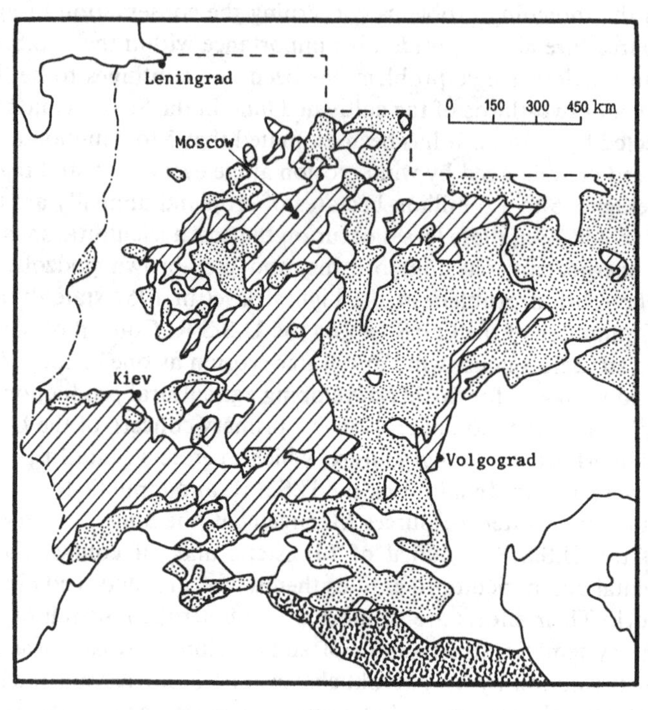

Figure 3.2. Extent of gully erosion in European Russia. From N. I. Sus (1949) p. 43.

estimates remain in the same range. Of these 50 to 70 million hectares, 30 to 35 million, or at least half, are on arable land.[17]

Both natural conditions (high local relief, light soils, unpredictable rainfall) and poor agricultural practices (plowing or grazing on steep slopes, neglect of incipient gullies, improper cropping practices, down-slope plowing, etc.) have contributed to the present state of water erosion on the croplands of the Soviet Union.

It is believed that really serious erosion problems began in Russia during the last century, particularly following the 1861 emancipation, when freed serfs farmed on steep slopes and in commune strips that ignored surface configurations. Although the problem was serious at the time of the revolution, little attention seems to have been devoted to it by the Soviet regime until the late 1930s.[18] Even then, the priority of maximum agricultural output limited work on the problem to little more than studies, at least until the advent of the short-lived 'Great Plan for the Transformation of Nature' in the late 1940s. Since Stalin's death, official concern for implementing erosion control measures has been high, yet the problem remains (Figures 3.3 and 3.4).

Figure 3.3. Ravine gullies in Armenia.

Figure 3.4. Hillside erosion along the Tsimlyansk Reservoir near Volgograd.

Land and soil resources

The Institute of Geography of the U.S.S.R. Academy of Sciences has recently completed a survey in which all lands of the Soviet Union are classified according to their incidence of, and susceptibility to, water erosion. Based on this study, it was estimated that almost two-thirds of all arable land on the country's state and collective farms lies in the zones categorized as suffering at least to some degree from water erosion, and over 40 per cent has 'significant' erosion problems.[19] The results of this survey are presented in Table 3.3.

Table 3.3 *Classification of U.S.S.R. state and collective farm land according to incidence of water erosion*

Class of land[a]	All state and collective farm land (per cent)	All arable land (per cent)	Gardens, orchards, vineyards, etc. (per cent)	Natural pasture lands (per cent)	Forests and scrublands (per cent)
1	38.0	35.0	27.2	42.0	31.2
2	27.0	1.5	—	27.0	53.0
3	2.2	0.6	0.3	4.0	0.5
4	2.2	1.8	—	3.0	1.6
5	11.3	20.0	8.0	9.8	8.5
6	8.0	13.5	7.0	7.5	2.6
7	10.0	25.8	44.5	5.4	2.2
8	1.3	1.8	13.0	1.3	0.4
	100.0	100.0	100.0	100.0	100.0

[a] Explanation of classes: 1 = no actual or potential erosion; 2 = no actual erosion but area is susceptible to it; 3 = areas with very little actual water erosion; 4 = areas with light (*slabyy*) erosion; 5 = areas with moderate (*umerennyy*) water erosion; 6 = areas with significant (*znachitel'nyy*) erosion; 7 = areas with heavy (*sil'nyy*) erosion; 8 = very heavily eroded areas.

From Sil'vestrov (1963) pp. 121–6.

Restoring heavily eroded lands is a tedious and costly process, and while some resources are being assigned to this end, the main thrust of the anti-erosion campaign is being directed toward preventing any widespread future erosion. The fight to ensure that no more of the nation's valuable farm land is unnecessarily lost in this way is considered to be of the highest priority. The 1967 study of the erosion problem referred to previously recommended that increased attention be given to the following practices:

Planting of agricultural crops at right angles to slopes; contour plowing...soil protecting crop rotation; the strip arrangement of agricultural crops; the sowing of steep slopes to grass; the cultivation of forest belts; the afforestation of gullies,

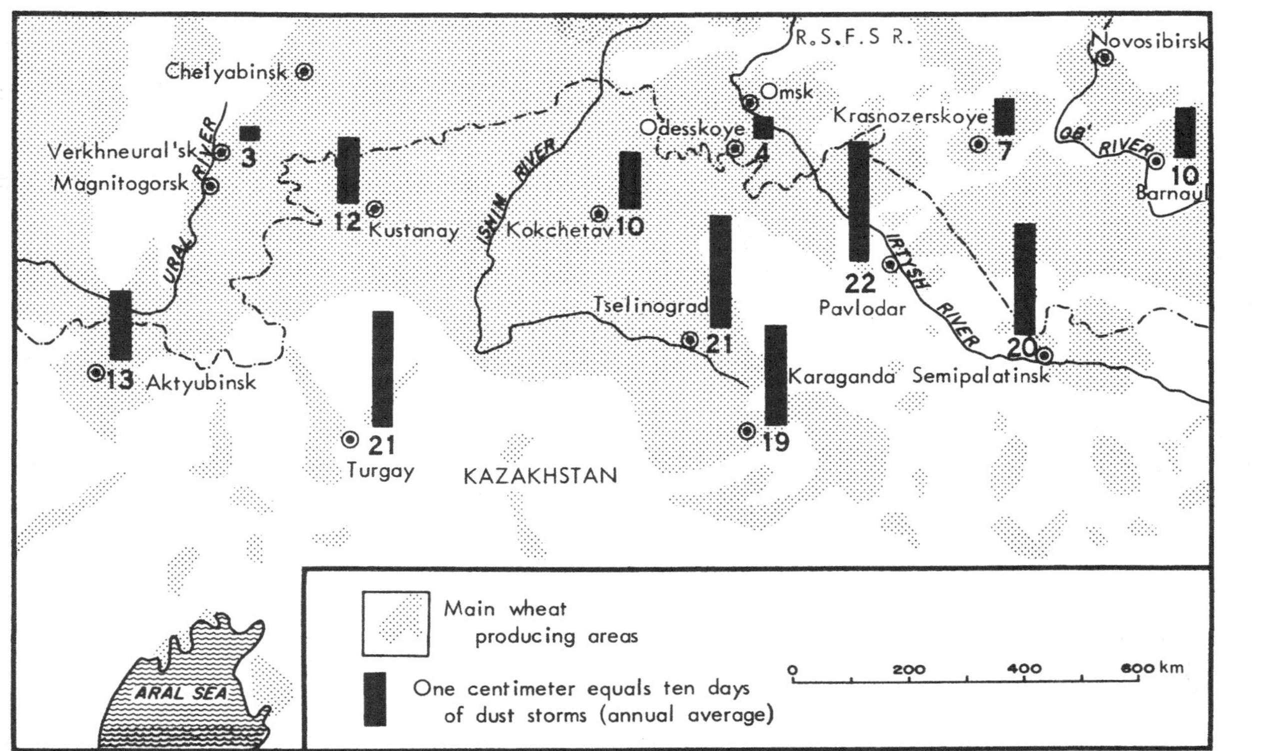

Figure 3.5. Frequency of dust storms in Kazakhstan and Western Siberia. From Table 3.4 and *Atlas sel'skogo khozyaystva S.S.S.R.* (Moskva, 1960) p. 106.

ravines, deserts, and the banks of rivers and reservoirs; and the construction of erosion-control hydrotechnical installations...[20]

To achieve these ends, for the period 1968–70, the resolution authorizes work to plant shelter belts on 324,000 hectares; to arrest and afforest gullies, ravines, and sand dunes over 827,000 hectares; to terrace steep slopes on 89,000 hectares; and to build erosion-control hydrotechnical and flood control structures costing an estimated 188 million rubles.

Whereas gullying acts slowly to decrease the amount of available arable

Table 3.4 *Incidence of dust storms in Western Siberia and Northern Kazakhstan*

	Number of days with dust storms			
Station	Average for April–October	Maximum, April–October in any one year	Average in maximum month	Month of maximum (statistical average)
Verkhneural'sk	2.7	13	0.9	June
Odesskoye	4.1	18	1.4	May
Krasnozerskoye	6.8	19	2.4	May
Barnaul	9.5	31	3.0	May
Kokchetav	10.3	37	2.9	May
Kustanay	12.0	37	3.0	May
Aktyubinsk	12.6	34	2.8	June
Karaganda	18.5	38	4.0	June
Semipalatinsk	20.1	37	4.7	June
Tselinograd	20.5	53	5.0	July
Turgay	20.9	56	3.7	July
Pavlodar	21.6	49	6.4	May

From K. F. Zhirkov, 'O pyl'nykh buryakh v stepyakh Zapadnoy Sibiri i Kazakhstana', *Izvestiya Akademii nauk S.S.S.R. seriya geograficheskaya* (1963) No. 6, p. 52.

land year by year, wind erosion poses an immediate threat to vast areas of crops every year in the Soviet Union.

Almost annually in the Kazakh Republic, the steppe areas of Western and Eastern Siberia, the Southern Ukraine, the North Caucasus, and a number of other regions of the country, dust storms blow away the fertile layer of the soil, damage and destroy crops over substantial areas, and, in some areas, cover up irrigated land, irrigation canals, and water sources.[21]

It has been estimated that the area of cropland subject to deflation – the process of surface soil being blown away by the wind – is in the order of

20–25 million hectares. An additional 30–35 million hectares of grazing land is in this category, for a total agricultural area of 50–60 million hectares endangered by deflation. In addition, no less than 51 per cent of the arable land of the U.S.S.R. lies in areas where the soils are at least potentially vulnerable to wind erosion.[22]

Among the areas most highly susceptible to dust storms are the agricultural lands of Western Siberia and Northern Kazakhstan, the same area that was developed in the 'Virgin Lands Program' of the 1950s. Certain portions of this area experience, on the average, more than twenty days

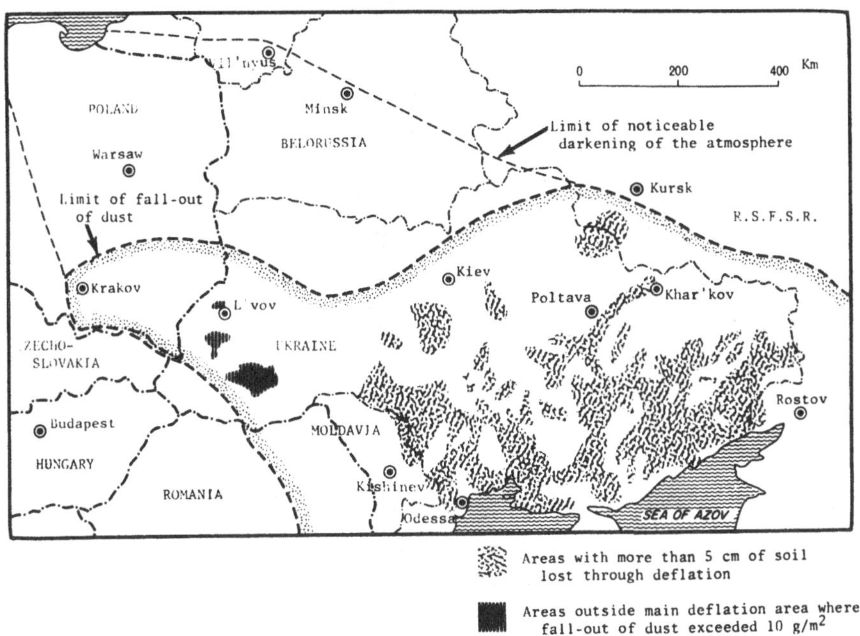

Figure 3.6. Area affected by the dust storm of April 26 and 27, 1928. From N. I. Sus (1949) p. 272.

with dust storms per year (Figure 3.5). The data in Figure 3.5 are based on twenty-five years of observations, and are elaborated upon in Table 3.4. All of the stations lie in or immediately adjacent to important grain-growing regions.

The drier steppe regions of the Ukraine and the North Caucasus are also very prone to dust storms, and particularly devastating ones hit here in 1928 and 1960. The 1928 storm removed at least 5 centimeters (2 inches) of soil from approximately a quarter of the entire land surface of the Ukraine, and this was redeposited over a vast area extending well into Poland (Figure 3.6).

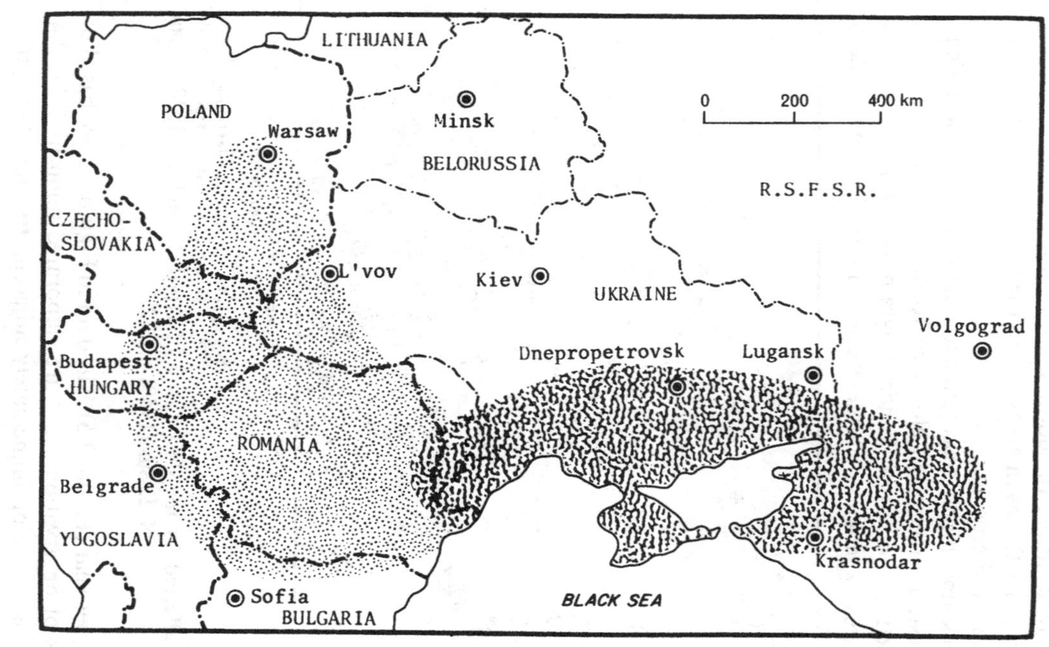

Figure 3.7. Areal extent of the dust storm beginning April 7, 1960. From I. V. Kravchenko (1961) p. 68.

The 1960 storm hit in the North Caucasus as well as the Ukraine, and damaged more than four million hectares of crops, of which 612,000 hectares were completely destroyed.[23] The effects of this storm at one collective farm in Krasnodar Kray, where dust reached 1 to 1.5 meters (3 to 5 feet) deep and covered some of the smaller buildings, was dramatically written up in the government press.[24] The fallout of dust from this storm was felt as far away as Warsaw, Budapest, and Belgrade (Figure 3.7).

Other dust storms occurred in the Kuban'–southeastern Ukraine region in 1964, 1965, and 1969. The 1969 storm in the Kuban' was fairly severe, stripping some areas of 20 cm (8 inches) of topsoil. In the downwind region, 4.5 million cubic meters of dust fell on the Armavir State Farm alone, burying 375 hectares of its plowland.[25] Less severe dust storms were reported in the same area in the spring of 1970.

As with water erosion, dust storms are the product of natural conditions which are often worsened by agricultural practices. Arid conditions, light soils, occasional drought, and the *sukhovey* combine to produce periodic dust storms even in areas undisturbed by man. The *sukhovey* is an extremely desiccating weather condition characterized by a steep rise in temperature and a sharp drop in relative humidity. These conditions may last for days, and can, in the most severe cases, completely kill young crops if artificial watering is not available. Moderate to strong winds usually accompany them.[26] Under such natural conditions, the plowing of the thick native grasses of the steppe could not help but contribute to the formation of dust storms.

Many of the measures recommended for reducing the frequency, severity, and effects of dust storms have been mentioned already. These include the planting of shelter belts, fallowing, and the use of crop rotations and sowing schedules which will reduce exposed soil and conserve soil moisture. Steps such as these taken to date have reportedly decreased the frequency of dust storms in the Virgin Lands region of Kazakhstan in recent years, although the role that the cyclical vagaries of climate might have played in this is unknown. Other anti-deflation measures which are still needed are the planting of heavily eroded areas and shifting sands to grass, the decreased use of moldboard plowing, the increased use of supplemental irrigation, and the use of buffer zones of perennial grasses.[27]

It might well be, however, that the only really satisfactory measure to prevent serious future deflation in such regions as the Virgin Lands of Kazakhstan would be to remove the more marginal fields from cultivation entirely. Bitter a pill as this might be for Soviet planners to swallow in the short-run, it could prove to be a long-term step worthy of serious consideration.

Wind and water erosion are among the most urgent contemporary conservation problems facing Soviet planners. The control of these debilitating processes, like most other aspects of improving the agricultural situation in the U.S.S.R., will probably require greater capital outlays and more man-hours of effort than the state has been willing to allocate in the past. But with the acceptance of the realization that there is really little other choice, it is probable that increased attention will henceforth be given to these fundamental problems.

The question of evaluating Soviet land resources

The assumption of Marxian economics that only labor produces value has had many unfavorable repercussions on natural resource conservation in the Soviet Union. This assumption leads Marxists to the logical conclusion that under their economic system land, water, and all their associated resources, as they exist in their natural state, should be considered as 'free' inputs into production. It is recognized, of course, that these resources vary greatly as they are found in nature, and both quantitative and qualitative inventorying is carefully carried out. However, no price per ton or per hectare is assigned to a resource in its natural state, since no one has to buy or lease these resources before they can be developed.

The consequences of this assumption can be easily foreseen. For example, if the land above a mineral deposit has no intrinsic ruble value, there is little economic reason to rehabilitate it after the mineral has been extracted. Likewise, there is no incentive to conserve low grade ore, or to avoid waste in timber harvesting, if the lost portion of the resource has no *in situ* value to the persons developing it.

Another aspect of this problem arises when land is withdrawn from agricultural uses for construction. In many cases, this is very productive land. N. V. Mel'nikov, chairman of the Commission for the Study of Productive Forces and Natural Resources, states that in 1965 alone 129,000 hectares of arable land were used for construction.[28] For 1967, the amount of arable land thus withdrawn from agriculture was reported to be almost three times as great – 335,000 hectares.[29] As much as a million hectares has been suggested as the total amount annually withdrawn from the agricultural and forestry sectors of the economy for construction purposes. Another major cause of lost agricultural land are the extensive reservoirs which fill behind large hydroelectric projects, drowning in some cases thousands of square kilometers of prime arable land (Table 7.1). Although similar losses of agricultural land occur in the United States, this land must at least be purchased at a price which reflects its present usefulness in the economy by those who believe its non-agricultural use will result in still greater economic gains. Such is not the case in the Soviet Union, where there exists at present no effective means for demonstrating a probable higher economic return for an alternate use of land.

In recent years, this problem has been increasingly discussed by Soviet economists, geographers, and other specialists in natural resource utilization. The normal avenue of approach is to stress the need for a complete, detailed national land cadastre, which would classify and evaluate in terms of agricultural potential the entire land surface of the country. However, the manner in which this economic evaluation would be expressed would not be as a price for the land, but rather as a 100-point scale, showing the relative agricultural potential of a given tract of land to any other.[30] Other suggestions have been made for evaluating all natural resources on such a 100-point scale.[31] While useful for certain purposes, such point-scale evaluations still do not assign any cost to wasted natural resources, and suggest no monetary penalties for the unnecessary 'depreciation' of the natural landscape.

It is possible that this problem could be resolved without resorting to the pricing of land, a solution that would be more acceptable to Marxist political theorists. It could be done, for example, by a very thorough set of resource management regulations, enforced (rigidly) by a much more severe schedule of fines and penalties than exist at present. More comprehensive land use planning would be required, as well.

However, the weaknesses in the basic assumption that land and other natural resources are 'free' are apparently considered to be so strong that serious debate is being conducted in the Soviet Union with regard to altering this assumption. It was suggested at a recent discussion on natural resource conservation that

> our economists have not yet abandoned one thesis which bars the way to legislation for the protection of nature and prevents us from adopting the attitude of a thrifty householder in our dealings with nature.
>
> This is the thesis concerning the estimation of natural resources in terms of value.
>
> As long as they are regarded as free and are not incorporated under the heading of the country's national wealth, the wastage of those resources will continue with impunity.[32]

The idea of charging some form of fee for the use of land has been debated vigorously in the Soviet press over the past few years, and numerous spokesmen have declared their support for this concept.[33] Nevertheless, proposals for making the users of land resources pay a fee for this privilege, although given consideration at the highest levels, continue to be rejected on ideological grounds. Thus, the 1968 U.S.S.R. Land Legislation Act explicitly reaffirms in Article 8 that the use of land in the Soviet Union is granted free of charge to state and collective farms, other enterprises, and individuals (Appendix 5).

This being the case, the use of regulations and penalties will remain as the basic means of enforcing proper conservation practices in the U.S.S.R. The need to raise these means to a higher level of effectiveness than exists at present (a need documented in subsequent chapters of this study) would seem to be mandatory whether a charge is made for the use of land and other natural resources or not. The fact that land is priced in the United States has not prevented blight, scarred landscapes, and pollution, and pricing by itself would seem unlikely to prevent them in the Soviet Union. But the pricing of Soviet natural resources *would* be useful as a monetary measure of wastage, and as a basis for more accurately levying penalties for such wastage. Since increased personal profits (except in the form of bonuses) cannot serve as the motivation for conserving natural resources in the U.S.S.R., the threat of sanctions must. But such sanctions need to be strict enough to be effective. The pricing of Soviet natural resources might have represented a way of accomplishing this, but since this method has been ruled out, other means of insuring the effectiveness of the penalties must now be devised. This represents a very basic problem which the Soviet Union must resolve if it is to guarantee the most efficient long-term utilization and conservation of its diverse natural resources.

4 *Zapovedniki* in the Soviet Union

...territories forever withdrawn from economic utilization...

In the fifty years since the Soviet government assumed power in Russia, it has developed a nation-wide network of natural areas devoted to the study and preservation of biotic resources. These areas are known as *zapovedniki* (singular: *zapovednik*, which literally means a forbidden or restricted area) and are somewhat similar to American national parks but place much less emphasis on tourism. In 1966 there were sixty-eight of them, embracing a total area of approximately 4,300,000 hectares, or about 10,600,000 acres (Figure 4.1 and Appendix 7). More have reportedly been created since then. They represent the most important territorial entities employed in the effort towards nature conservation in the Soviet Union, and as such are worthy of close examination.

The nature and history of *zapovedniki*

Soviet *zapovedniki*, or nature preserves, are basically protected natural areas set aside for research in the natural sciences. For this reason it is not possible to make a precise analogy between them and any particular type of American land management institution; some most closely resemble the American concept of a national monument, others are more like wildlife refuges, and a few resemble our national parks. Hunting, fishing, and timber harvesting are forbidden in all of them, and many have large areas of *de facto* wilderness. A list of activities which are prohibited in Soviet *zapovedniki* is presented as Appendix 8. Almost all have scientific research facilities connected with them, some of which are quite extensive.

This latter characteristic represents the main distinguishing feature of a Soviet *zapovednik*, as compared to a national park or a national wildlife refuge in the United States. The laws authorizing the establishment of *zapovedniki*, in defining them, underscore their importance as research facilities. These areas are specified as being 'forever withdrawn from economic utilization, for scientific research and cultural-educational purposes' (see Appendix 9). Typical fields of research include various aspects of zoology, botany, forestry, pedology, ichthyology, ornithology, parasitology, and many other more specialized areas.

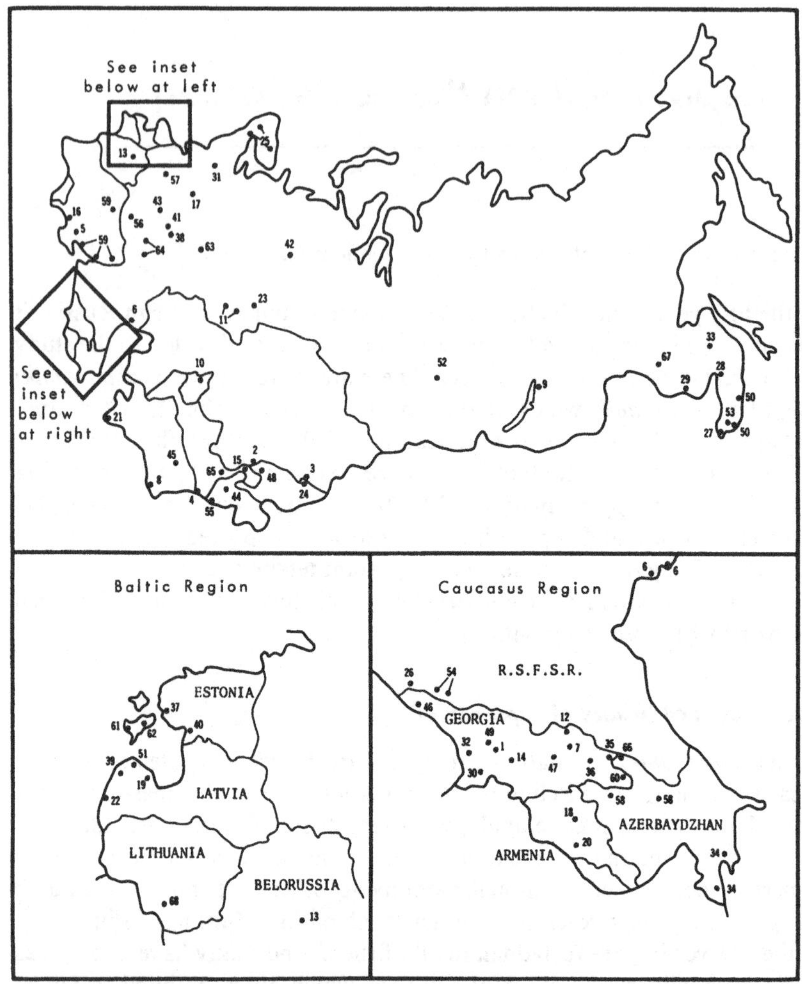

Figure 4.1. Distribution of *zapovedniki* in the U.S.S.R. (1966). From Bannikov (1966). Numbers correspond to *zapovedniki* as listed in Appendix 7 (Part 1).

Not only do most *zapovedniki* have their own research laboratories, scientific staff, and natural history museums, but specialists from the Academy of Sciences, universities, and other scientific institutions often come to work at these preserves as well. Many of them publish their own research journals, and some are well known for their scientific literature.

The use of these preserves for biological and ecological research thus represents one of the main justifications for withdrawing such areas from economic development. It goes without saying, however, that a good portion of the research carried out in these preserves has direct applications in the national economy. Examples of this would include the improving of techniques for breeding and disseminating economically useful animals, the control of predators, research on plant adaptation, the control of infestations, and many other studies of a similar nature.

Many *zapovedniki*, however, have been created primarily to preserve some unique or threatened species of plant or animal life, and the basic role of all of them in the preservation of natural objects and landscapes should not be understated.

Soviet writers like to describe *zapovedniki* as 'standards of nature' (*etalony prirody*), to which similar areas subject to changes brought about by man's activities upon them can be compared. Thus, these preserves can be thought of in a sense as the control portion of a controlled experiment – the effect of man upon the natural environment.

The following represent some of the most important functions of *zapovedniki*, though not necessarily in any particular order of priority:

(1) to propagate species of animals that are rare or threatened with extinction, such as the European bison, sable, flamingo, beaver, fur seal, kulan, goral, desman, spotted deer, and other wildlife (undertaken, for example, at the Kavkaz, Belovezhsk, Prioksko-Terrasny, Badkhyz, Tigrovaya Balka, and many other preserves);

(2) to protect and study rare, unusual, or useful trees or other vegetation (Turianchay, Adzhameti, Khosta exclave of Kavkaz, Batsarskiy, Badkhyz, and many other preserves);

(3) to protect and study nesting or wintering areas of birds (Matsalu, Astrakhan', Kyzyl-Agach, Kandalaksha, Gasan-Kuli, Zhuvintas, and other preserves);

(4) to preserve as 'standards' and to study landscapes typical of the various natural zones of the country (Sikhote-Alin, Ukrainian Steppe, Central Chernozem, Repetek, Barsa-Kelmes, Pechora-Ilych, and other preserves);

(5) to study the total or partial ecology of a given area (the Darvin preserve was established primarily for this purpose; the Barsa-Kelmes, Astrakhan', Kandalaksha, and many other preserves are engaged to a large degree in ecological research);

(6) to preserve unique geological or archaeological features (Stolby, Il'men, Satapliyskiy, Bashkir, Garni, Borzhomi, and others. This function is less common than it is in the case of American national parks and monuments);

(7) to breed or improve useful species of animal or fish life, often for transplantation elsewhere (Barguzin, Berezina, Voronezh, Pechora-Ilych, Askaniya-Nova, and other preserves);

(8) to accommodate tourism; these preserves are referred to as 'open' *zapovedniki* (Kavkaz, Il'men, Ritsa, Teberda, Stolby, Askaniya-Nova, and a limited number of other preserves).

Any particular *zapovednik* may have any one, or a combination, of the above as its primary function(s), with others as a secondary purpose. Only a few are presently open to large-scale tourism, but many have museums which can be visted by the general public, and most are open to special delegations, students engaged in research, and other special categories of users.

Zapovedniki are established by means of a decision of the Council of Ministers of either the U.S.S.R. or one of the Union republics. Their administration is not centralized, either. Of the sixty-six preserves existing in 1964, twelve were administered by the Central Directorate for Conservation, Preserves, and Hunting of the U.S.S.R. Ministry of Agriculture; forty-three were the responsibility of some agency under the Council of Ministers of the republic in which they were found; ten were controlled by the Academy of Sciences of the U.S.S.R., one of its regional affiliates, or the Academy of Sciences of a Union republic; and one (the Kivach preserve) by the Autonomous Republic in which it is located.

Proposals for new *zapovedniki* are occasionally brought before the Council of Ministers by influential groups. A noteworthy example of this occurred in 1957, when the Commission for the Conservation of Nature of the U.S.S.R. Academy of Sciences compiled a detailed long-range plan for the establishment of new preserves throughout the Soviet Union.[1] This plan called for a total of ninety-nine preserves which would have embraced about 16 to 17 million hectares. This would have nearly doubled the then existing number of preserves, and increased the protected area by more than ten times. Many of the new preserves would have been in parts of Siberia and the steppe areas of Kazakhstan and European Russia where there are now no preserved sections of the natural landscape. Although this body was perhaps the most influential one existing at that time which could have made such a proposal, its recommendations have received little action in the past decade. Despite the fact that many new preserves have been created (and abolished) since 1957, only a very few of this report's recommended new areas have actually been established.

Restricted natural areas have existed on the territory of the Soviet Union over a period of many centuries. Often, their establishment was not for the conservation of nature, but rather reflected other objectives, such as national defense against mounted invasion. In other cases, forest areas were placed off-limits to unrestricted hunting, although many of these areas were used as private hunting reserves for the aristocracy, especially in the mixed and deciduous forests of European Russia and in the Caucasus.

Some, such as Belovezhskaya Pushcha in western Belorussia, have been maintained as some classification of forest preserve for many hundreds of years. Other areas, such as large church-owned estates, existed in a basically protected status as well.

The impetus for setting aside natural areas for scientific purposes came towards the end of the nineteenth century. By this time the inevitable results of centuries of exploiting Russia's natural resources (particularly soil, game animals, and forests) were becoming quite evident. Prominent men in the natural sciences, such as V. V. Dokuchayev, G. A. Kozhevnikov, and I. P. Borodin, were among the first to express the need for setting aside a network of unspoiled natural areas in perpetuity. At the Twelfth Congress of Naturalists in 1909, upon the urging of Borodin, a list of suggested areas for establishing nature preserves was drawn up. In 1915 a similar report was prepared by the Nature Conservation Commission of the Russian Geographical Society, and on October 2, 1917, the noted Russian explorer and geographer V. P. Semenov-Tyan-Shanskiy presented a paper entitled 'On the types of locales in which it is necessary to establish *zapovedniki* analogous to American national parks', in which he suggested suitable areas in forty-six different geographic zones of the country.[2]

Some results were realized, and by the time of the revolution, several preserves had been either established or authorized. Among them were the Barguzin Sable Nature Preserve on Lake Baykal, created to revive the exhausted sable industry (1916); the Lagodekhi preserve in the mountains of eastern Georgia, organized by the Academy of Sciences in 1912; the Suputinka (1911) and 'Kedrovaya Pad' (1916) preserves, both near Vladivostok in the Far East; a Sayan preserve in Central Siberia (no longer in existence); Askaniya-Nova in southern Ukraine (established as a private zoological garden in 1874); and Moritssala in Latvia. In addition, the Kuban' and Ramon (Voronezh) reserves, Belovezhskaya Pushcha, and other tsarist hunting reserves, which are now either *zapovedniki* or state hunting reserves, were in existence prior to 1917. However, due to boundary changes, Belovezhskaya Pushcha and Moritssala were not on the territory of the Soviet Union between 1918 and World War II. In 1916, the same year that the National Park Service was established, the first law in Russian history authorizing the creation of *zapovedniki* was enacted, largely as a result of efforts by the Russian Geographical Society.

Shortly after the revolution, two important new preserves were created. Despite the prevailing internal upheaval, the Astrakhan' preserve in the Volga delta was established in 1919, and the Il'men preserve in the Urals in 1920. In 1921 Lenin signed a decree calling for a Baykal Natural Reserve, which would have been located in the same general area as the present Barguzin preserve, but would have been much more extensive in area than the one actually created in 1926. In 1921, the Central Forestry Department of the Agricultural Commissariat placed large areas of forests off-limits to logging, calling them 'territories of future national parks and monuments of nature'.[3]

Such actions may have been due in part to the fact that in these early days of the Soviet Union conservationists had friends in high places. For example, P. G. Smidovich, who was responsible for the creation and direction of many of the early preserves, was chairman of the Moscow Soviet, and later was a member of the Presidium of the All-Union Central Executive Committee (VTsIK). In addition, Lenin himself seems to have had an appreciation of the need for protected natural areas, and for conservation in general.

On September 16, 1921, Lenin signed a decree 'On the Preservation of Natural Monuments, Gardens, and Parks', thus giving official status to the emerging system of *zapovedniki* (Appendix 10). In 1924 they were given a more formal legal status through concurrent action by the Council of People's Commissars and the Presidium of the All-Union Central Executive Committee (VTsIK), at that time the highest legislative body of the country.[4] This act established penal procedures for violations of the protected features of *zapovedniki.* Another decree taken the following year further elaborated upon the establishment of these natural preserves. Their protected status was again emphasized in Article 40 of the 1968 U.S.S.R. Land Legislation Act (Appendix 11).

Until 1933 the protection of natural areas and objects was carried out by an 'Interdepartmental State Committee on Conservation'. In 1933 this body was replaced by the 'Committee on *Zapovedniki* of the Presidium of the VTsIK', which after 1938 fell under the Council of the National Economy of the Russian Republic, rather than under the All-Union VTsIK. To compliment this move, in 1939 similar agencies were established in the other republics. Following the 1951 reorganization of the preserves, twenty-eight of the remaining forty were being administered by the 'Central Directorate for *Zapovedniki*' under the U.S.S.R. Council of Ministers, eleven were being run by the Academy of Sciences, and one by a separate research institute. Subsequently, their management has fallen increasingly to agencies under the Councils of Ministers of the Union republics, as noted above.

Within this framework, the number of *zapovedniki* in the Soviet Union grew rapidly. In 1937 there were 37, in 1947 91, and by 1951 the system had grown to 128 preserves covering an area of about 12,500,000 hectares. In 1951, however, a major revision of the system was carried out, and over two-thirds of the preserves and seven-eighths of the protected area was abolished (Table 4.1).

The reasons for this large reduction are not known. One possibility is that some of these areas, particularly the larger among them, may have been wanted for economic development during the Fifth Five Year Plan (1951–5). In any event, several of the most extensive preserves, such as the Sayan (about 1,200,000 hectares, an area larger than Yellowstone), were terminated at this time, and others, such as the Sikhote-Alin, were greatly reduced in size.

The number of *zapovedniki* gradually increased again during the 1950s, until by the end of 1960 there were ninety-three with a combined area of 6,360,000 hectares. In 1957, the Commission on Conservation of Nature of the U.S.S.R. Academy of Sciences put forth their proposal for 'rounding out' the nationwide system of preserves. However, as noted above, very few of their recommendations have been realized.

Table 4.1 *Number and area of* zapovedniki

Year or date	Number	Area (hectares)	Area (acres)
1920	about 7	?	?
April 1937	37	7,138,300	17,638,739
1947	91	*ca.* 12,000,000	*ca.* 30,000,000
1951	128	*ca.* 12,500,000	*ca.* 31,000,000
1952	40	1,465,668	3,621,665
May 1 1957	52	1,554,000	3,840,000
Sept. 1 1959	76	3,296,000	8,140,000
1960	90	5,764,500	14,240,000
Jan. 1 1961	93	6,360,000	15,700,000
Mar. 1 1964	66	4,267,400	10,544,745
Jan. 1 1966	68	4,621,600	11,420,000
mid-1968	79[a]	*ca.* 6,400,000	*ca.* 15,800,000

[a] Includes the 68 shown in Figure 4.1, plus five areas reclassified from other types of preserved status, four former *zapovedniki* which have been reactivated and two entirely new ones (see Appendix 7, Part 3, and Figure 4.2).

Sources: 1920: text; 1937: Brouwer, 1938; 1947: Makarov, 1947; 1951: Lavrenko *et al.*, 1958; 1952: *Bol'shaya Sovetskaya Entsiklopediya*, 2nd edn, 1952; 1957: Lavrenko *et al.*, 1958; 1959: Kirikov, 1960; 1960: *Kratkaya geograficheskaya entsiklopediya*, 1961; 1961: *List of National Parks and Equivalent Reserves*, 1962; 1964: *Zapovedniki S.S.S.R.*, 1964; 1966: Bannikov, 1966; 1968: Bannikov, 1969 and Belousova *et al.*, 1969.

On June 10, 1961, a resolution by the Council of Ministers of the U.S.S.R. called for another reorganization of the network of *zapovedniki*. Although fewer preserves were abolished at this time than in 1951, two of the ones that were – the Altay and Kronotskiy preserves – were the two largest then in existence. With a combined area of about 1,880,000 hectares between them, they alone accounted for almost 90 per cent of the reduction in total area that took place. These two preserves, together with the Kemeri and Gumista preserves, had all been abolished in the 1951 reduction, and had all been newly re-established for less than five years when the decision to abolish them again was made. The disposition of the then existing ninety-three preserves as a result of the 1961 reorganization is presented in Table 4.2, and those undergoing a change in status are listed in Appendix 7, Part 2.

Table 4.2 *The 1961 reorganization of Soviet* zapovedniki

Category	*Zapovedniki* existing before 1961	*Zapovedniki* existing in 1964
Remaining as *zapovedniki* after 1961	61	61
Combined into other existing *zapovedniki*	6	0
Combined into a new *zapovedniki*	4	1
Reclassified as hunting reserves	5	0
Apparently dropped as *zapovedniki*[a]	17	0
Created in 1963	0	4
Total, as in Table 4.1	93	66

[a] It is possible that some of these may have been reclassified as *zakazniki* (temporary or partially restricted natural areas), or as local (republic or oblast level) *zapovedniki*. Some have since been reactivated as *zapovedniki*.

From *Zapovedniki S.S.S.R.* (1964); Bannikov, *Po zapovednikam Sovetskogo Soyuza* (1966); *Kratkaya geograficheskaya entsiklopediya* (1961).

Two works which appeared in 1969 indicated that several new additions to the system have been made.[5] Eleven new areas were apparently added between 1966 and 1969, although only two of these had never been *zapovedniki* before. Of the others, five were elevated to full *zapovedniki* status after having been parts of other preserves or in a different category of preserve, and four were former *zapovedniki* which had been abolished in 1961. In addition, there were in 1969 four state hunting preserves (the Russian term for these is *zapovedno-okhotnich'ye khozyaystvo*) which in the past had been *zapovedniki*, and ten other areas which were under consideration for *zapovednik* status (Appendix 7, Part 3, and Figure 4.2).

The combining of two or more non-contiguous preserves, such as took place in 1961, is not uncommon. Although the United States has examples of non-continguous preserves in its national park system (Saguaro National

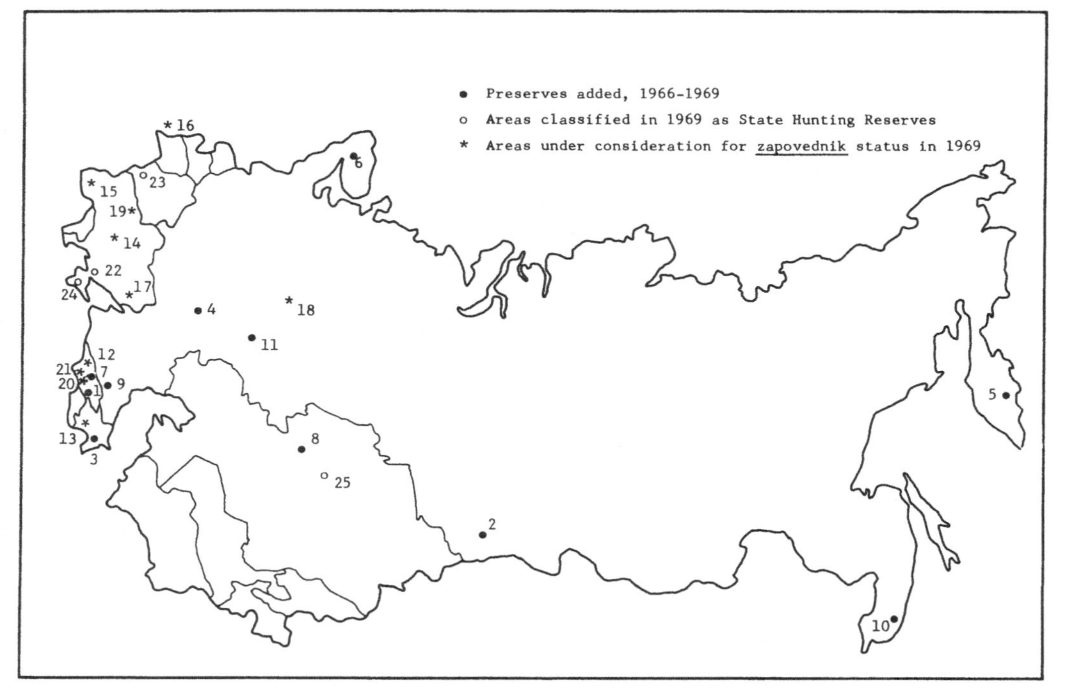

Figure 4.2. *Zapovedniki* added, or under consideration, 1966–9. Data from Appendix 7 (Part 3), to which the numbers refer.

Monument, North Cascades National Park, etc.), such units in the Soviet system may be quite far apart. For example, two sections of the new Ukrainian Steppe preserve are separated by a distance of nearly 500 kilometers.

In addition to the 1957 plan, other proposals have been put forth for new *zapovedniki*. For example, in 1963, S. M. Uspenskiy, a leading researcher in arctic conservation problems, suggested that:

In the Soviet arctic very valuable game preserves could be organized in the regions of the Indigirka delta and in the central Taymyr, since these regions possess a very wide variety of landscapes and an extremely interesting and varied fauna and flora. It is essential to establish a number of state nature preserves in the arctic, especially in areas where the lairs of polar bears and the nests of snow geese are most numerous. One such preserve should be planned for Wrangel Island. Another nature preserve should be organized on Tit-Ary Island.[6]

Profile of a prominent *zapovednik* – the Kavkaz

In order to elaborate upon the nature and functions of *zapovedniki*, a closer examination of the natural environment and operation of a major Soviet preserve might be of value. For this purpose, the Kavkaz (Caucasus) preserve, situated in the northwest portion of the mountain range after which it is named, will serve well.

The Kavkaz *zapovednik* is located in southern Krasnodar Kray, almost adjacent to the northwest corner of the Georgian Republic (Figure 4.3). A small (300 hectares) grove of old yew and box trees near the resort town of Khosta is a non-contiguous administrative unit (*filial*) of the preserve.

The preserve is located within an area established in 1888 as the private Kuban' hunting reserve. In 1909 it was recommended by the Academy of Sciences for development as a *zapovednik*, but not until May 12, 1924 was this protected status granted to it. In 1930 the Khosta grove was also set aside as a preserve. The purpose of this *zapovednik* is to 'protect, study, properly utilize, and renew the natural riches of the northwest Caucasus'.[7]

Both the boundaries and the area of the Kavkaz preserve have undergone revisions. In 1933 its area was given as 350,000 hectares, in 1947 as 338,000, in 1952 as 99,600, in 1960 as 251,800, in 1964 as 262,500, and in 1966 as 266,000 hectares.[8] Several major changes in its boundaries seem to have taken place (Figure 4.3).

The preserve straddles the Caucasus divide, with the major portion of it lying within the lateral ridges extending north from the high divide. The relief varies from about 800 meters up to 3,360 at Smidovich Peak (2,600–11,000 feet). Many of the higher peaks are glacier-covered.

The climate is quite diverse, even if vertical zonation is disregarded. The Caucasus mark the boundary between the warm, moist, sub-tropical Black

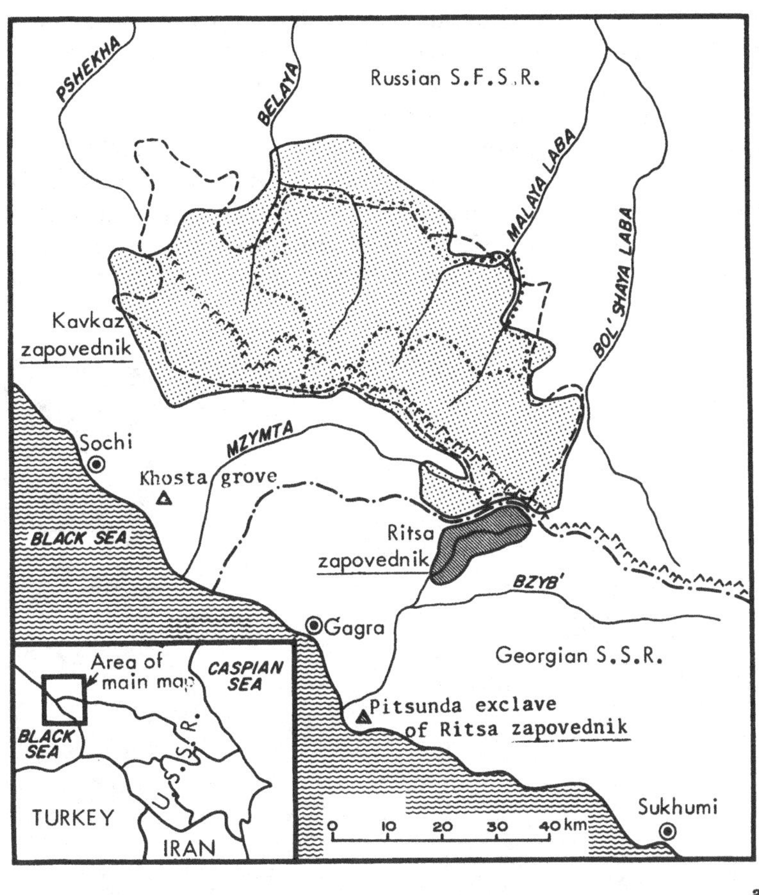

Boundaries of Kavkaz zapovednik as shown in 1967[a]
Probable 1952-1961 boundary of Kavkaz zapovednik[b]
1951 boundary of Kavkaz zapovednik[c]
Boundary between the Russian and Georgian republics
Caucasus drainage divide

Figure 4.3. The Kavkaz *zapovednik*. [a]From *Geografiya v shkole* (1967) No. 2, p. 95. [b]From *Atlas S.S.S.R.*, p. 20. [c]From A. I. Solov'yev, *Zapovedniki S.S.S.R.* (1951) p. 377.

Sea coast to the south, and the drier, cooler Kuban' area to the north, which has a continental climatic regime. Precipitation is moderate to heavy the year around.

The above characteristics, combined with the effects of vertical zonation, give the Kavkaz preserve an extremely rich and varied natural vegetation cover. Over 3000 species of flora have been counted. The large virgin forests consist mainly of beech, fir, spruce, and hornbeam, while many groves of less common or relict vegetation can be found along the river valleys and in the Khosta exclave (Figures 4.4 and 4.5). The preserve is also noted for its wild rhododendrons.

The wildlife of the preserve is equally varied, and includes 58 types of mammals and 192 species of birds, of which 121 nest there.[9] Among the more interesting endemic species are the West Caucasus mountain goats (*tur*, of which there are about 12,000), black grouse (*teterev*), vole (*polevka*), mountain turkey (*ular*), and Caucasus adders. Other important wildlife

Figure 4.4. Entrance road to the Khosta grove of the Kavkaz *zapovednik*.

include wisent (about 400), chamois (about 3500), Caucasian deer (about 6500), roe deer, wild boars, Caucasian bears, wolves, lynx, marten, wildcats, otters, foxes, badgers, weasels, vultures, and excellent trout. A very few snow leopards and mink can be found within the preserve.

The scientific research carried out in the preserve has been considerable. Almost 700 papers have been prepared on the basis of work conducted there. One of the main tasks of the preserve has been the re-establishment of the nearly extinct European bison (wisent); this effort is described in more detail in Chapter 5. In addition to the work with the bison, other areas of investigation have included:

(1) inventories and ecological relationships of various mammals (especially deer, chamois, mountain goats and marten);
(2) inventories, classification and mapping of flora;
(3) studies of seeds and vegetation reproduction;
(4) the microclimatic role of different types of forests;
(5) watershed properties of different types of forests;
(6) mammalian pathology;
(7) restoring high mountain pastures;
(8) classification, properties, and erosion prevention of mountain soils;
(9) large scale cartographic representation of the area's natural properties;
(10) control of predators (especially wolves);
(11) acclimatization of exotic species of flora and fauna.

Figure 4.5. A yew tree posted as being about 1000 years old, along a self-guiding nature trail in the Khosta grove of the Kavkaz *zapovednik*.

As a more explicit example of the type of studies carried out in the Kavkaz preserve, the table of contents of one of its research digests has been translated as Appendix 12.

The Kavkaz is also a leader among Soviet preserves in the accommodation of tourists, who primarily engage in hiking, camping, and alpinism. As of 1964, approximately 200,000 tourists from the Soviet Union and abroad were visiting the area each year, and it is probable that this total will increase annually.[10] The Kavkaz preserve is somewhat unique in the Soviet Union in the extent to which it combines both scientific research and extensive tourism. As a natural preserve in a very scenic setting receiving heavy tourist use, of all the Soviet *zapovedniki* it probably most closely resembles an American national park.

Problem areas in the administration of *zapovedniki*

Several types of problems are evident in the current manner by which Soviet *zapovedniki* are established and managed. These problems include some of an administrative and some of a policy nature.

The first category of problems results from the centralized and production-oriented manner in which decisions regarding the development of the country's natural resources are made. Although *zapovedniki* can be created with perhaps more political ease than can national parks in the United States, one of the main shortcomings of the Soviet system is the similar ease with which they can also be abolished. The two major reductions of 1951 and 1961 have been noted above. It should be pointed out that although the 1961 reorganization was ordered by a decree of the Council of Ministers of the U.S.S.R., it involved preserves which had been established by the governments of the Union republics as well as those created by the U.S.S.R. government.

For similar reasons, changes in the size and configuration of *zapovedniki* are not uncommon. Fluctuations in the size of the Kavkaz preserve have already been mentioned; many others have experienced the same thing. For example, the Sikhote-Alin preserve encompassed about 1,000,000 hectares in 1933, 1,700,000 in 1947, 1,105,900 in 1959, 599,600 in 1961, and 310,000 in 1964; the Barguzin contained over 200,000 hectares in 1933, 570,900 in 1951, 52,400 in 1952, and 248,180 since the late fifties. Similar situations could be cited for many of the other preserves as well.

Associated with the above is the resultant problem of conducting any long-term biological or ecological research in areas subject to such major changes in protected status. Mention has already been made of the Altay and Kronotskiy preserves, for example, which have both been abolished, re-established, abolished again, and re-established a third time within the past twenty years, a situation openly lamented by many Soviet conservationists.[11]

A second major source of problems stems from the fact that there is no equivalent of the National Park Service in the Soviet Union; that is, there is no one agency coordinating the activities of all of the nation's *zapovedniki*. They can be managed by agencies of the governments of the U.S.S.R., the Union republics, or even autonomous republics, or they can be managed by the Academy of Sciences of a Union republic, the Academy of Sciences of the U.S.S.R., or one of its regional affiliates. As might be expected, this situation precludes the establishment of any uniform set of administrative policies and, as a result, policy disagreements sometimes arise. In order to achieve better coordination, national conferences of *zapovednik* directors are occasionally convened. For example, one such conference was held on May 22–4, 1963, at the Voronezh preserve, where were discussed current problems associated with the scientific work of the preserves, and ways of making this work more effective in the national economy.[12]

One major area of policy disagreement concerns the areal extent to which experimental work in the preserves should be permitted. While granting that the primary function of a preserve is research, much of it experimental, some Soviet conservationists nevertheless feel that a portion of the preserves should be kept as 'natural plots' (what we might call wilderness areas), where no experimental work would be allowed and where no exotic plants or animals would be introduced.[13] It should be noted that although many preserves have large areas of *de facto* wilderness, there is apparently no provision for statutory wilderness areas in the U.S.S.R. such as exists under our own 1964 Wilderness Act.

Although *zapovedniki* are defined as 'areas forever withdrawn from economic utilization', they are nevertheless occasionally used for economic ends, such as pasturing, hay production, and timber harvesting.[14] If the economic potential of a preserve is great enough, it may simply be abolished. This was the case with the unique Kronotskiy preserve on Kamchatka. Located around one of the world's four great geothermal areas and containing many large geysers, it was abolished in 1951 to permit prospecting for oil (which was unsuccessful), and then again in 1961 to allow the construction of the 12,500 kW Pauzhetka geothermal power station in the geyser area. Additional geothermal stations have also been planned for the area. It is not known whether these power stations are located within the current boundaries of the newly reactivated *zapovednik*.

A problem akin to that of unauthorized economic use is that of poaching on the preserves. Although neither hunting nor fishing is allowed in Soviet *zapovedniki*, illegal trespass to engage in these and other prohibited activities is not uncommon, even by state officials. Poaching is common, for example, on the Kyzyl-Agach preserve in Azerbaydzhan, and in conjunction with an extensive loss of natural habitat, has resulted in a decrease in the number of waterfowl wintering there from 6,500,000 in

1958 to 1,500,000 in 1968.[15] Adequate personnel to control poachers appears to be lacking in many of the large preserves.

Another major policy question concerns the admittance of tourists to the preserves. Rapidly growing tourism in the Soviet Union is producing strong pressures to open more of these areas to public use, although there is an equal amount of pressure for continuing to direct the tourist traffic in other directions.[16] Despite the fact that existing interpretive facilities in the preserves are receiving greatly increased use, no general movement towards opening more of the preserves to widespread tourist use appears imminent. The ones that are now open, therefore, are bracing themselves for greatly expanded visitations. For example, the Stolby Preserve near the East Siberian city of Krasnoyarsk experienced an increase in visitors from 18,000 in 1965 to 100,000 in 1969. Many of these newcomers are without experience in how to behave in protected areas, and illegal tree cutting, vandalism, and even abandoned campfires resulting in forest fires have been reported.[17]

A third category of problems is those of an ecological nature. It goes without saying that the more active man is in an area, the more difficult it is for the natural ecological relationships there to maintain themselves. This presents an especially complex problem in those preserves which are charged both with preserving the natural landscape and with carrying out extensive scientific experimentation (Pechora-Ilych, Kavkaz, etc.).

In other cases the area set aside as a *zapovednik* may be too small for the territory to maintain its own natural community. In the Streletskaya section of the Ukrainian Steppe preserve, for example, a decline in the number of birds of prey in the surrounding area allowed marmots to increase far beyond their normal numbers. As a result, in places they have dug up the virgin steppe, allowing weeds to grow on the transformed land.[18]

The management of predators is a normal problem in any protected natural area, and the *zapovedniki* of the Soviet Union are no exception. However, unlike the situation in American parks where predators are only controlled, not eliminated, the Soviet Union appears to be committed to a program of complete extermination of wolves from certain preserves, such as the Kavkaz, where they are normally found. Although only about twenty remain in the Kavkaz preserve, the effort to eliminate them by shooting, trapping, and poisoning continues.[19] This contrasts sharply with the carefully protected status of wolves in our own Isle Royale National Park.

As a final consideration, many Soviet conservationists are not satisfied with the total area presently to be found within the nationwide network of preserves, nor with its regional composition. They very frequently point out that only 0.3 per cent of the total area of the country is in nature preserves, while the figure for Japan is 4 per cent, for Czechoslovakia

1.5 per cent, and for the United States 1 per cent. However, the figure of 1 per cent for the United States includes national parks and national monuments only; to correspond to the Soviet system of *zapovedniki* it should also include wilderness areas and wildlife refuges (Table 4.3).

Table 4.3 represents a rough comparison only. The Soviet figures should ideally include local (republic and oblast) *zapovedniki*, but comprehensive figures for these are not available. Their total area is probably less than that for the national *zapovedniki*, and many of them preserve such things as architectural monuments in cities, etc. Also, large state parks and state forests in the United States, which would be somewhat analogous (such as the Adirondak Forest Preserve in New York and the Anza-Borrego Desert State Park in California), are not included in the American total.

Table 4.3 *Federally protected areas in the United States and the Soviet Union*

Type of area	United States (1000 hectares)	Soviet Union (1000 hectares)
National Parks and Monuments (Jan. 1, 1968)	9,368	
Zapovedniki (mid-1968)		*ca.* 6,400
Forest Service Wilderness Areas (June 1, 1968)	5,867	
State Hunting Preserves (mid-1968)		128
National Wildlife Refuges (June 30, 1968)	11,824	
Total	27,059	*ca.* 6,500
Area of country	936,333	2,240,220
Per cent of country in federally protected natural areas	2.89	0.29

Sources: U.S.S.R.: Belousova *et al.*, p. 20; U.S.A.: Information brochures, dated as above, as published by the respective agencies.

Regarding the types of land included as *zapovedniki*, Soviet conservationists regret that there are still many varieties of natural landscapes which are not represented in the system of preserves, especially arctic areas and steppe areas outside the Ukraine. Even so, they are far ahead of the United States in this respect, as our own country has never made a systematic effort to include representative samples of all major North American natural zones in its national park system.

By way of summation, it would appear that the Soviet system of *zapovedniki* stands in need of a strong, centralized parent organization, one that can create a uniform set of administrative policies for all of the country's preserves. At the same time, it should also be one that could exert enough influence to increase the area under protection, while simultaneously insuring it from periodic reorganizations and depletions in the future.

National parks and other types of protected areas in the U.S.S.R.

In the years since the first national park was established around the Yellowstone geyser basin in 1872, a great many countries of the world have similarly set aside some of their most scenic areas for public enjoyment and recreation. In the past decade or so, interest has arisen in the Soviet Union on this subject as well. The first significant discussion regarding the creation of national parks in the U.S.S.R. took place at the 4th All-Union Conference on the Conservation of Nature, September 16–19, 1961, in Novosibirsk. A major impetus for this has no doubt been the recent large increases in internal tourism, and the resultant pressures on the 'open' *zapovedniki.* Inasmuch as most conservationists in the Soviet Union with backgrounds in the natural sciences wish to see the *zapovedniki* preserved as outdoor laboratories and kept untrampled by tourists, other outlets for mass tourism have been sought. One such outlet would be the creation of national parks.

Although national parks have been authorized in the Soviet Union since 1921 (Appendix 10), none had actually been developed as of the late 1960s. In recent years, however, considerable planning and discussion has taken place, and the creation of such parks has been promised in the press. Two in particular have received wide publicity: a 'Russian Forest' park south of Moscow, and a Lake Baykal National Park in Eastern Siberia. In addition, the Central Laboratory for the Conservation of Nature (a division of the U.S.S.R. Ministry of Agriculture) has compiled a list of thirteen other sites throughout the country suitable for the creation of national parks; these areas are briefly described in Appendix 13.

The features and functions of areas designated as 'national parks' differ greatly from country to country. Those proposed for the Soviet Union would be somewhat different in function from their American counterparts, in that tourism would be their primary responsibility, with the preservation of scenic or unique landscapes having in general only a secondary importance. Thus they might actually be more analogous to our national recreation areas. Soviet writers seem to be encouraging a different kind of semantic distinction, however, in that some would favor referring to their proposed areas as 'natural parks' (*prirodnyy park*), rather than 'national parks'. Bannikov explains the preference for 'natural park' by noting that since all land is nationalized in the Soviet Union, the use of the distinguishing adjective 'national' is inapplicable and unnecessary in the Soviet case.[20] This would seem to be only a partly valid argument, since we also use the term to distinguish our federal parks from their state and local counterparts, and the Soviet Union also has similar areas under republic and province management.

It has been recommended that Soviet natural parks be at least 10,000 hectares in size (only four American national parks are smaller than this).[21] Although the proposed parks are located in areas with scenic and varied natural landscapes, no effort has been made to include in their numbers the most scenic or geologically unique areas of the country; this represents one major conceptual difference between Soviet and American parks. One major similarity that has been recommended would be the division of each natural park into developed and undeveloped sections. The undeveloped 'central' area would be kept in a semi-wilderness state, having only trails and designated camping areas. The 'peripheral' zone would be the developed area, containing the park's hotels, stores, museums, administrative buildings, automobile-accessible campgrounds, and other improvements.[22] It has been suggested that the campgrounds be periodically rotated, so as to prevent permanent site loss through soil compaction, damage to vegetation, and other 'human erosion'.

Like American parks, a major purpose of Soviet natural parks would be to educate the visitors in the concepts of conservation. Unlike American parks, a very limited amount of tightly controlled economic activities might be permitted within the boundaries of the parks (such as grazing, hay growing, hunting, orchards, and also the non-commercial collecting of mushrooms, berries, and nuts by the park's visitors).[23] In addition to the purchase of entrance tickets at a park's boundaries, such tickets might also be distributed at factories or other places of employment, as a bonus to outstanding workers.

The Baykal and Russian Forest parks would have somewhat broader functions than the Central Laboratory's proposed thirteen 'natural parks'. Interestingly enough, although both have been discussed for years, neither was mentioned at all in the chapter on natural parks in the work cited above (preceding three notes).

The Baykal park has largely arisen out of a dispute over the threat of pollution to Lake Baykal's waters from new pulp mills being built near its shores (Chapter 8). As a result, the 'park' proposed for portions of Lake Baykal's shoreline would have, in addition to campgrounds and other tourist facilities, a supervisory authority over all regional activities that could endanger the ecology of the lake or the purity of its waters:

The purpose of the park is to utilize and protect Baykal's natural resources. It will also exercise sanitary and technical control over the operations of the chemical, fishing and lumber enterprises in the lake basin, as well as over water transport. To this end there are provisions for the organization of a science center as well as protection and inspection services, departments of mass activities, and a park administration.[24]

The park will initially take in about 1.3 million hectares; this would eventually be increased to around 4 million. The first areas to be developed for tourism will include Olkhon Island and parts of the adjacent shoreline, Cape Svyatoy Nos and sections of the Khamar-Daban and Barguzin

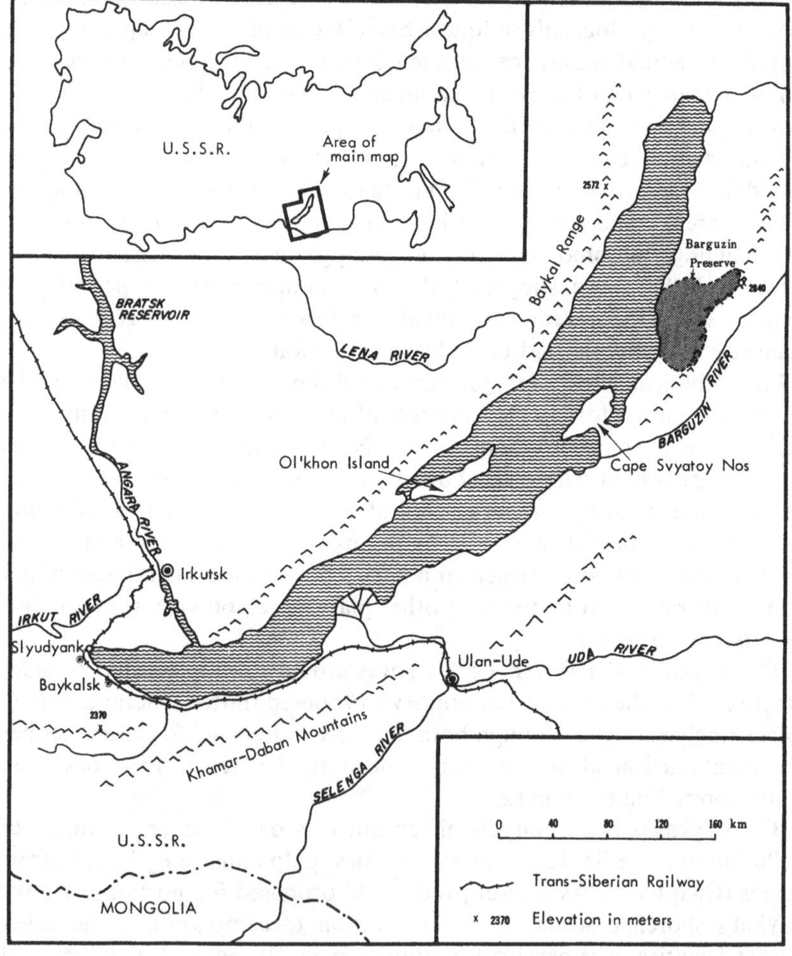

Figure 4.6. Lake Baykal and environs.

mountain ranges (Figure 4.6). Whether the existing Barguzin *zapovednik* would be incorporated into the park or not has not been made clear, but it is likely that it would remain as an independent, restricted research institution. In its capacity as a tourist facility the proposed park would be able to accommodate half a million visitors a year, and campgrounds, motels and guest houses will be established for their use.

The 'Russian Forest' park was first proposed as early as 1956. It is to be located on the left bank of the Oka River between the towns of Serpukhov and Stupino, about 100 km south of Moscow. In size it will be around 50,000 hectares. Extensive facilities are planned for the accommodation of tourists: five tourist centers will have rooms for 15,000 guests, and six artificial reservoirs and beaches and 135 km of hiking trails are being built.[25] The first phase of construction is scheduled for completion in the early 1970s. Reportedly, only biological methods, rather than chemical pesticides, will be employed for insect control in this park.[26]

The special feature of this park which distinguishes it from the proposed Soviet 'natural parks' is that it will serve both as a forest preserve and as a demonstration forest. The latter may even include a demonstration timber harvesting area.

From the above, it is clear that a firm organizational policy for national, or natural, parks in the Soviet Union has not yet been developed. However, it is probable that this type of recreational institution will come into reality in the U.S.S.R. within the next few years, in order to relieve the mounting visitor pressure on the supposedly restricted *zapovedniki*.

In addition to *zapovedniki* (and the proposed national parks), other categories of protected natural areas exist in the Soviet Union at present. A second important type is known as a *zakaznik*. There are two types of *zakazniki*, temporary and permanent.

Temporary *zakazniki* are normally established only for a predetermined period of time, usually not more than ten years. In addition, they are normally concerned only with animal resources, and generally with only one or a few of the wildlife resources found within them. For example, the hunting of a particular species of animal might be controlled over a period of years in such an area if its numbers or condition were observed to be deteriorating. Current statistical data on *zakazniki* are lacking, but in 1960 there were over 1000 of them, taking in a total area of over 10,000,000 hectares.[27]

Permanent *zakazniki*, like the temporary ones, protect only a portion of the natural resources within them, but are not so limited to wildlife resources. As their name would imply, it is intended that they should be maintained indefinitely. Occasionally they are termed 'natural monuments' (*pamyatniki prirody*), rather than 'permanent *zakazniki*'.

A permanent *zakaznik* might be employed, for example, to protect wintering or nesting birds. However, on the same territory, fishing might be permitted, or artificial water impoundments for urban or agricultural needs might be constructed, etc. Another type protects spawning areas of important varieties of fish; this in some cases might involve prohibiting fishing in certain portions of lakes, rivers, or reservoirs.

Another type, termed 'permanent landscape *zakaznik*', is used in many areas of the country, most extensively in the Baltic republics, to preserve scientifically significant, unusual, or picturesque portions of the natural landscape. In these, any activity which would alter the general appearance or attractiveness of the area is forbidden. These areas are usually not too large in size and are generally intended for tourist use. They would be somewhat analogous to certain American national monuments, or to areas on the National Park Service's Registry of Natural Landmarks.

A compilation of natural areas in the Soviet Union thought deserving of permanent *zakaznik* status, and deemed to be of national importance, was carried out by the Central Laboratory for the Conservation of Nature, and was published in 1967.[28] It contained 370 recommended sites which were grouped under five general categories of *zakazniki*: botanical, geologic, botanical-geologic, lacustrine, and zoological.

Hunting reserves (*zapovedno-okhotnich'ye khozyaystvo*) have been mentioned above in passing. In 1967 there were five of national significance, all of which were listed as *zapovedniki* in 1961 (Appendix 7, Part 2). They appear to be still in a general protected status, except that controlled public hunting is allowed in them. In 1968 one of them, Naurzum in Kazakhstan, was reportedly reclassified as a *zapovednik*.

All of the forms of protected areas discussed above, which are of national importance, can be duplicated at the republic level, or in some cases even lower (province, autonomous republic, territory, etc.). Many republics and provinces (*oblast*) have developed extensive systems of protected areas and objects. Depending on the primary functions of the particular areas and the extent of protection involved, these systems may include locally designated *zapovedniki*, *zakazniki*, hunting preserves, parks and gardens, natural monuments, historical monuments, and architectural monuments.

To cite some examples, Zakarpatskaya Oblast has a total of 103 areas in some type of officially preserved status, ranging from individual buildings and gardens to *zakazniki* of 4600 hectares. L'vov Oblast has forty natural areas alone, of which four are of republic (all-Ukrainian) significance, including the 52,000 hectare 'Maydan' hunting preserve.[29] Sverdlovsk Oblast has eighty-eight areas established as natural monuments, and Chelyabinsk sixty-one.[30] Estonia has made a very complete inventory of its natural monuments, and has no less than 653 such areas under legal protection; some of them are individual glacial erratics, and some are only 1.5 hectares in area.[31]

As in the United States, there exist natural areas in the Soviet countryside belonging to universities and other higher educational institutions, and which are used by them for biological and ecological research. These areas are not designed to be preserves, however, and experimental research is the dominant theme within them. Another major category of natural area – urban parks and green belts – will be dealt with in a later section on urban environment.

5 The management of fisheries and wildlife

Unfortunately, fishing remains the stepchild of hydrotechnical construction...

The preservation of Soviet fish and wildlife resources is marked by some very successful efforts, as well as by many other areas in which the record is decidedly poorer. The preservation of the U.S.S.R.'s internal fisheries, for example, has not been as successful as might be desired, for reasons almost identical to those affecting the freshwater fisheries of the United States. Fish and wildlife conservation in the U.S.S.R. is supported by an impressive amount of research in those biological sciences which deal with the life cycles, zoogeography, ecology, and pathology of mammals, birds, fishes, and other categories of animal life. This research notwithstanding, wildlife and particularly fisheries conservation in the U.S.S.R. has encountered numerous difficulties, many of which can be seen to be the result of pragmatic decisions made by economic planners concerning resource use priorities.

The regulation of hunting and wildlife preservation

There is no uniform code governing wildlife management and hunting throughout the whole of the U.S.S.R. Rather, the preservation of wildlife stocks, as well as the regulation of both commercial and sport hunting, is the resopnsibility of special wildlife and hunting directorates within each of the Union republics.[1] In the Russian Republic, Article 11 of the 1960 R.S.F.S.R. conservation law pertains to wildlife preservation (Appendix 14).

Specific regulations regarding hunting (seasons, areas, registration of hunters and guns, limits, etc.) are issued at a still more local level, that is, by provinces and autonomous republics, where local wildlife conditions are better known and can be more effectively managed. However, the inter-republic supervision and coordination of hunting activities, and wildlife conservation for the country as a whole, is the responsibility of the Central Directorate for the Conservation of Nature, Preserves, and the Hunting Economy, which operates within the U.S.S.R. Ministry of Agriculture.

Table 5.1 *State purchases of pelts by major types*

Species	Per cent of total value of state purchases	
	1926–9	1956–9
Squirrel	34.5	15.8
Muskrat	? (1.8)	13.7
Suslik	5.1	12.1
Fox	9.9	11.1
Sable	3.4	11.0
Arctic fox	6.5	9.5
Water rat	3.3	5.9
Mole	1.8	5.4
Hare (all types)	9.4	3.6
Marten	1.8	3.4
Ermine	5.4	1.8
Other	18.9	6.7

From K. N. Blagosklonov, A. A. Inozemtsev and V. N. Tikhomirov, *Okhrana prirody* (Moskva: Izdatel'stvo 'Vysshaya shkola', 1967) p. 302.

Hunting in the Soviet Union may be divided into two broad categories: professional (i.e. commercial), and sport.

Professional hunting and trapping takes place as a state industry, mainly on collective and state farms in the tayga and tundra zones. Here valuable

Table 5.2 *Reserves and annual extraction of wild ungulates in the U.S.S.R.*

Species	Reserves (thousands)	Extraction (thousands)
Sayga	2500–3000	350–400
Roe deer	1000–1200	60–65
Elk	700–800	35–40
Reindeer	450–600	10–15
Siberian mountain goat	400–550	20–25
Musk deer (*Kabarga*)	350–500	6–8
Persian gazelle (*Dzheyran*)	90–100	—
Mountain sheep (*Argali*)	60–70	2–3
Snowy sheep (*Chubuk*)	30–40	*ca.* 1
Red deer	25–30	1.5
Mountain goat (*Tur*)	25–30	*ca.* 1
Chamois	25–30	*ca.* 1

From Blagosklonov *et al.* (1967) p. 304.

fur animals live in abundance, and in these zones there are about 150 such enterprises engaging in commercial hunting and trapping. These farms employ some 100,000 special professional hunters, who are on leave from their regular jobs during the hunting seasons to work for the state and collective farms. Their take is turned over to the government, and is used either for domestic commodities or for export (mainly furs). The composition of government procurement of pelts is shown in Table 5.1.

Of the above government procurements, the price paid for some, such as certain species of suslik (a rodent considered an agricultural pest), should probably be considered more as a form of bounty.

The absolute catch by species varies greatly from year to year, and, for certain species (such as sable), figures are not released. However, for some of the above species it averages approximately as follows: fox – 165,000; silver fox – 120,000; marten – 80,000; ermine – 200,000; hare – 2,000,000; muskrat – 5,000,000; water rat – 10,000,000; mole – 18,000,000; and suslik – 30,000,000.[2]

The annual allowable kill (both commercial and sport) is determined by local authorities and is based on the current strength of reserves and reproductive capacity of the particular species. For the main types of ungulates, recent data are presented in Table 5.2.

A widespread problem which hinders the efficient management of wild game is the very uneven distribution of the activities of both commercial and sport hunters. Particularly in the more remote areas of the country, they tend to concentrate near rivers and other transportation routes, so that the game in these easily accessible corridors may be over-hunted, while in the more remote back-country the numbers of some species may exceed the carrying capacity of the land.

Sport hunting is very popular in the Soviet Union, and numerous local hunting organizations exist throughout the country. Specific areas for hunting are allocated to them, and these organizations are responsible for supervising the hunting that takes place on their allotted territory, and for maintaining proper conservation practices. A hunting license may be obtained by anyone over the age of eighteen, but he is obliged to belong to one of the local hunting organizations and to remain in good standing within it. There are about two million sport hunters in the U.S.S.R.

A short (ten day) season for game birds exists in the spring; the fall season for game birds is about three months long. The primary season for rabbit, squirrel, and fox begins in the fall, and the length of the open season is fixed independently for each of these and other species. The more valuable fur-bearing animals and ungulates require a special permit in those areas where they may be hunted at all, and the rarer species (polar bears, tigers, wisent, swans, and others) are unconditionally protected.

The hunting organizations provide game wardens and are responsible for enforcing regulations.

Whether the number of game wardens is too small, or whether there are other basic causes, poaching represents a very serious problem in Soviet game management. Hunters who exceed limits, hunt out of season or outside posted areas, or kill restricted species appear to be extremely numerous in all parts of the country. Complaints of illegal hunting are frequently seen in the press, and one Soviet article states that 'poaching has assumed the proportions of a national calamity in our country'.[3]

One publicized case concerned the slaughter of thousands of wild antelope on the steppes of Kazakhstan. It was reported in part as follows:

> In recent years the herd of antelope in Kazakhstan has been on the increase. So it was decided to shoot them under licenses from the hunters' unions. Tens of teams of hunters rushed to the places where the antelope live. The pitiless destruction of antelopes began.
>
> Most often antelope shooting is done at night, 'by headlights'. Finding themselves in the bright beam of light from the spotlights mounted on cars, the animals are bewildered and come to a standstill. And then the hunters open fire on the herd with ten-round automatics. They fire point blank.
>
> In one burst they would get 150 head each...the leader of one of the teams boasted to us.[4]

Jacking game is prohibited by Kazakh laws, but the officials in charge consider the situation not simply as sport hunting, but as antelope meat procurement. And since it is regarded as procurement, the article continued, exception to the laws are permitted.

Illegal killing of elk is widespread in many northern areas of the country. Some publicized cases have even involved officials ignoring or sheltering influential persons caught in the act of poaching, and other reports of 'ordinary' poaching are common. In an effort to bring the situation under control, the Supreme Soviet of the Russian Republic (R.S.F.S.R.) increased the penalties for hunting violations in October of 1963. Currently, illegal hunting is punishable by a year of corrective labor or by fines of up to a maximum of 500 rubles. Nevertheless, Soviet wildlife conservationists contend that poaching in all of its forms is still far too widespread, and that even stricter enforcement measures may be necessary.

Species of wildlife which are threatened with extinction or which are unduly decreasing in number are the concern of various scientific bodies, particularly the zoological institutes of the Academy of Sciences and the research institutes at those *zapovedniki* where such animals are protected or bred. As was noted in the preceding chapter, the research facilities at *zapovedniki* conduct studies on all species found on their territories, not just endangered ones.

Considerable success has been realized in the preservation of threatened species in the past fifty years. Soviet writers assert that no major species of wildlife has become extinct over the territory of the Soviet Union since the tarpan was exterminated from the Russian steppes about a hundred years ago. Some species which were greatly depleted in the first quarter of the century, such as the beaver, sable, sayga, otter, and pine marten, have been assisted in making a strong recovery.

Figure 5.1. The European bison (or wisent), an endangered species making a comeback.

Many other animals and birds have been threatened with extinction, however, and a number are still in some danger. As elsewhere, the main causes of wildlife depopulation have been over-hunting and the loss of natural habitat. The extensive clearing of forests and plowing up of virgin steppe and prairie have greatly reduced the former range of numerous important wild animals, such as the elk and sayga.

Among the species whose numbers are considered to be low enough to require special attention are the wisent, or European bison, the kulan (a steppe animal resembling a donkey), the desman (*vykhukhol'*), the walrus, the polar bear, the Ussuri tiger, the sea otter (*kalan*), the fur seal (*kotik*), the white heron, and native swans. In addition, the beaver, European elk (*los'*),[5] musk deer (*kabarga*), chamois, pine marten, goral, and many others, although not considered as presently threatened with extinction, are nevertheless being carefully studied. Some of the more threatened species, such as the wisent and the kulan, are today almost unknown outside of *zapovedniki* and zoos. The case of the wisent is worthy of special comment.

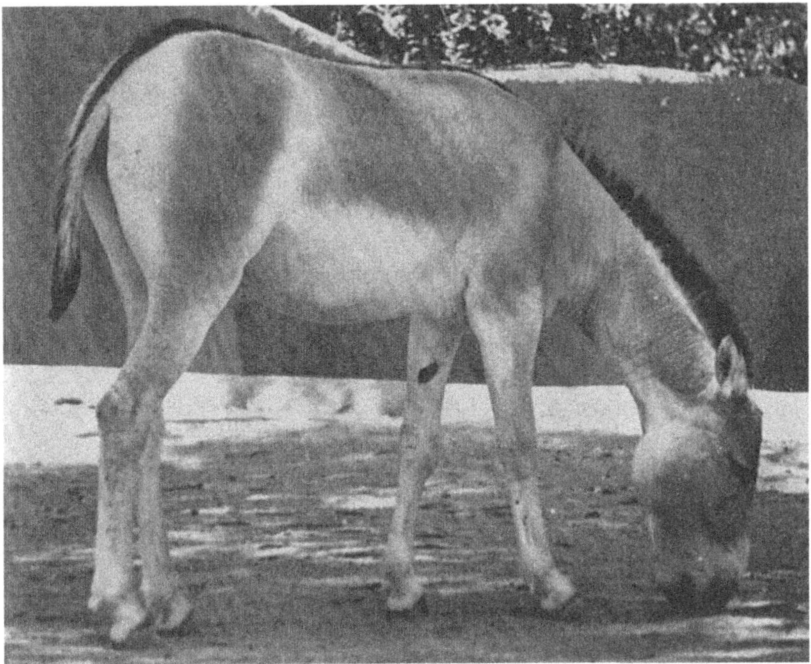

Figure 5.2. The Kulan, an endangered ungulate of the Central Asian steppes.

The saving of the European bison, or wisent (*Bison bonasus*), from extinction within the U.S.S.R. represents one of the most determined efforts of Soviet wildlife conservationists. Originally this species, not too different in appearance from the American bison, ranged over most of Western and Central Europe, European Russia, and the Caucasus. Today, its total numbers are in the hundreds, and it is found only in zoos and in natural preserves in Poland, Russia, and a few other countries. There are two sub-species of the European bison, the Belovezh of Eastern Europe, and the 'mountain' or Caucasian bison of the Caucasus Range. They are

Figure 5.3. The Siberian (or Amur or Ussuri) tiger, an endangered species under strict protection.

Figure 5.4. Sika or spotted deer, an endangered species under protection in the U.S.S.R.

quite similar, and a distinction between them in counting the herds is often not made in Soviet sources, and will not be made here.

The animal was approaching extinction in Russia in the late nineteenth century, and efforts to revive the species began before the turn of the century by importing wisents from other countries. Although some small success was realized, it all went for nothing as the animal was almost totally exterminated in Russia during World War I and the civil war which followed. In 1921 there were only fifty-three wisents in the world, and in 1927 only forty-eight.[6]

In the Soviet period the herd has increased relatively rapidly, considering its low starting base. However, the herd now consists of two components – pure-blooded wisents, and those cross-bred with American

Table 5.3 *The distribution of wisents, bison, and their hybrids within the Soviet Union (as of Jan. 1, 1956)*

Area	Wisent	Bison	Wisent-bison	Complex hybrids	Total
Kavkaz preserve	4	0	86	0	90
Belovezhskaya Pushcha	25	0	0	0	25
Prioksko-Terrasnyy preserve	18	2	0	17	37
Other preserves	4	10	1	3	18
Soviet zoos	6	7	10	15	38
Totals	57	19	97	35	208

From Zablotskiy (1962) p. 47.

bison (termed wisent-bison). It was necessary to use bison and second generation hybrids for breeding in many cases because of the difficulty in obtaining pure wisents. No females at all young enough for breeding existed on the territory of the Soviet Union after the civil war, and therefore about thirty have been imported for breeding purposes. This breeding has been carried out mainly in three *zapovedniki* of the U.S.S.R. – Belovezhskaya Pushcha, Kavkaz, and Prioksko-Terrasnyy. The results to 1956 of the breeding of both the pure-blooded and hybrid strains are shown in Table 5.3.

The success of the breeding program can be seen in the fact that nine years later, on January 1, 1965, there were no less than 231 full-blooded wisents in the U.S.S.R. (and around 1000 of all kinds). It is expected that the number will continue to double about every five years.

Many other European countries are also breeding full-blooded wisents, and Poland stands even ahead of the U.S.S.R. in this respect. In 1962, when there were 160 wisents in Russia, there were 176 in Poland, 50 in West Germany, 28 in Sweden, and about 95 elsewhere in the world (almost all in Europe), for a world total of slightly over 500.[7]

Two categories of wildlife which have experienced widespread depletion in the Soviet period are land and water game birds. Although the annual take of game birds reaches perhaps 100,000,000, the primary cause of depletion has not been over-hunting, but rather the loss of the species' normal natural habitat. This phenomenon, stemming from man's ability to alter the natural environment over large areas, represents one of the common problems resulting from economic priorities in the competition for land resources.

In the Soviet Union, the main causes of the problem have been the clearing of woods and thickets, the plowing up of prairies and meadows for agricultural purposes, the reforestation of cut-over lands with monoculture stands, and the construction of dams and drainage projects which eliminate natural marshes and ponds.

The hazel grouse is the leading commercial species in the Soviet Union, and used to be hunted in the mixed forest regions of European Russia. With the cutting of these forests, it is now found in large numbers only in the coniferous forests to the north and east. The same situation applies for the wood grouse, although it is a less important commercial species. The grey partridge is being forced farther north in smaller numbers as the forest-steppe is increasingly brought into cultivation, and the great and little bustard have been forced from their former habitats in the virgin steppe areas.[8] On the other hand, shelter belt plantations in the steppes, where successful, have been a small positive factor in providing a new habitat for game birds. Among steppe mammals, the steppe marmot (*baybak*) has also been disappearing before the plow.

The situation regarding waterfowl is very bad. In Azerbaydzhan alone, reclamation and the lowering level of the Caspian Sea have reduced waterfowl wintering grounds by more than ten times. The shallow Gasan-Kuli Gulf on the Caspian has been reduced from 1,030,000 hectares in 1888 to 20,000 in 1932 to almost nothing today. In addition, about 20,000 to 25,000 birds perish annually due to oil pollution in Azerbaydzhan alone.[9] In the more northern and eastern portions of the Soviet Union waterfowl habitats have not been so markedly affected, but nevertheless a gradual decrease in the quantity of many species has been observed (e.g. eider duck, Brent goose, white goose). A basic problem in conserving waterfowl anywhere in the country is the lack of an accurate count of the number killed by hunters each year.

As a means towards preserving some of the more important remaining wildlife habitats, the suggestion has been made to establish more and larger *zakazniki*, patterned after the system of wildlife refuges which exist in the United States. Also suggested were more effective inventorying of wildlife reserves, more thorough hunting regulations and counts of kill, and stricter enforcement of such regulations.[10]

Soviet wildlife managers are somewhat less tolerant than their American counterparts in their attitude towards the more destructive predator mammals. Whereas the management goal in the United States is normally to control the numbers of predators so as to maintain a reasonable ecological balance, Soviet zoologists often call for the elimination of certain of the more obnoxious species, phrasing it in such a manner as, '[only] those species whose harmfulness is indisputable may be totally exterminated'.[11] The main predator to which this applies is the wolf, in which the Russians have historically seen no redeeming qualities whatsoever.[12] A change in this attitude may be occurring at the present time, however. Although not the subject of an overt extermination campaign, there is also a no-limit, year-round open season on the Nepal marten, lynx, and wolverine. Most other predators, including birds of prey, are only controlled, not exterminated. All-out campaigns are also waged against certain agricultural pests, such as the suslik, hamster, and water rat, but their very numbers will probably prevent their total eradication. Only occasionally are protests heard against the generally accepted theory that some biological species are unconditionally expendable, and the attendant implication that man already knows all the biological information these species might be able to reveal.

Another interesting facet of Soviet wildlife management has been their efforts at domesticating certain species of wild animals. Two in particular should be mentioned; the reindeer and the European elk. The reindeer is bred on collective farms by the native population in the northern regions of the Russian Republic, and for them represents a very versatile and basic natural resource. The European elk is being domesticated at the Pechoro-Ilych preserve; it is hoped that it can be developed as a work animal, and for the use of its rich milk.

Soviet conservationists are aware of the dangers to fish and wildlife posed by chemical pesticides, herbicides, and fertilizers, and an increasing number of articles and publications have appeared on this subject in recent years, as greater and greater quantities of these agricultural aids are being applied throughout the U.S.S.R. Although their use is still at a relatively low level, unfortunately incidents have already occurred. In one instance, the death of elk in a pesticide-treated area of Yaroslovl' Oblast was reported in late 1966.[13]

Management of fisheries and wildlife

In 1970, the common practice of applying zinc phosphide-treated grain by aerial spraying to control field rodents led to two kills of beneficial wildlife in the steppe region near Rostov. In one incident, the counted kill included 50 cranes, 11 grey geese, and 5 black geese; in the other several dozen foxes and about 200 rare great bustards were destroyed.[14] As has often been the case in the United States, it is uncertain that the pesticide even had its intended effect on the gophers which, of course, are food for the foxes that were killed. The manager of the collective farm where the incident occurred was fined only 150 rubles, because the spraying was done 'in the interests of the harvest'.

The problem of pesticides was discussed by the newspaper *Komsomol'skaya pravda* in March, 1967, in which it was suggested that Rachel Carson's *Silent Spring* be translated into Russian.[15] Although Soviet agronomists call for increased production of insecticides, the U.S.S.R. may be heeding Miss Carson's advice, as it has been reported that the Soviet Union has stopped the manufacture of DDT. In addition, considerable interest has developed in recent years concerning the use of biological controls over agricultural pests, rather than chemical controls. An All-Union Research Institute for Biological Methods of Plant Protection has been established.[16] Soviet scientists are doing considerable research on the effects of chemical poisons on animal life, and a collection of papers on this subject was recently published by the Central Laboratory for the Conservation of Nature.[17] With the current planning emphasis on the 'massive chemicalization of agriculture', many additional studies and translations on this subject would appear to be very timely for the interests of wildlife conservation in the Soviet Union.

The conservation of Soviet fisheries

A somewhat different set of problems surrounds the preservation of fish resources than is the case for the protection of terrestrial wildlife. There is less danger of extinction from over-extraction of the resource, as it is much harder to catch the last few existing fish of a species than it is to kill the last few survivors of an animal type. Of course, improvident fishing can easily reduce a stock to where it will no longer be profitable to continue using the resource, but the danger of total depletion is less.

On the other hand, man's ability to destroy an individual fish run by changing the environment is often greater, and more final, than his ability to alter a natural wildlife habitat. By means of damming, diverting, polluting, channelling, or draining, man can very quickly eliminate an established fish habitat or spawning area. All of these alterations of nature have taken place in the Soviet Union, in some cases with very adverse effects on important fisheries resources.

The supervision of fishing activities is the responsibility of fishing ministries existing in each of the fifteen Union republics, as well as of a Ministry of Fishing of the U.S.S.R. The latter has five regional divisions, one each for the Far East, Northern, Western, Azov–Black, and Caspian–Volga fisheries. These regions were to account for a total planned 1967 catch of six million tons as follows: Far East – 36 per cent, Western – 28 per cent, Northern – 17 per cent, Azov–Black – 11 per cent, and Caspian–Volga – 7 per cent. The ministries maintain staffs of conservation inspectors, which are supplemented at the local level by inspectors from public fishing societies.

The various republic conservation laws (e.g. Appendix 14) contain the basic guidelines for fisheries conservation in the Soviet Union, but enforcement regulations have been enacted at the national level as well. For example, on March 27, 1964, the U.S.S.R. Supreme Soviet passed a law establishing administrative fines for violations of fish conservation regulations. Amateur fishing is regulated by the use of licences issued by local sporting societies for particular bodies of water. Receipts from the sale of the licences go into stocking and maintaining these fisheries.

The concealed and fugitive nature of the resource requires specialized research facilities, and for this purpose there has been established within the Ministry of Fisheries an All-Union Scientific Research Institute for Marine Fishing and Oceanography (acronym: VNIRO), which operates several subordinate regional institutes. An Ichthyological Commission exists under the State Committee for Fisheries of the U.S.S.R. Council of Ministers, and a research institute for fisheries in lakes and rivers is found within the Ministry of Fisheries, and one for pond fisheries within the Ministry of Agriculture. In addition, the U.S.S.R. Academy of Sciences operates several institutes of importance to fisheries research, and certain *zapovedniki*, such as the Darwin and Pechoro-Ilych, also conduct fisheries investigations.[18]

Currently, only about 20 per cent of the total Soviet catch of fish is from internal waters, as compared to more than 80 per cent prior to the revolution and nearly 60 per cent in the 1940s. This reflects the intensified efforts of the Soviet Union over the last two decades in the field of open seas fishing. A summary of the quantity of Soviet fish extractions by species is presented in Appendix 15.

Conservation problems which currently surround Soviet fisheries include determining maximum sustained yields, stocking new or depleted water bodies, poaching, pollution, and conflicts with dams and irrigation projects.

The determination of maximum sustained yields is a basic problem of all marine biologists, and Soviet specialists are well aware of the necessity for understanding the population dynamics of their domestic fisheries. For certain major fishing regions, such as the Volga–Caspian, it has been acknowledged that optimum annual catches have not yet been satisfactorily determined.[19] Over-fishing of Karelian salmon, Lake Sevan trout, and Lake Baykal *omul* has also been reported.[20]

Considerable work is being done in the Soviet Union, as in the United States, on the artificial breeding of valuable species of fish and the stocking of ponds and streams. In 1961 it was reported that there were seventy-three hatcheries in the U.S.S.R. for salmon and sturgeon alone, which produced about 800 million young fish a year.[21] About thirty of these hatcheries are in the Far East, and are used for rearing chum, pink, and red salmon. Sturgeon hatcheries exist on the lower reaches of the Volga, Don, and Kura rivers.

The Seven-Year Plan (1959–65) called for the construction of forty-four new hatcheries and the improving of 537,000 hectares of spawning grounds. However, in 1966 it was reported that since 1961 only one-seventh of the plan's goals for hatchery construction had been fulfilled.[22]

On the other hand, in many areas new spawning grounds have been created or natural ones improved. This is especially important in the case of the Volga–Caspian sturgeon, since many of its natural spawning places along the Volga have been inundated by the series of reservoirs which have been constructed on the river, while others around the Volga estuary have dried up as the Caspian Sea level has continued to lower.

The stocking of inland bodies of water with both fish and fish nutriments is receiving great emphasis, particularly with regard to the many new reservoirs behind hydroelectric projects, and with regard to the numerous small ponds which dot the countryside. The Tsimlyansk Reservoir on the Don has been made into an especially productive fishing ground, and the string of reservoirs on the Volga and Dnepr are also receiving considerable attention. Among the leading reservoirs in terms of their fish productivity in 1964 were the Tsimlyansk with 104,000 metric tons of fish extracted, Kakhovka with 83,300, Kuybyshev with 53,000, Kremenchug with 50,600, Rybinsk with 41,900, and Volgograd with 27,300 metric tons.[23] Collective farms are encouraged to stock and harvest fish in ponds on their lands. It would appear that there is still much undeveloped potential for utilizing small inland bodies of water for fish cultivation, and the development of these resources is currently being stressed.

As was the case with the supervision of hunting, poaching is a major problem in the regulation of Soviet fisheries. From time to time, the names of persons who are caught poaching or officials who aid or ignore poachers are spotlighted in the Soviet press, but apparently to little avail. Enforcement seems to be very lax, as one such article reported that the Urals State

Fisheries and Water Administration in 1961 registered 1137 instances of poaching, but during that year the courts heard only three cases involving illegal fishing.[24] Similarly, in Azerbaydzhan, the instances of illegal hunting and fishing rose from 1300 in 1965 to 1900 in 1966, but in a year's time only twenty cases of illegal fishing and not one case of poaching by hunters were examined by the courts.[25]

Fines for poaching in the Soviet Union are fairly stiff. Typical Russian Republic fines for poaching important commercial species run (per fish) from $11 for Pacific salmon to $111 (100 rubles) for large varieties of sturgeon, and imprisonment for up to four years is possible. In the past, few poachers were brought to trial, and the more successful ones could easily afford even such fines as these. Recently, however, heavier penalties, including imprisonment, have been meted out for the more flagrant violations.[26]

Water pollution is today a serious threat to fish runs in many places in the Soviet Union, and represents a major area of conflict between economic planners and conservationists. Industrial wastes which are emptied into rivers and lakes represent the main source of contamination, and sections of several major rivers, such as the Don, Volga, Kama, Northern Dvina, Belaya, Ural, and others are heavily polluted. The extensive loss of fish stocks is an inevitable result. A 1965 review of the fulfillment of the 1960 conservation law in the Russian Republic stated:

> Because of the pollution of rivers, lakes, and reservoirs with sewage and their obstruction with timber and debris from log-floating operations, hundreds of rivers and bodies of water that used to have large fish reserves have completely lost their importance for the fishing industry. Reserves of valuable types of fish in the republic, particularly whitefish, salmon, sturgeon, and fish caught in small-mesh nets have decreased considerably. There are instances of mass deaths of fish.[27]

One reported occurrence of the mass destruction of fish occurred on the Lysva River in 1967, when 1050 tons of spawning bream were killed by effluent from the Lysva Metallurgical Plant.[28] Fish reserves in parts of the Caspian Sea have decreased greatly in recent years, due to oil pollution of the surface water and along the shoreline. Sharp declines in the catch of the prized Baykal *omul* have occurred due to the loss of spawning grounds as a result of over-fishing, lumbering operations and the pollution of tributary streams. Many other examples could be cited. The problem generally is an unwillingness on the part of administrators to enforce existing anti-pollution laws, partly through fear that to do so would risk having enterprises under their supervision not meet their planned production or profit targets. Calls for a correction to this imbalance in priorities are frequently made, but progress is slow at best.

Management of fisheries and wildlife

The problem of the loss of fish runs and spawning grounds due to the construction of dams for navigation, power, and irrigation is as significant in the Soviet Union as it has been in the United States. This is another common conflict area, and those who speak in the interests of Soviet fisheries conservation consider themselves the underdogs:

'Unfortunately, today fishing remains the stepchild of hydraulic construction. It has long been a bad habit to save funds in hydrotechnical construction chiefly at the expense of facilities necessary for hatching fish. Such installations are usually provided for in designs, but later they are often left out. Sometimes they are built in spite of everything, but most often it is toward the end of construction, after the river has been stopped by a dam.[29]

Figure 5.5. Volgograd hydroelectric dam, with fish elevator facility in center.

This was true at the large Saratov Dam on the middle Volga, which was completed in 1968 but without the planned fish passage facilities having first been built. Consequently, no Volga sturgeon arrived at spawning grounds above the Saratov Dam during the 1968 season.[30]

Although there are fish passage facilities at some of the major dams in the Soviet Union (for example, at Volgograd and Tsimlyansk, where fish elevators rather than ladders are employed – see Figure 5.5), they are not always as successful as might be desired. It has been stated that 'the Volgograd hydroelectric plant almost completely blocks the access of beluga sturgeon to their spawning grounds, and to a significant degree cuts off the spawning grounds of [other] sturgeons'.[31] The effect of the new dams on fisheries resources in some of the major river systems of the U.S.S.R. is shown in Table 5.4.

Other major rivers have been cut in half by large hydroelectric facilities. As examples, the Angara and Lake Baykal are separated by the dam at Bratsk; the Vakhsh by the dam at Nurek; and the Yenisey by the ones upstream from Krasnoyarsk. These dams also prevent nutrient material from reaching the fisheries in the downstream regions, and in the case of the Volga–Caspian fishery it has been proposed to compensate for this by

Table 5.4 *Effects of dams on some major river fisheries*

			Effect of dams on fisheries	
River basin	Pre-dam fish catch (1000 metric tons)	Number of dams	Increase in fish catch as a result of the reservoir (1000 metric tons)	Decrease in catch of migratory fish due to dams (1000 metric tons)
Volga	275	9	62.5	116.7
Kura	9	3	1.2	6.0
Don	21	1	11.0	15.4
Dnepr	7	4	4.1	23.4[a]
Ural	5	1	1.0	—

[a] An explanation of how this figure could be greater than the pre-dam catch is not given. If no typographical error is involved, then possibly the last column refers to potential fish harvests, or includes salt water catches of migratory species which spawn in the river.

From A. B. Avakyan (1966) p. 29.

the addition of three thousand tons of superphosphates directly into the Caspian Sea.[32]

A related problem exists in preventing young fish from accidentally entering irrigation headworks and the tail-races of dams. Many fish are lost annually in this way, and satisfactory solutions have not yet been discovered.

Interestingly enough, the American problem of the sovereignty of the individual states has not always been completely avoided in the Soviet Union, either. In the U.S.S.R., Union and autonomous republics have sometimes failed to coordinate adequately with one another. For example, *Pravda* complained in 1962 that fishermen of the Astrakhan' Economic Council, and of Turkmenia, Azerbaydzhan, Kazakhstan, Dagestan, and Kalmykia carry on fishing separately on the Caspian, for the same fish, practically on the same fishing grounds. They all try to take as much as possible from the sea and are not concerned with the preservation and reproduction of the fish reserves...There is no single agency to look after fish reserves and regulate the catch on the Caspian.[33]

This particular problem may have been alleviated by the abolishment of the regional economic councils in 1965 and the corresponding restrengthening of the ministries system. Nevertheless, in 1970 it was reported that from 20 to 40 per cent of all fish caught in the Caspian Sea were taken in violation of existing fishing regulations.[34]

The Caspian Sea fisheries have been further depleted by the lowering of the surface level of the Sea in recent years, which has eliminated many

Table 5.5 *Fish catch in the Caspian Sea*

	1000 metric tons					
Species	1913	1930	1940	1950	1956	1959
Sturgeon (*Osetr*)	28.3	13.3	8.9	12.9	13.1	11.8
Herring (*Sel'd'*)	329.5	134.1	124.5	54.4	43.5	52.2
Roach (*Vobla*)	142.5	263.4	52.3	55.3	63.9	64.9
Zander (*Sudak*)	—	90.9	34.8	31.4	20.8	14.8
Bream (*Leshch*)	—	37.4	61.2	75.4	27.7	31.9
Carp (*Sazan*)	—	13.9	19.5	33.8	14.4	8.5
Sprats (*Kil'ka*)	—	3.9	9.6	21.7	187.0	152.4

From Berdichevskiy, *op. cit.*, p. 28.

spawning and feeding areas along its northern shores. By the start of the 1960s, the combined effect of over-fishing, loss of habitat, and pollution had been a marked reduction in the catch of almost all major varieties of commercial fish in the Caspian Sea (Table 5.5). In addition, other spawning areas in the lower Volga (below the Volgograd dam) are being transformed by reclamation projects, and a channel-control dam on the lower Volga has been seen as preventing 90 per cent of the fish that reach it from continuing on to their spawning grounds.[35]

A serious problem is also developing in the Sea of Azov, where development projects on the Don (and projected ones on the Kuban River) could greatly reduce (by up to 40 per cent) the amount of fresh water flowing into that Sea. This would destroy a natural flushing action and allow water of much greater salinity from the Black Sea to enter the Sea of Azov. In turn, this could severely harm or possibly even destroy the rich fish resources in this latter body of water. This consequence was foreseen as long ago as the 1930s, but no measures have been carried out to avert it.

A similar problem is arising in the Aral Sea, where the withdrawal of freshwater from the Amu-Darya and the Syr-Darya for irrigation purposes has led to a lowering of the surface area of the Sea and an increase in its salinity. As a result, the catch of fish from the Sea has dropped markedly, and exotic salt-water species are being introduced.

On the whole, it would appear that in planning the use of the Soviet Union's natural resources, power, irrigation, and navigation projects have generally received greater priority than have the country's internal fisheries. For the resolution of the many conflicts arising out of the development of the Soviet Union's rivers and lakes, a supra-ministerial planning agency would seem advisable. Suggestions to this effect have been put forward, and will be discussed further in Chapter 7. Meanwhile, several of the more valuable internal fish runs of the U.S.S.R. appear to be facing an uncertain future.

The acclimatization of fish and wildlife

Considerable work has been carried out in the Soviet Union on the introduction of valuable fish and wildlife species into new habitats. These acclimatization efforts may be divided into three categories: the re-introduction of native species into areas where they once lived but from which they have since disappeared, the introduction of native species into entirely new areas, and the introduction of exotic species into suitable environments within the U.S.S.R.

Work of the first type represents the dominant effort. The intrusion of man into the forests, steppes, and other natural habitats of wild animals have forced many of them into much smaller ranges than they previously enjoyed. The cutting of forests has reduced the natural breeding area of many valuable fur-bearing animals, and the use of the forest-steppe and steppe for crop and livestock production has greatly diminished the natural range of many wild ungulates. Efforts to re-establish a number of species, such as the beaver, sable, elk, red deer, sayga, and others into areas where they formerly roamed have been conducted for many years.

Diffusion of the valuable Siberian sable has been a central research project of the Barguzin Preserve on Lake Baykal, which lies in a major natural habitat of this animal. Thousands of sables have been introduced into new locales, with the Urals as one of the primary receiving areas. The results of work with the Siberian sable thus far have been encouraging.

Very successful results have been obtained from re-introducing the river beaver into suitable habitats over much of European Russia. Almost extinct in the U.S.S.R. in the early part of the century (less than 1000 remained), its numbers have increased many tens of times. The main centers for the re-acclimatization of river beavers have been the Voronezh and Berezina preserves.

Red deer were formally found in almost all forested areas within the present territory of the U.S.S.R., but this majestic animal is now limited to the Caucasus, the Baltic region, and a few other scattered areas. In the 1930s and 40s, various areas in European Russia were successfully populated with the Siberian variety of the red deer (*maral*), and the species as a whole is making a comeback.

In addition, the habitats of the European elk and roe deer have increased in recent decades, although this has been achieved more through hunting bans and proper husbandry than through artificial transplantation. The same is true of the sayga and Persian gazelle (*dzheyran*), although some of these species were transplanted to the island preserve of Barsa-Kelmes in the Aral Sea.

Species which have been introduced into entirely new areas of the country include the brown hare, sandy suslik, Teleut squirrel, raccoon dog, and arctic fox. Various degrees of success have been realized.[36]

The brown hare and sandy suslik (the only type of suslik not considered a pest, due to its valuable fur) were also introduced on Barsa-Kelmes island. A remarkable degree of success was realized in the case of the sandy suslik; only moderate results were achieved with the hares. Both have subsequently been introduced elsewhere.

Efforts with the Teleut squirrel and raccoon dog have had mixed results, mostly negative. Both took well to the areas into which they were transplanted, but the Teleut squirrel in its new Crimean habitat underwent a noticeable decrease in the quality of its fur. The raccoon dog has been suspected of destroying game birds and has even been associated with rabies epizootics which took place in the 1940s and 50s.[37]

Efforts to introduce the arctic fox into the coastal islands in the north and east of the country have not been successful, due in part to the interruption of its normal migration habits.

Animals which have been experimentally transported into new areas in lesser numbers include skunks, foxes (other than arctic), steppe marmots, stone marten, kolinsky minks, and sea otters.

Three species of animals are dominant among those which have been brought into the Soviet Union from other countries. These are the muskrat, the American mink, and the nutria. The muskrat, first introduced in 1927, has done so spectacularly well that it is now one of the leading commercial animals of the country (Table 5.1). Several million pelts are taken annually. However, its presence and competition have been a major cause of the decline in the numbers of the native desman (*vykhukhol'*).

The American mink, first released in 1933, has successfully adapted to the mountainous regions of central and eastern Siberia, and in the Tatar A.S.S.R. In some areas it is now numerous enough to be used as a commercial species.

Nutria (or coypu), first brought to the Soviet Union from Argentina in 1930, have been introduced in many of the southern republics, but only in Transcaucasus have they done well. North American raccoons and rabbits from Switzerland have been introduced in specific locations in small numbers. Recent efforts to acclimatize South American chinchillas into Central Asia have thus far proven unsuccessful.

In specialized game farms, such as Askaniya-Nova in the Ukraine, animals from all over the world are kept in the open, but they are not intended for dissemination elsewhere in the country.

The emphasis of Soviet specialists on the acclimatization of wildlife has been questioned by a leading Soviet biologist, V. G. Geptner, and led to a discussion as to whether this approach, or one which would try to increase the numbers of game animals in their natural habitats, was the more advisable method.[38]

Special studies are also made of the entrance of animals into habitats newly created by man, such as shelter belts and reservoirs, and of the zoocoenoses which are created there. Soviet biologists, like their American counterparts, also keep track of the movements of birds and animals by the process of banding, which is carried out at many of the *zapovedniki* and research institutes throughout the country.

Efforts have also been made to introduce fish runs into new areas. One of the major effects has been to establish runs of pink salmon in the White and Barents Seas. An early attempt in the 1930s had little success, but a new effort was initiated in 1956. Over 61 million eggs were transplanted between 1956 and 1960, the year of the first appreciable returns. Not only were tens of thousands of pinks observed in the rivers of northwestern Russia in 1960, but they were also found in the rivers of Finland, Norway, Scotland and Ireland.[39] Results since 1960 have been variable, but the project appears to have long-term promise.

In other cases of acclimatization, the Black Sea mullet (*kefal'*) has been successfully introduced into the Caspian and Aral Seas, sevruga sturgeon have been transplanted from the Caspian to the Aral, and Baltic sprats have also been brought to the Aral.

A great deal of work has been done in stocking the many new freshwater reservoirs which have been created throughout the country. The types of fish which have been introduced vary greatly depending on the local natural conditions. As one example, *omul* from Lake Baykal have been transplanted into the Bukhtarminsk Reservoir. Some reservoirs, such as the Tsimlyansk, have been turned into highly productive fisheries. The stocks of existing lakes have been augmented, too. For example, bream, carp, and zander have been introduced into Lake Balkhash.

The Soviet Union is also experimenting with the introduction of certain types of carp into some of its large irrigation canals, such as the Kara-Kum, in the hope that they will assist in reducing the growth of nuisance vegetation in and along such systems.

In addition to the work with fish, initial efforts have also been made to transplant king crab from the Pacific to the Barents Sea, and to introduce Pacific shrimp into the Black Sea.

International cooperation in fish and wildlife conservation

Many migratory species of wildlife can only be properly protected through international cooperation. Other species which are not migratory but which are threatened with extinction can be preserved only if the countries in which they remain are willing to take joint action towards this end. The Soviet Union is a representative to many international regulatory commissions, and has entered into numerous bilateral and multilateral agreements for regulating and conserving a wide range of natural resources, particularly in the last two decades.[40] Some of the most important of these have dealt with the conservation of fauna resources.

Soviet participation in the international effort to restore the wisent has been noted above. It might be mentioned that one of the preserves in which this work is being conducted, Belovezhskaya Pushcha, lies on the border with Poland immediately adjacent to a Polish national park of a similar name which is also working on the preservation of these animals. Trans-border movements of wisents between the two preserves have been noted.

Another area of wildlife preservation in which the U.S.S.R. has taken perhaps stronger conservation measures than any other country is the preservation of the polar bear. All hunting of this animal has been prohibited in the Soviet Union since 1956, and at international conferences on the preservation of the polar bear in 1965, 1968, and 1970 the U.S.S.R. urged that its arctic neighbors restrict the rate of hunting of this species. A permanent committee of international scientists was created at the 1968 meeting to supervise research on the polar bear, with S. M. Uspenskiy of the Soviet Union as its first chairman. The Soviet Union is also party to a 1964 agreement on the preservation of the fauna of Antarctica.

Although the highly threatened Przhevalskiy horse (*Equus caballus przewalskii*) no longer has a natural habitat within the Soviet Union, the

U.S.S.R. is assisting in efforts towards its restoration through a hybridization program at the Askaniya-Nova preserve in the Ukraine.

With respect to maritime mammals, the Soviet Union has entered into conventions with the United States, Canada, and Japan for the conservation of the fur seal (1957 and 1963), and with Norway regarding the harp seal, hooded seal and walrus (1957). The basic compact agreed to in 1957 concerning the fur seal differed little from one established in 1911 between tsarist Russia, Canada, Japan and the United States. As a member of the International Whaling Commission, the Soviet Union is party to several multilateral agreements regulating the extraction of whales (1946 and later years). However, the whaling practices of the U.S.S.R.

Figure 5.6. Przhevalskiy horses, an endangered Asian species now found only in zoos.

and some of the other Commission members have come under considerable criticism, and some specialists fear depletion of certain species such as the sperm whale.

The Soviet Union, however, is very active in the preservation of its own maritime animals. For example, in 1970 the I.U.C.N. reported that a new regulation

> bans private hunting of seals, Kamchatka otter, and other marine animals...A thirty mile zone where fishing and work of any kind are banned, including the erection of navigation signs, has been set up around the seal colonies on the Komandorskiye Islands. The area is out of bounds to ships.[41]

High seas fishing, especially along continental shelves, is an area in which international cooperation is absolutely essential if the resources are to be preserved. The U.S.S.R. is party to various agreements covering fishing in both the North Atlantic and the North Pacific. In the Atlantic, the U.S.S.R. has been a member of the International Commission for Northwest Atlantic Fisheries since 1958, and joined the Northeast Atlantic

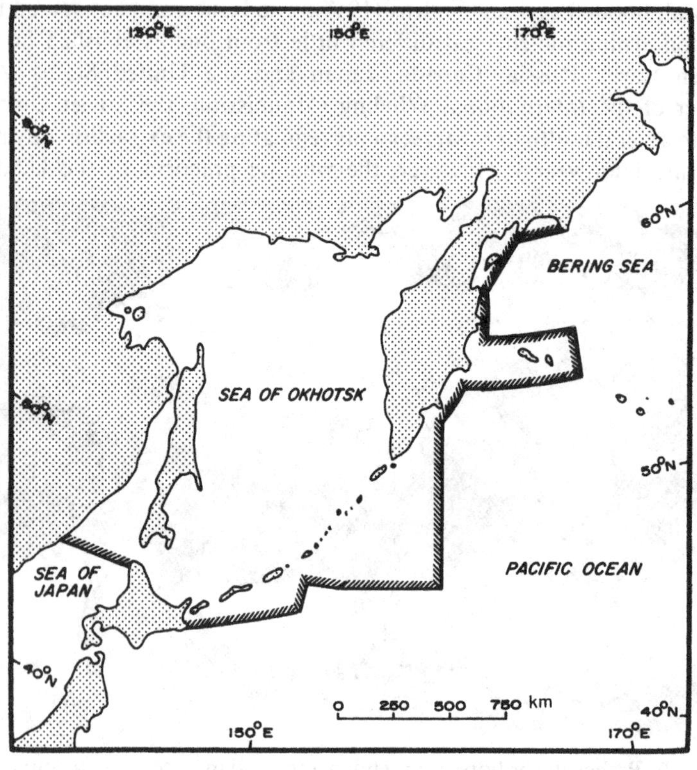

Outer limit of areas closed to open seas fishing for salmon according to provisions of the 1956 Soviet - Japanese treaty.

Figure 5.7. Prohibited areas for salmon fishing in the Soviet Far East. From *Washington Law Review*, **43**, No. 1 (October 1967) p. 74.

Fisheries Commission in 1959; it has been included in all subsequent conventions of both organizations. In addition, supplemental bilateral agreements are in effect with several countries, most notably Norway. The Soviet Union is also party to multilateral treaties signed in 1958 and 1959 for preserving fish resources of the Danube River and Black Sea, respectively.

In the Pacific, the Soviet Union has two major agreements, one with Japan (1956), and one with the United States (1964), both of which are subject to periodic review. The 1956 convention with Japan represents an effort to resolve a major fisheries conservation problem. In the mid-1950s, Japan began intensive high-seas fishing for pink salmon, most of which spawned in Soviet rivers. The Soviet Union, like the United States, does not fish the high seas for salmon, only along coastal waters. By 1958, the Japanese fishing practices began to take a heavy toll on the normal Soviet catch of pink salmon; between 1953 and 1958 the average annual Japanese catch increased by over four times and the average annual Soviet catch was reduced to about one-third its former size. The 1956 convention placed the entire Sea of Okhotsk and large parts of the Sea of Japan and the Bering Sea off limits to Japanese high-seas salmon fishing (Figure 5.7). In addition, the regulation of king crab and Pacific herring was added to the scope of the convention in 1958. King crab agreements with the United States were signed in 1965 and 1969.

In 1956, another Pacific fisheries commission was established by representatives of the Soviet Union, mainland China, North Korea, and North Vietnam.

The Soviet Union participates actively in the work of the International Union for Conservation of Nature and Natural Resources (I.U.C.N.). The I.U.C.N. acts in the capacity of a coordinating agency for international research and conferences on all aspects of the conservation of natural resources, and leading Soviet scientists are active on its commissions and executive boards (Appendix 16).

Through their activities in the I.U.C.N., in various agencies of the United Nations such as U.N.E.S.C.O., and in other international organizations (such as the International Council for the Exploration of the Sea), specialists from the Soviet Union and other countries have continuing opportunities to discuss conservation problems and to seek to develop coordinated approaches towards resolving these problems. With time and the realization of more stable international relationships, it is possible that the initial achievements realized to date could broaden into significant accomplishments on larger and more complex conservation issues, and eventually lead to the creation of those global approaches necessary for assuring a sustainable high level of environmental quality for the entire earth.

6 The extraction and conservation of timber and mineral resources

...we accept wholesale (the) pernicious...assertion about 'inexhaustible natural resources'

In the management and extraction of a nation's forest and mineral resources, the interrelationships and interdependencies which characterize most communities of natural resources come quite clearly into view. Economic activities which interrupt these normal interrelationships frequently bring about many adverse secondary consequences which can manifest themselves over wide areas.

The cutting of forests, for example, can cause soil erosion, eliminate wildlife habitats, increase stream sediment, damage fish runs, and alter local microclimatic conditions. Surface mining activities may not only destroy forests or agricultural lands which overlie the deposit, but often seriously pollute nearby streams, induce soil erosion, and generally result in some degree of aesthetic disutility. Clearly, these resources demand the most careful planning prior to their use, both for their own sake and for the sake of other natural resources associated with them.

The conservation record of the Soviet Union in the development of its forests and mineral deposits has fallen considerably short of what even Soviet specialists consider desirable, and warrants an examination of some of the problems that have arisen and their causes.

The organization of Soviet forestry

The vast forest expanses of the Soviet Union, one quarter of the world's total, are one of the U.S.S.R.'s greatest resources, but their very abundance has historically produced a rather indifferent attitude toward their management and conservation. Unfortunately, this is a trend that has not been fully checked even to the present day.

The problem is not a lack of appreciation of the important natural benefits of forests. Soviet forestry experts are as cognizant of the role of forests in providing watersheds, regulating run-off, conserving soils, sheltering wildlife, and modifying climate as are their counterparts in other countries. The difficulty in the U.S.S.R., as elsewhere, seems to be in bringing forestry management specialists and logging and wood products specialists together, and coordinating their activities and goals.

A management institution for federal forest lands with the scope, efficiency, and authority of the U.S. Forest Service has never been brought into existence in the Soviet Union. Since World War II, the agency responsible for forestry practices has been changed or reorganized on an average of once every four years. In all cases, though, the loggers have generally been the dominant force and the foresters generally have lacked the administrative muscle to ensure and enforce proper forest conservation practices.

During the period of decentralized economic management (1957–65), forestry and wood processing activities were the responsibility of regional economic councils. After the removal of Khrushchev economic functions were again centralized, and from 1965 until 1968 the responsible agency was the U.S.S.R. Ministry of the Lumber, Pulp and Paper, and Wood Processing Industry, and corresponding ministries in each of the fifteen Union republics. On June 11, 1968, the U.S.S.R. ministry was abolished as such and two separate ministries were formed: a U.S.S.R. Ministry of the Lumber and Wood-Processing Industry (a Union-republic ministry) and a U.S.S.R. Ministry of the Pulp-and-Paper Industry (an all-Union ministry). At present, the agency charged with overseeing the management and conservation of the nation's forests is the 'State Forestry Committee', first established under a different name in 1962. This Committee apparently superseded in function the 'Central Forestry and Shelter Belt Administration' under the U.S.S.R. Ministry of Agriculture, which existed in the late 1950s. Thus far, however, the State Forestry Committee has had little success in correcting the many inefficient timber harvesting practices still prevalent in the country. In 1970, committees of the U.S.S.R. Supreme Soviet were reportedly preparing a draft of basic forestry management and conservation legislation for the whole of the Soviet Union. Portions of the existing Russian Republic conservation law pertaining to forestry are presented in Appendix 18.

A brief description of the resource itself would be desirable at this point. The forests of the Soviet Union collectively constitute what is termed the State Forest Reserve (*Goslesfond*). The State Reserve takes in 1238 million hectares, or more than half of the entire land area of the country. As is the case with the National Forests of the United States, not all this land is actually covered with forests. The actual forested area of the Soviet Union is 738.2 million hectares, and, if burned and unreforested logged areas are included, this figure becomes 910 million hectares.[1]

Of the forested area, 36.4 million hectares, or about 5 per cent, belongs to collective farms (and therefore is the responsibility of the Ministry of

Agriculture), and 15.0 million hectares is attached to other Ministries or organizations. All the rest (686.8 million hectares) is the responsibility of the State Forestry Committee. It is worth pointing out, too, that over 93 per cent of all the forests of the U.S.S.R. lie in the Russian Republic (R.S.F.S.R.); the other fourteen republics account for less than 7 per cent.

The total timber reserves of the U.S.S.R. are estimated at about 76 billion cubic meters. The volume cut annually has held fairly steady over the past decade, amounting to between 350 and 400 million m³ each year;

Table 6.1 *Composition of Group I forests of the Soviet Union*

Use of forest	Total area (hectares)	Timber reserves (million m³)
Urban green belts and parks	7,806,000	770
Field and soil protective belts	9,912,000	1090
Health resort forests	932,000	198
Forested belts along highways	778,000	87
Forested belts along railways	843,000	84
Protected belts along rivers[a]	20,009,000	2477
Zoned for nut-tree utilization	1,938,000	332
Other restricted forests	2,028,000	234
Total U.S.S.R. Group I forests:	44,246,000[b]	5272

[a] The protected belt along rivers, established by law in 1936, is up to 20 km wide along portions of such rivers as the Don, Volga, Ob′, and Oka, and from 1 to 6 km wide elsewhere.

[b] This figure does not agree with the total for Group I forests given in Table 6.2 (40.3 million hectares) because the above total includes forested areas in Group II and Group III forests which are managed according to Group I specifications. Subsequently, this total has been greatly enlarged, mainly by the addition of 72,600,000 hectares of sparse forests bordering the tundra.

From V. P. Tseplyayev, *Lesa S.S.S.R.* (Moskva: Sel′khozgiz, 1961) p. 26 of the English translation. Data based on 1956 forest inventory.

the cut in 1965 was 378.1 million m³. The composition of the forest reserves of the Soviet Union by species is summarized in Appendix 19.[2]

The forests of the Soviet Union have been divided since 1943 into three management and conservation categories, based on their location or special function, and the degree to which they can be economically exploited or, conversely, must be protected for particular purposes. These three categories are:

Group I. This group takes in forests requiring the highest degree of preservation and protection. In it are included forests located in urban green belts, in parks and preserves, in shelter belts, protective forests along rivers and coastlines, those astride highways and railroads, and others of a similar nature (Table 6.1). Between 1956 and 1961, the area of Group I forests was increased almost four-fold (from 44 to 170 million hectares), mainly as the result of including large areas of scrub forests bordering on the tundra in this category. Only maintenance timber cutting is permitted in these forests, although the volume of this cutting has increased sharply since the mid-1950s.

Group II. This group takes in forests whose watersheds are of special significance, and most of the woodlands in the lightly forested regions of the central and south European portion of the Soviet Union. Some limited

Table 6.2 *Distribution of forests by management categories*[a]

	Area in State Forest Reserve		Forest covered area			
Category of forest	Million hectares	Per cent of total	Million hectares	Million hectares in R.S.F.S.R.	Per cent in R.S.F.S.R.	Volume of timber, billion m^3
Group I	60.8	5.6	40.3	32.7	81	4.7
Group II	76.2	7.0	56.1	26.7	48	4.1
Group III	947.3	87.4	584.5	580.7	99	66.3
Totals	1084.3	100.0	680.9	640.1	94[b]	75.1

[a] Excluding forests belonging to collective farms and to other ministries and organizations (totaling about 41 million forest-covered hectares).

[b] Overall per cent of forest area in R.S.F.S.R.

From A. I. Mukhin (ed.), *Les-nashe bogatstvo* (Moskva: Goslesbumizdat, 1962) p. 52. Data based on 1956 forest inventory.

timber harvesting is permitted in these forests under special regulations, and expansion of the forested area is often a management goal. In no case should the annual cut in a given area exceed the annual growth increment. Most of these forests lie in the mixed forest zone of European Russia.

Group III. Included in this group are almost all of the densely forested areas of northern European Russia, the Urals, Siberia, and the Far East. These are the commercial forests, subject to intensive timber harvest and all other appropriate uses. They lie almost exclusively in the Russian Republic (Table 6.2). The designation of a given forest tract to one of these three categories is not permanent, but may be changed administratively as circumstances might seem to warrant.

Timber and mineral resources

Problem areas in Soviet forestry

As indicated above, serious conservation problems exist in the management of Soviet forests, stemming primarily from the ascendancy of the timber harvesting interests over the forestry management organizations. The problems encountered may be categorized under the general headings of poor timber harvesting practices, waste or inefficient use of the harvested resource, and fires and forest pests.

Of the timber harvesting practices which adversely affect the long-range conservation of the resource, perhaps the one most strongly criticized by Soviet writers is the lack of permanent logging bases and effective sustained-yield harvesting. According to a 1963 report dealing with problems of the timber industry, the average volume of timber on the tracts assigned to the 2300 logging bases of the U.S.S.R. was 4,304,000 cubic meters, and the average procurement quota was 300,000 m^3. Therefore, the average life span of a lumbering operation in a given area is around fourteen years, at which time its activities must be terminated or relocated.[3] It should be noted that 2300 times 300,000 m^3 comes to almost double the annual timber harvest in the Soviet Union. Perhaps there is a problem of what type of averages are being used. In any event, even if half the given procurement quota were the correct 'average', the resources of the average base would last only 28 years. Since this is still much less than the length of time needed to regenerate a commercial coniferous forest, the need for nomadic harvesting practices would seem to be unaffected. Since the loggers can assume that the base will be abandoned and that they themselves will not have to harvest again on the same area, there is little incentive to spend time and money on extensive conservation measures. This nomadic system of harvesting creates other inefficiencies, too, such as the closing or relocating of local sawmills, long hauls to the under-utilized major processing combines, and difficulties in attracting workers with families.

Because logging enterprises do not have permanent responsibility for a given tract of land, incentive is low for carrying out proper rehabilitation work. Although reforestation measures are today usually carried out, often this involves only scarification, and in many cases this proves to be ineffective. As a result, in the early 1960s it was stated that 'only 30–35% of the total annual area of logged forests, comprising three million hectares, is replenished with the principal economically needed varieties of timber, half the area is grown over with deciduous varieties, and 10–15% is not renewed'.[4] It was estimated in 1960 that around 10 million hectares which had been cut in the preceding decade alone were not undergoing reforestation (and another 42 million unreplenished hectares existed as the result of forest fires).[5] Even in the mid-1960s, 300,000 to 400,000 hectares annually were not being properly cleared and reseeded. As a result, between 1963 and 1968 the reserves of coniferous timber in the U.S.S.R. decreased by 3,300,000,000 m^3 or approximately 5 per cent of the total stand. By way of similar evidence, that portion of the forested area of the State Forest Reserve which is under exploitation and which is classed as 'not covered by forests' increased from 61 to 94 million hectares between 1956 and 1961.[6] Although some areas are replanted with nursery stock or direct-seeded, as in the United States a considerable area is left to reseed itself by means of residual seed trees. For the Russian Republic (R.S.F.S.R.), where almost all of the commercial timber is located, the average amount of the clear-cut area which was directly planted or seeded equalled only 14.3 per cent over the period 1946–60. The failure rate of the plantings and seedings equalled about 16 per cent.[7] In these areas of the Soviet Union, the regeneration is quite frequently with aspen and birch, rather than with the more valuable coniferous species which were extracted. It has been recommended that clear-cut timber harvesting practices be used less often, and that the annual sown area be increased to about three times what it was in the early 1960s to insure proper regeneration of the necessary species.[8] However, no early transfer to a universal system of guaranteed sustained yield forestry appears to be planned:

> It would seem that by the start of the twenty-first century it will be necessary to bring most of even the densely forested regions of European Russia and the Urals fully under proper forestry management practices. Cutting 'on loan' [i.e. without proper reforestation] in these parts of the country will, in all probability, be retained only in the least cut-over sections of Arkhangel'sk Oblast and the Komi A.S.S.R. In Siberia, this manner of cutting can continue to be practiced in remote areas, but in the more settled regions proper forestry management procedures will also have to be adopted by this time. The regeneration of forests will start to be engaged in everywhere.[9]

Thus it would appear that it will be another thirty years or more before assured reforestation of the commercial species being cut will be standard practice in the U.S.S.R.

Another significant syndrome running through Soviet forestry is the normal tendency to extract the most convenient timber, rather than the most desirable timber from the standpoint of forest conservation. For example, the extraction of timber is still heavy in the moderately and lightly forested zones of the country; it equaled 42 per cent of the total cut in 1959. This is particularly true in the southern half of the European part of the U.S.S.R., where there is a great demand for all types of wood products. For example, logging activity in 1959 was over twice the 1940 level in lightly forested Penza, Tula, and Voronezh Oblasts, and was more than four times greater in the very sparsely forested steppes of Rostov and Astrakhan' Oblasts.[10] In 1960 it was estimated that the reserves of mature timber in Moscow, Ryazan, Bryansk, Lipetsk, Tambov, and Kursk Oblasts would be exhausted in four to seven years at the existing rate of extraction.[11]

Large clear areas are also appearing in other easily accessible locations, such as along rivers, roads, and railways, and even around towns and cities in the heavily forested areas of Siberia. In the dense Group III forests of the northern and eastern regions of the country, there are 56 billion m³ of timber (over 70 per cent of all the timber in the U.S.S.R.) that are classified as either mature or over-age; yet, because of the appeal of these low-cost harvesting practices, much of this goes unused while 'convenience' cutting around cities and transportation routes, and in Group II forests, continues. Larch represents another case in point; although it is the most common species in Eastern Siberia, it is little harvested by the logging concerns simply because it is too dense to be conveniently floated via rivers to the mills. In a similar way, potentially valuable broadleaf species are often passed over in the effort to harvest the more attractive coniferous stands.

Despite the vast reserves of mature timber in northern European Russia and Siberia, not only has increased cutting been tolerated in the moderately tree-covered Group II forests of central European Russia, but much overcutting as well. In 1959 such overcutting amounted to 4.1 million m³, and overcutting in these areas has also been permitted in some subsequent years. For example, in 1961 the cut in Estonia was 112.5 per cent of the average annual growth, and in Latvia it was 120.7 per cent. Even in the Group III forests, cutting over the established quotas is not uncommon. In 1964, the volume cut in Perm' Oblast amounted to 197 per cent of the quota; in Gor'kiy Oblast – 178 per cent; and in Kostroma and Vologda Oblasts and Udmurt A.S.S.R. – 150 to 155 per cent.[12] Cutting has even taken place in Group I forests around Lake Baykal, where serious soil erosion and water pollution is threatened.

However, the inefficient approach of the timber industry to its natural resource does not stop with the manner in which it is cut. Exceptionally heavy waste of the harvested resource takes place in the course of its transportation, processing, and use. The complaint has been frequently raised in the press that up to a third of all timber cut in the Soviet Union is wasted (Appendix 20). An article in *Izvestiya* in 1958 noted that this included 50 million m³ annually wasted at the cutting areas (burnt, abandoned, etc.), and a similar amount wasted during milling. It was stated, for example, that 30 per cent of the timber cut in four Ukrainian oblasts between 1948 and 1958 had simply been abandoned at the site.[13]

In the United States, most harvested timber is transported to mills by trucks, but in the Soviet Union most is still floated down rivers. This practice has resulted in the following type of complaint:

The floating of unstripped logs singly, not in rafts, litters the rivers with sunken logs and poisons the water with the bark that comes off the logs and rots on the bottom. Between 1958 and 1961 approximately 825,000 m^3 of timber sank in the Kama alone, and in some places the river bottom was covered with a wooden roadway several rows deep.[14]

The press also occasionally complains of the inefficient and wasteful use of wood in industry and construction. In addition, much wood is still used for fuel in the Soviet Union. The amount of wood used in this manner (about 50 million m^3 annually) has decreased very little since 1940 and is three times the 1913 level.

A schedule of fines for enterprises that do not follow prescribed cutting practices exists, but its effects are minimal. No manager objects to paying a few hundred or thousand rubles of the firm's money in fines, if the extra timber gained enables his firm to overfulfill its quota, thereby bringing in ten times the amount in bonuses payable to the individual employees themselves. A solution to this obviously fundamental problem has apparently not yet been found.

The fight against forest fires is as much of a problem in the Soviet Union as it is in the United States. In part, this is an inevitable result of the U.S.S.R. having vast areas of undeveloped and almost inaccessible forest lands in the northern and eastern portions of the country. Here, forest fires often range uncontrolled over large areas, even as they do in remote portions of Alaska and Canada.

One of the largest forest fires in recorded history burned within a general expanse of 160 million hectares in 1915, an area over twice the size of Texas. Within this area, stretching from the upper Lena to the Ob' – Irtysh confluence, over 12 million hectares of forests were burned away.[15]

In August 1957, an aerial survey observed fifteen to twenty unchecked forest fires in just the area between the Lena and Angara rivers in Eastern Siberia.[16] As much as a million hectares are consumed by fire annually, despite a network of observation towers and aerial surveillance. The figure of 42 million hectares was cited earlier as the estimated total area in the country which has been denuded by forest fires. Although Soviet foresters are giving increased attention to fire prevention measures, until such time as greater access is available into the vast stands of coniferous timber in Siberia and the Far East, losses from forest fires will probably remain high.

Damage from insects and other forest pests is another major cause of timber losses in the Soviet Union, as it is elsewhere. Extensive research is being conducted into this problem, and aerial spraying is being used in the fight against this type of despoilation. As noted in the preceding chapter, the potential of biological controls over insect pests is also being investigated.

Another source of forfeited forest land is the creation of large reservoirs behind hydroelectric projects. Besides the area directly flooded by the reservoirs, much additional land is lost through waterlogging, as many of the large reservoirs are situated on fairly level ground. Nearly 18 million hectares of forests have been lost in this manner[17] (see Table 7.1).

The forests on the collective farms, totaling 37 million hectares, are, if anything, managed even more poorly than are the Group III forests under the State Forestry Committee. Much wood is extracted from these forests and used mainly for firewood. The selling of timber from the collective farm forests is often a convenient way of supplementing a farm's annual income. Not only are the forests on many collectives heavily overcut, but reforestation is the exception rather than the rule, despite the fact that all collective farm forests are classified as either Group I or Group II forests. Some collectives have considerably more forest land than they do plow land, and it has been urged that excessive tracts of forests on collective farm land be transferred to the state reserve.

As noted above, the Soviet Union does not have at present an administrative agency with the enforcement capabilities of the U.S. Forest Service. It has been frequently suggested that the inefficient use of the U.S.S.R.'s forests be ended by the creation of such a comprehensive forest management ministry, one which would have responsibility for the scientific management of all the nation's forests, and for overseeing the proper commercial extraction of their timber resources. Such a ministry existed and was reported to have worked reasonably well during the period 1947–53, and calls for the re-establishment of something similar have been made for over a decade.[18] The abolition of the regional economic-administrative councils in 1965 was no doubt a useful step in this regard, since a more coordinated, national administration of forestry practices became possible. However, it goes without saying that the controlling force in any such unified ministry would have to be the forest managers, rather than the economic functionaries, if such an agency were to be an effective answer to the present low level of resource conservation which still characterizes too much of Soviet forestry.

Conservation in the extraction of mineral resources

Of all the nations of the world, the Soviet Union comes closest to being self-sufficient in its reserves of mineral resources. It has tremendous reserves of most of the main industrial raw materials, in particular coal, oil, and the ferrous group of metals.[19] Unfortunately, as was the case with its vast forest resources, the ready availability of mineral resources has generally not encouraged their efficient development or the utilization of proper conservation procedures.

Extraction of mineral resources

The prospecting for new mineral wealth is carried out by the U.S.S.R. Ministry of Geology, which, prior to a 1965 reorganization, was known as the Ministry of Geology and Conservation of Mineral Resources. Responsibility for the conservation of mineral wealth during its extraction and basic processing belongs to State Committees for Mining Supervision and Safety (acronym: *Gosgortekhnadzor*) which operate in each of the Union republics. As was the case with forestry, a new fundamental mining law was reportedly being drafted in 1970.

Unlike most other natural resources, mineral resources generally have no value in their natural state; they take on value only when exploited (it should be emphasized, however, that the land overlying these mineral resources might have considerable value even while the minerals themselves lie undeveloped). Also, they are nonrenewable in their natural state, at least in the time span we are interested in. Therefore, in the case of mineral resources, conservation problems usually center around questions of the rate of their use, of the avoidance of waste in their use, and the possibility of finding substitutes. There are also other associated problems, such as landscape renewal, and competing uses for the land over the mineral deposits.

The attitude of a nation toward the rate of use of its mineral wealth is always an interesting study. Generally speaking, there are three approaches: (1) since mineral resources exist in fixed and limited amounts, they must be used cautiously, possibly even rationed; (2) they can be used as needed, as exploration will turn up new reserves at a rate at least equal to the rate of use; or (3) they can be used as needed, as economic laws will bring substitute materials into use if and when a mineral resource begins to become scarce.

The first attitude is most often encountered in the early stages of a nation's industrialization, when only limited geologic exploration has been conducted and often only very high grade resources are being employed. Naturally, it is more common in small or densely populated countries than in those occupying vast territories. It is rarely voiced in either the United States or the Soviet Union today.

Of the latter two approaches, the second seems to be more common in the U.S.S.R., the third more common in the United States. Perhaps this is a natural reflection of the huge area of the former and level of technology of the latter.

However, the enormous mineral reserves of the Soviet Union have encouraged the popular development of the notion of 'unlimited mineral resources' to a greater degree than probably existed even in our own country. Although most Soviet scholars employ the term 'vast' mineral resources, in popular or propaganda writings the phrases 'unlimited' and 'inexhaustible' are still often encountered.[20] This idea has led to rather poor conservation practices in the extraction and use of Russia's mineral wealth, not only on the part of the workers themselves, but on the part of governmental supervisory agencies as well. Examples of this will be shown below.

A turning point may be approaching, however. An article in the January 27, 1968, issue of *Izvestiya* started out as follows: 'Among the many customary sayings that we accept wholesale and as a matter of course, one is erroneous and pernicious in its consequences. We mean the oft-repeated assertion about "inexhaustible natural resources".'[21] It is likely that this attack on the inexhaustibility concept will gain momentum as the rate of exploitation of these resources increases and as many lower grade ores are necessarily brought into use.

This is not to imply that the Soviet Union is not actively seeking substitutes for the products of many of its mineral resources. On the contrary, it is doing considerable work on the development of such power sources as breeder reactors, thermionic converters, fuel cells, controlled fusion reactors, solar batteries, semi-conductors, and so forth. Oil shale deposits are being developed, new concentrating plants are processing many lower grade ores, and a greatly increased emphasis has been placed on the chemical and plastics industries in the past decade. For the immediate future, it seems clear that steadily increasing amounts of conventional mineral resources will continue to be processed for the mandatory expansion of the Soviet economy.

The waste and inefficient use of mineral resources has been under attack in the Soviet press for many years, but from the unabated reappearances of such articles, it would appear that progress to date has been slow at best.

The most common target of these articles is the widespread practice of burning off natural gas which occurs in conjunction with the extraction of crude petroleum. Complaints against the practice have been raised for over ten years, but tremendous losses of burned-off natural gases are still being tolerated. In 1957, four billion m^3 of gas were burned off or escaped into the atmosphere; this wastage equaled about one-sixth of the entire national production for that year. In both 1962 and 1963 three billion m^3 were burned off in the Volga–Ural oil fields alone. In the mid-1960s about six to seven billion m^3 were being burned off each year, and in 1969 the total was stated to have reached ten billion m^3, the equivalent of several million tons of coal.[22] Including that gas which escapes into the atmosphere without being burned, the total lost can be assumed to be well in excess of ten billion m^3 annually. An unusual approach to extinguishing a wildcat natural gas fire in Central Asia was reportedly used in 1966, when an underground nuclear blast effectively closed off a large burning gas vent which had resisted control by conventional techniques.[23]

Crude oil may be seen lying on the ground in many producing areas. Large oil losses into the Caspian Sea, both from seaside refineries and off-shore exploratory and drilling operations have resulted not only in a substantial wastage of the mineral resource itself, but also in widespread water pollution and much loss of valuable fish life.[24] Ruptures in oil pipelines have also been reported.

Another particularly wasteful phenomenon has been the occurrence of underground fires in coal mines. Numerous such fires break out every year, sometimes causing losses of up to 400–500 tons of coal a day in a given mine. Often their cause, or the severity of losses, has been attributed to improper mining practices or conditions.[25] In addition, complaints are sounded from time to time about the inefficient use of coal, particularly the use of high grade coking coal in locomotives and thermal power plants, where lower grade coal could be employed. Tens of millions of tons of coking coal are used up in this way each year. The efficiency of extraction appears to be low as well. At some mines in the Karaganda and Kuznetsk coal fields almost 50 per cent of the coal suitable for mining is left in the ground; the average for the industry as a whole is about 28 per cent not extracted.

Objections have also been frequently voiced to certain practices in the mining of ferrous and non-ferrous ores and minerals. These include the mining of only the highest grade ores, while other ores of lesser but still exploitable quality are left in the ground, the extraction of only a portion of the different elements out of polymetallic ores, and the loss of much of even the primary metal in ores during processing. For example, at the Soviet Union's main iron ore deposits at Krivoy Rog in the Ukraine, ore with an iron content of less than 46 per cent is not used, as this would require a concentrating plant. However, it has been calculated that such a plant would pay for itself in just three years. At the Solnechnyy tin deposit, copper, lead, and zinc are left on the waste heaps, and at the Vysokaya Gora and Sokolovka-Sarbay iron ore deposits, the iron is used but copper and cobalt are lost.[26] At another metallurgical plant, the material in one of the waste dumps, which had been collecting for thirty years, was found to contain 10–15 per cent zinc.[27] In some cases, the concentration of metal in the waste product piles is several times higher than in the original ore which the plant is processing. At the Karabash metallurgical plant, only four of the fifteen mineral components of the copper pyrite ore were reported as being extracted.[28] Unfortunately, the new economic reforms, under which enterprises work to produce profits as well as to fulfill quotas, encourages the use of lowest-cost methods and hence the abandonment of lower grade or supplementary ores.

Timber and mineral resources

Large quantities of minerals are also lost into the atmosphere. One report has stated that the amount of sulfur discarded into the air as a component of industrial waste gases is considerably greater than the total industrial demand for this basic raw material.[29]

A common source of waste of basic ores is brought about by planners assigning a mine a norm for the metal content of its extracted ore which is much higher than the average concentration to be found in the deposit. As a result, only the highest grade ore is extracted and the rest is left behind. It has been stated that at the important Dzhezkazgan copper mines in Kazakhstan a third of all the copper reserves are being lost in this way.[30] In a similar way, incorrect planning targets and other inefficiencies have been cited as causing more than 80 per cent of the Muscovite mica at the Mama-Chuya deposit in East Siberia to be wasted.[31]

Unfortunately, such wastage does not end with the processing of the ore. Lack of adequate loading, unloading, and storage facilities often causes further losses during the shipment and storage of many processed minerals and mineral products, such as coal, cement, and particularly fertilizers.

With regard to the restoration of areas affected by open pit, strip, hydraulic, and other forms of surface mining, it appears that in general the necessary capital to perform this type of work is not being provided. It is known that a very large area is so affected. Currently, for example, about 70 per cent of all iron ore is extracted from surface mines, and about 25 per cent of all coal. For the future, surface mining is talked of as the preferred form of mineral extraction wherever feasible. For example, in the huge Kursk Magnetic Anomaly, a major future iron ore development area, it is thought that it may be possible to conduct open-pit mining from a rolling plain to a depth of more than 200 meters. There is controversy over this point, however. Such huge depressions could adversely affect agriculture by lowering the surrounding water table, as well as by dust blown from the overburdened piles. In addition, acid water pumped out of the pit could pollute nearby streams.[32] Some underground mining would probably be necessary in any event, as some of the high grade ore lies at depths of 400 to 450 meters.

Despite the fact that such surface transformations cover a very large area, despite the fact that the desirability of restoring surface mines to agricultural uses and quarries to productive lakes is readily admitted, and despite the fact that both the 1968 U.S.S.R. Land Law and the 1960 R.S.F.S.R. Conservation Law call for such work (Appendix 5, Article 11, and Appendix 6, Article 2), the necessary expenditures to reclaim and redevelop these lands are not being allotted. The Chairman of the Commission for the Study of Productive Forces and Natural Resources of the U.S.S.R. Academy of Sciences recently pointed this out when he stated that 'there are hardly any examples now of the reclamation of land destroyed in open cut mining – there are tens of thousands of hectares of such land'.[33] Another source suggests 30,000 to 35,000 hectares per year as the annual increment to lands affected by surface mining operations.[34] The small degree of rehabilitation of these lands which takes place is suggested by the following experience in the Georgian republic:

> In the past 20 years more than 4000 hectares of collective farm and state farm land, including more than 1000 hectares of plowland and land planted to perennial crops, has been allocated for open-cut mining. More than 70% of the area allocated for open-cut mining has been worked out, and this land is no longer suitable for further mining. However, to this day only 100 hectares, i.e., 3% of the worked-out area, has been levelled and returned to the users of the land.[35]

The amount of stream pollution which may be resulting from unrehabilitated mining operations is likewise hard to estimate, but based on experiences in other countries, including the United States, it would be surprising if it were not substantial.

Conflicts between underground mineral deposits and the desirability of keeping the overlying land inviolate have inevitably occurred, and while the mechanism for resolving such conflicts in the Soviet Union can only be surmised, there is evidence that the outcome is often weighted in favor of the economic developers. Instances of such conflicts have come to light, and in several cases outstanding *zapovedniki* have been abolished to allow geologic exploration and development, as was noted in Chapter 4.

Another area of direct relevance to the exploitation of mineral resources is oceanography. Vast reserves of many minerals exist in the ocean water itself, in concretions on the ocean floor, and in deposits lying under the world's oceans and seas. Current Soviet endeavors in studying and developing the natural resources of the sea are noted in the following chapter on water resources.

With regard to both forestry and mineral extraction, one final point which was brought out earlier in Chapter 3 should again be noted. That point centers around the current discussion in the U.S.S.R. concerning pricing and charging for the use of land resources, a suggestion that applies equally as well for timber and mineral resources. Although a lack of coordination between geologists, mining enterprises, and those who plan economic targets accounts for much of the inefficiency in Soviet mineral extraction, the lack of a precise monetary value for the resource itself is also cited as a prime cause of the waste.[36]

Both the Soviet and American experiences have shown that so long as forests and mineral deposits are available as 'free goods' and have no inherent value to those who develop them for use in the economy, there will continue to be little built-in incentive to guard against waste during their extraction and processing. To a Soviet mining engineer, for example, there is little reason to avoid wasteful practices in the extraction of ores if the losses are not reflected in anyone's economic accounting. Even in the Soviet system, the monetary value of the ore must be at least equal to the cost of geologically exploring for it, but the mining enterprises are not held responsible even for this measurable value of the worth of the deposit. Similarly, if trees are assigned no *in situ* value, then whose concern is it if a few 'extra' are cut and then burned or left to rot during the course of logging operations? Even if fines are levied against such practices, how can it be assured that they will be at all commensurate with the losses involved if there is no ruble value placed on the wasted resource?

The suggestion by Soviet specialists that an economic value be assigned to biotic and mineral resources as they exist in their natural state is designed in part to act as a cost-accounting check on developmental extravagances. However, as noted in Chapter 3, such suggestions have apparently been overruled for the foreseeable future. This makes it all the more mandatory to train those who extract the U.S.S.R.'s natural resources in the most advanced concepts of conservation and ecology, and to back up this training with adequately stringent and rigorously enforced penalties. Otherwise, a striving for the minimum short-term operating cost, an innate desire of unenlightened natural resource users under any form of economic system, will act to perpetuate the existing inefficient use of the Soviet Union's forest and mineral resources.

7 Water resource utilization and conservation

...these projects have caused serious, largely unforeseen difficulties...

The development of a nation's water resources has traditionally been hailed as one of the forms of economic investment most bountiful in its returns. In most cases, the social benefits from such projects are indeed very great. It is also true, however, that virtually any form of hydrotechnical development will likewise involve the loss either of other resources, or of some of the alternative ways in which the original water resource itself might have been used.

Unfortunately, it is all too common for planners to overlook, either intentionally or through oversight, many of the very real losses encountered as the result of the construction of new hydroelectric, irrigation, and flood control facilities. Many projects are authorized simply because they have been 'proven' to be economically feasible, but this feasibility normally takes into account only the cost of the more direct and immediately apparent losses – relocations, flooded forests and farms, fish passage facilities, and so forth. More subtle losses to the surrounding environment and resources, and to those who use them (the so-called external effects, or 'externalities'), usually do not appear among the costs of the project. Nevertheless, these external losses inevitably come to the surface, and their emergence has caused considerable second guessing on more than one project.

This phenomenon appears to be equally common in both socialist and capitalist economies, and the frequency and magnitude of these unforeseen losses form the basis of the conservationist's continuing concern with plans for new water development projects. The record of the Soviet Union in this respect is open to some serious questioning, and an effort to introduce greater ecological sophistication into Soviet hydroelectric project planning has emerged over the past decade.

The planning and management of the U.S.S.R.'s water resources

The management of the vast water resources of a country as large as the Soviet Union is a complex matter, and, as in the United States, more than one agency shares in this responsibility. One of the primary administrative agencies is the Ministry of Reclamation and Water Resources Management, and its counterpart ministries in each of the fifteen Union republics. In addition, there exist national and republic ministries of Power and Electrification, and, as noted in earlier chapters, of Fisheries and of Agriculture, all of which are directly concerned with the management of the country's water resources. Departments dealing with water resources exist within both the state planning agency (*Gosplan*) and the State Economic Council, and state committees on water development exist under the Council of Ministers of the R.S.F.S.R. and other Union republics.

The nature of the relationship between these agencies, the numerous research institutes which deal with water resources, and the Communist Party itself in deciding the basic directions of river developments, major irrigation schemes, etc., is somewhat unclear. It is widely believed that at various times in the history of the Soviet Union, particularly under Stalin, many large scale projects were initiated primarily at the insistence of the Party, and possibly at the insistence of Stalin himself, with insufficient attention given to their effects on the surrounding environment. Such one-sided approaches are in disrepute today, but the political mechanism for deciding what will actually be undertaken from among those projects deemed economically feasible remains uncertain. It can be assumed, however, that a major role in setting forth overall trends is played by the planning sessions of the Party Congress immediately preceding each new five-year plan.

The actual designing of new projects is carried out by various technical-research organizations which exist at both the national and republic levels. Among these is the S. Ya. Zhuk All-Union Design and Research Institute for Hydrotechnical Installations (*Gidroproyekt*) which has designed many of the Soviet Union's largest hydroelectric facilities. Similar institutes (termed *Giprovodkhozi*) are associated with the U.S.S.R. and republic Ministries of Reclamation and Water Resources Management; these are more concerned with reclamation and the agricultural uses of water. In addition, there are numerous other research institutes connected with governmental ministries and the Academy of Sciences. Other specialized agencies exist for both the construction and the operation of hydrotechnical facilities.

About three-quarters of the water resources of the U.S.S.R. are found in the Russian Republic (R.S.F.S.R.). Article 4 of the 1960 R.S.F.S.R. Law on the Conservation of Nature deals with the utilization of water resources, and has been included as Appendix 21. Just prior to the adoption of this law, on April 22, 1960, the U.S.S.R. Council of Ministers issued a decree outlining measures for regulating and conserving the water resources of the country as a whole. A decree governing the use and conservation of subsurface waters was enacted on September 4, 1959. A new statement of basic principles governing U.S.S.R. water legislation was enacted by the Supreme Soviet in 1970. These principles are presented as Appendix 22.

Laws such as these, however, tend to be very general, while conservation problems tend to be quite specific, and republic-level laws which would explicitly regulate the use of local water supplies are often inadequate. Many types of problems are common to all water development projects, but every project will also have a particular set of problems and externalities uniquely its own. This underscores the importance of project planners being guided by persons skilled in ecology, ichthyology, regional land use, pollution control, and numerous other such subjects, in addition to the usual specialists trained in hydrology, reclamation, power technology, and economics. The former have been all too often either missing, or their advice has been underestimated, when feasibility studies for Soviet projects have been carried out.

Benefit–cost analyses as used in the United States compare the expected annual benefits (monetary and other) from all aspects of a project with the amortized construction costs plus the annual operating expenses and other associated costs. The greater the ratio in favor of the benefits, the more desirable the project is considered to be.

This type of benefit–cost analysis (BCA) for Soviet hydrotechnical developments is generally not made public. However, in 1961, one such set of cost and return figures was published, albeit they were compiled eight years after the construction was completed. These figures are for the Mingechaur project, the major power and irrigation complex in the Transcaucasus.[1] This study is now several years old, and perhaps more sophisticated approaches have since been created, but at that time the benefit and cost figures cited were extremely generalized, and, to all appearances, most incomplete. The study shows an almost total disregard for secondary costs and benefits, and, in addition, it must be remembered that (*a*) there were no direct interest charges for the use of the capital which built the dam, and (*b*) no charge is made for the use of irrigation water. Obviously, these conditions will greatly restrict the use of any normal BCA procedures.

The system for determining project feasibility most frequently used in the U.S.S.R. is to compare the 'recoupment period' of two alternative projects to an arbitrary norm. The 'recoupment period' represents the length of time the more capital-intensive variant (but having the lower operating costs) would require to equal in total monetary outlays the second variant which would be cheaper to build but more costly to operate. The recoupment period (T) may be expressed as $T = (I_1 - I_2)/(C_2 - C_1)$, where I equals the initial cost of the two variants and C their operating costs. The benefits of the two variants are assumed equal (a somewhat questionable assumption, unless all secondary and intangible costs and benefits have been fully considered, which is always doubtful). If this period of time is less than a set norm, the capital-intensive project will be built; if not, the initially cheaper variant will be constructed. Obviously, arbitrarily setting the length of the recoupment period norm provides the planners with as much built-in flexibility as does the use of very low interest rates in American benefit–cost analyses.

Using the American benefit–cost approach for this particular project (Mingechaur), the benefits and costs, at least as stated in the article, do not seem to be very comprehensive. As stated, they do not seem to include such factors as the silting of the reservoir (located in a very arid area), the annual loss of agricultural production foregone by the flooding of 53,300 hectares of land, the changes in the delta ecology resulting from the altered stream flow regimen, the monetary benefits to navigation which are resulting from the project, the cost of measures to overcome bank collapse and shoreline erosion, or the future costs of averting salinization on lands affected by the project. The evaluation also attributes large returns to the project from flood control, but size-of-flood estimation procedures in the Soviet Union have also come under question.[2]

Another serious detriment to any meaningful BCA study of the Mingechaur data is that the electricity produced by the dam can be sold at arbitrarily set prices which, in this case at least, appear to be extremely high: the reported cost of producing one kWh of power is 0.2 kopecks, but this kWh is sold for about 1.75 kopecks.[3] The 0.2 kopecks kWh includes total amortization of the dam, power station, and irrigation network over about a 100-year period, or of the dam and power station in about fifty years. In view of the overlooked secondary costs, it might represent an unrealistically low figure. Although the real marginal cost of producing the power could theoretically be as high as 1.75 kopecks, the point here is that any hydroelectric project can be made to appear feasible in terms of a benefit–cost ratio if the power produced can be sold at arbitrarily set rates sufficiently high to cover whatever the cost of producing it might be.

The purpose of all of the foregoing is to illustrate that the Soviet Union, like the United States, apparently lacks any valid means of deciding whether a major water development project is truly beneficial from the standpoint of total, integrated resource management, rather than just from the standpoint of economic profitability. Present procedures in both countries would seem inadequate to commit with confidence, in all cases, large amounts of monetary and natural resources in such irreversible ways; and to preclude the alternative use of the capital elsewhere in the economy where it might actually give a greater net social return. In any event, anyone familiar with BCA procedures as used in the United States will find this Soviet study a most interesting article.

This tendency for Soviet planners to overlook hidden secondary costs and benefits has resulted in many undesirable consequences to associated natural resources. Damages to fish runs and wildlife habitats have already been noted in Chapter 5. Decreases in flood-plain fertility below some large reservoirs have been noticed, extensive waterlogging and other hydrologic changes are common, and large reservoirs in almost all cases have affected microclimatic conditions in the immediate environs. The seriousness of the problem is acknowledged:

The construction of systems of hydroelectric stations. . . has had great importance for the country's power development. At the same time, these projects have caused serious, largely unforeseen difficulties in the utilization of other natural resources. For example, reservoirs in the European part of the U.S.S.R. flooded more than 4 million hectares of low-lying land, much of which was agriculturally highly productive. Problems also arose in connection with the spoilage of additional land from a rise in the water table, beach erosion, disruption of transport lines, the inaccessibility of upstream reaches for migratory fish, [etc.]. All these kinds of damage have not been adequately taken into account in selecting dam sites; in fact, an economic comparison of alternative sites is impossible as long as we lack a system for evaluating the basic flooded object, namely the land.[4]

Difficulties such as these in dealing with the complexities inherent in water development planning have led to considerable discussions in recent years. Both national and republic-level laws which specifically define the relationships between water suppliers and the various categories of water users, and regulate and conserve local water supplies, have been reviewed and termed inadequate.[5] In the early 1960s, a study of the long-term integrated development of the country's water resources was initiated. The suggestion of encouraging more public participation in the planning of water resources has been made. There is also a need for an inter-ministry (or supra-ministry) commission to coordinate and direct the use of the nation's water resources.

Perhaps most significantly, the practice of the free availability of water, like the free use of land discussed previously, has come under question. Irrigation water which has been calculated to cost an average of 0.7 kopeck per cubic meter is not only provided free, the irrigated farms receiving it are not even required to keep account of it.[6] Naturally, this provides little incentive for its efficient use, and seriously undermines water conservation efforts in those dry areas where such conservation is needed most. A system under which irrigation water was sold was in effect from 1949 to 1956, and arguments for re-instating such a program appear to be very strong. However, as can be seen from Article 15 of Appendix 22, it is anticipated that water will continue to be provided free of charge.

Water resource utilization and conservation

Another possible cause of an improvident attitude towards the use of the Soviet Union's water resources was noted in Chapter 1, and merits repeating here. This is the tendency of Soviet natural resource specialists to classify water as an 'inexhaustible' resource. The danger in this lies not so much in whether this categorization is technically accurate or not, but rather in the psychological attitudes towards the use of water which it promotes. If water is popularly believed to be an inexhaustible resource, concern for preventing its waste and pollution will be that much harder to instill in the minds of its users. A change in this particular aspect of Soviet natural resource education might be a useful step in advancing water conservation in the U.S.S.R.

It has been the normal situation in the Soviet Union for major river development projects to be undertaken primarily for the benefit of the electric power industry, with insufficient consideration for the needs of other branches of the economy or for the possible adverse environmental consequences of such projects. Currently, Soviet planners appear to be orienting their thinking more towards regional and integrated water resource development planning, and towards a fuller calculation of the entire complex of consequences which result from major dam and reservoir construction. This should represent a useful step toward averting some of the problems connected with particular projects, reviewed below, that have arisen in the past.

Conservation problems associated with Soviet river development planning

The Soviet Union has the most extensive river network of any country in the world (Fig. 7.1 and Appendices 23 and 24).[7] Used for centuries as the most convenient form of cross-country transportation, it has in the past few decades begun to be developed for power, irrigation, and deep-draft navigation. Both irrigation systems and hydroelectric projects have enjoyed considerable priority in the Soviet period. Irrigation is needed to make productive large areas of what are potentially some of the country's best agricultural lands in the warm, but arid, southern portions of the country. The many large rivers have made possible the development of hydroelectric power at extremely low costs per kilowatt hour.

However, as was noted in the preceding section, the development of even a small, single-purpose river facility has very complex and far-reaching implications for the use and conservation of other natural resources. Occasionally, unfortunate consequences have come about as the result of Soviet river development planning which has not taken these implications sufficiently into consideration. It is well beyond the scope and intent of this study to review each river development project or cascade of

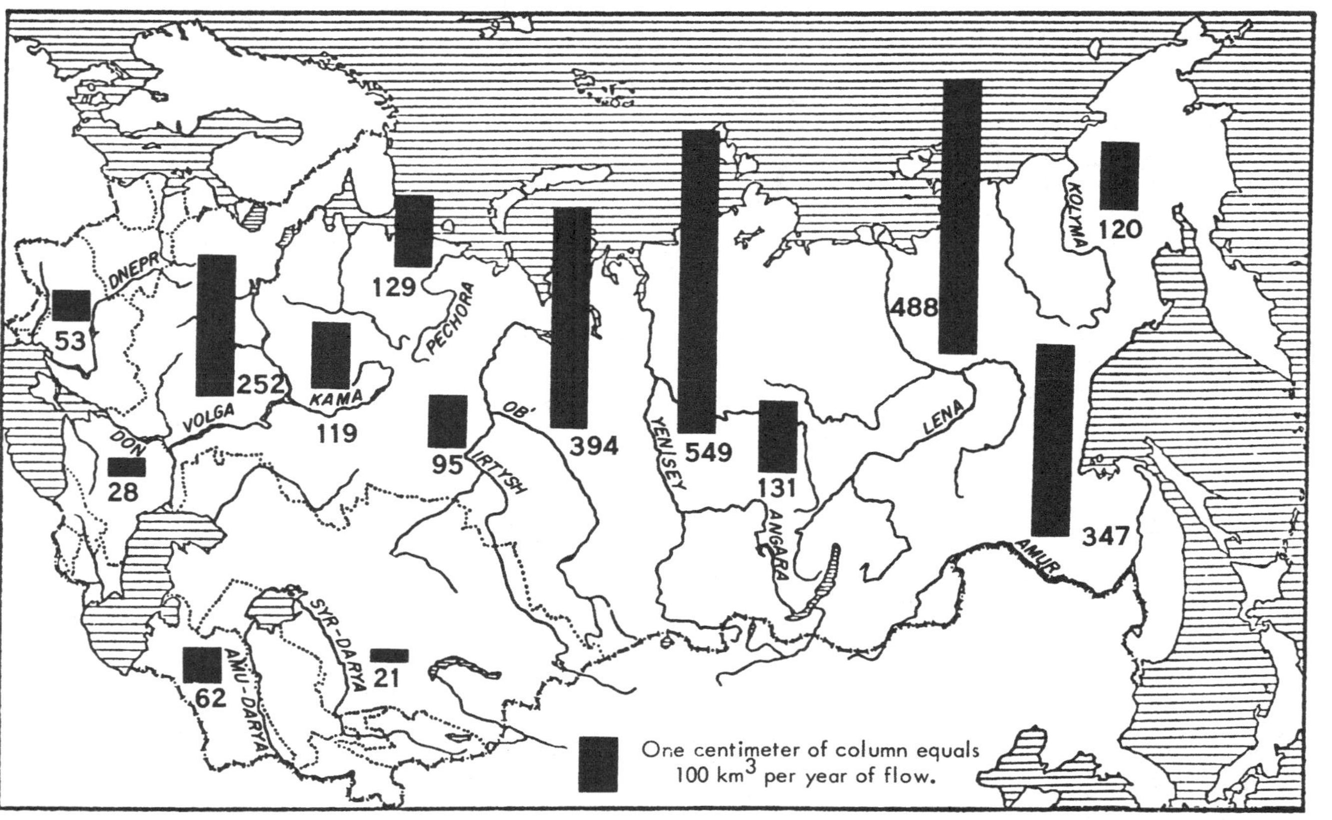

Figure 7.1. Flows of major rivers of the U.S.S.R. From L. K. Davydov, *Gidrografiya S.S.S.R.*, Vol. II (1955).

projects which the Soviet Union has undertaken. This section will therefore be limited to an examination of some which have had the greatest impact on other associated resources, or which have been the subject of prolonged debate regarding the wisest manner of their development.

Most of the largest river development projects in the U.S.S.R. have had power generation as one of the main goals of the project. At some of the large dams in Siberia, such as Bratsk and Krasnoyarsk, power is almost the

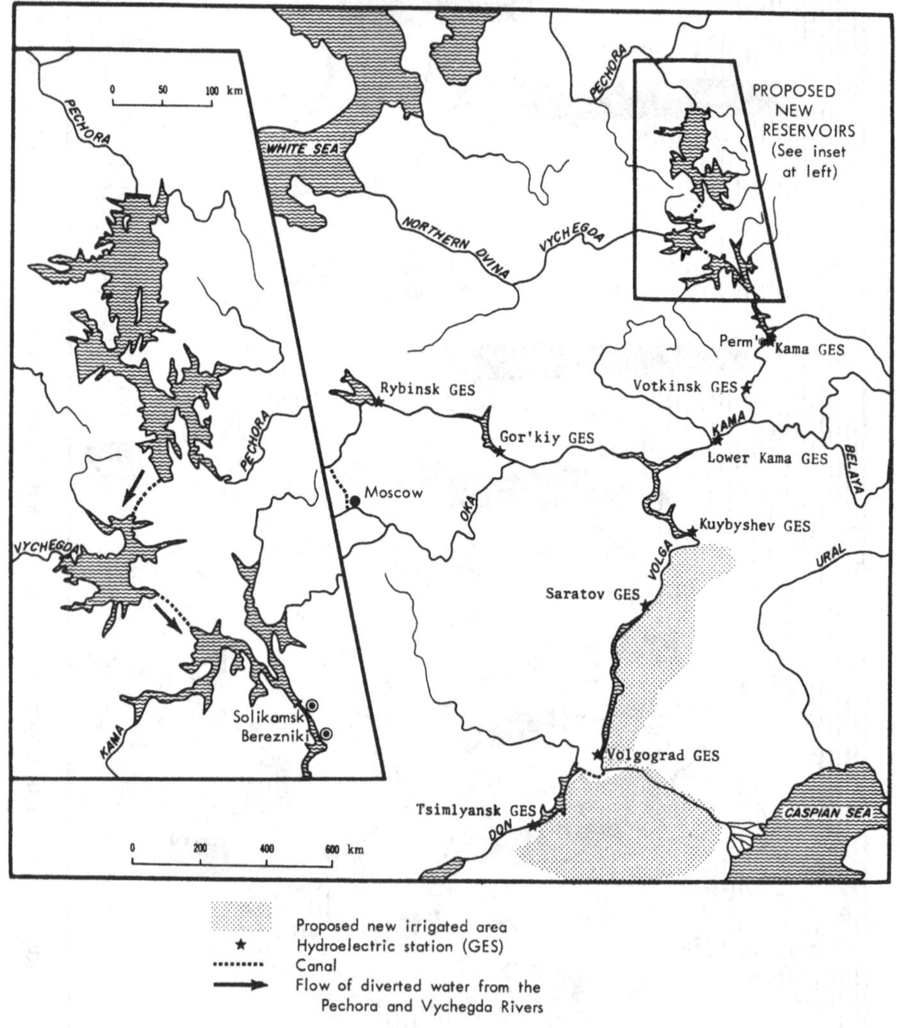

Figure 7.2. The proposed Pechora–Vychegda–Kama diversion. From *Gidro tekhnicheskoye stroitel'stvo* (1961) No. 7, p. 13.

sole consideration. In the European part of the country, however, the dams are frequently part of a multiple-purpose scheme involving power, navigation, flood control, and sometimes irrigation as well. The foremost examples of this are the cascades of dams on the Volga and Dnepr (Dneiper) rivers (Fig. 7.2 and Appendix 25).

An outstanding feature of Soviet hydro-developments, both in the western and Siberian portions of the country, has been the enormous size of the reservoirs created behind many of the larger dams. Some of these reservoirs have been very costly in terms of foregone forest and agriculture resources. The Kuybyshev Reservoir alone, for example, inundated approximately 277,000 hectares (684,000 acres) of agriculturally useful land (Table 7.1).

Table 7.1 *Agricultural and forest land flooded by large reservoirs*

Reservoir	Total area of reservoir (1000 ha.)[a]	Flooded agricultural land (1000 ha.)	Flooded forests and other wooded areas (1000 ha.)	Original water-course and other flooded land (1000 ha.)
Kuybyshev	600	277	165	158
Volgograd	326	110	65	151
Gor′kiy	175	107	41	27
Rybinsk	455	174	241	40
Tsimlyansk	270	195	30	45
Lower Kama (planned)	373[b]	176	110	87
Bratsk	550	166	357	27
Krasnoyarsk	213	115	37	61

[a] 1000 hectares equals 2471 acres. See Appendix 2.
[b] Other sources give 240,000 hectares.
From V. G. Grebenkin, 'O zatoplenii zemel′ pri obrazovanii vodokhranilishch GES', *Gidrotekhnika i melioratsiya* (1966) No. 3, p. 16.

For the U.S.S.R. as a whole, about 7,500,000 hectares of all categories of land have been flooded by hydroelectric station reservoirs, of which approximately a quarter has been agricultural land.

Although the reservoirs have permitted some other compensating lands to be brought under cultivation or to enjoy increased productivity, the areas irrigated with water from most of these large reservoirs is considerably less than the agricultural lands flooded by them. In addition, considerable moisture is lost from the reservoirs through evaporation, mounting to 5 km^3 annually from the Volga–Kama reservoirs and 4 km^3 from those on the Dnepr. These losses would be adequate to irrigate 3 to 4 million hectares of steppe lands.[8] Among the large reservoirs which exist at present in the U.S.S.R., only a few, such as Tsimlyansk, Kakhovka, and Mingechaur, are used extensively for irrigation. At present, most of the large irrigated areas of the country receive their water from diverted mountain rivers in the Caucasus and Central Asia. Only if the proposed Pechora–Vychegda diversion is carried out will the large reservoirs on the Volga be able to take on their originally planned irrigation role. The land flooded by these reservoirs is a factor of considerable importance in a country where good agricultural land is at a premium, and have undoubtedly been compounded by the lack of any reliable method for evaluating the economic usefulness of Soviet land resources.

Attempts have been made to intervene on behalf of conserving agricultural land, but generally the arguments of the hydropower proponents prevail. As an example, an effort was made to persuade the planners of the Ust-Ilim dam in Eastern Siberia to locate it above the confluence of the Ilim River, so as to preserve the towns and 20,000 hectares of agricultural land in the Ilim valley.[9] Despite persuasive arguments, the site chosen was at the location with the greater power-producing potential, below the confluence of the Ilim.

In areas where farm land has been inundated, the local towns and villages, and the transportation network which connects them, must be relocated as well. The larger the reservoir, the larger the relocation expenses involved; all the more so since rural settlements in Russia tended to develop along major rivers. Thus, the filling of the Tsimlyansk Reservoir necessitated entirely relocating 159 towns, villages, and hamlets, including the former *rayon* (district) administrative centers of Tsimlyanskaya and Verkhne-Kurmoyarskaya. The Rybinsk Reservoir necessitated moving over 600 settlements, including the town of Mologa. The Kuybyshev Reservoir required relocating about 300 towns and villages, including the city of Stavropol′; in all, an estimated 150,000 persons had to be moved from this one reservoir site. At Bratsk, 238 settlements, including the old town of Bratsk, were relocated.[10] The Bratsk dam also forced the rebuilding of a large section of the Tayshet–Lena River railroad, which had only recently been completed.

In addition to the towns which have been flooded, many others have had to be protected by dikes. The construction of the Gor′kiy Reservoir on the Volga, for example, required protective works around six cities and towns, and for the Kuybyshev dam, around eleven, including very extensive dikes, drainage canals, and pumping stations around the major industrial city of Kazan′.[11]

Fish losses are another cost attributable to the damming of rivers, and the unfortunate effects of the Volga dams on anadromous fish runs was discussed in Chapter 5. Table 5.4 shows the effect on fish catches brought about by some of the large dams. It should be noted that the nesting grounds of waterfowl and fish spawning grounds can also be adversely affected, and can in fact be completely wiped out by sudden releases of water from storage reservoirs. Inasmuch as the requirements of the electric power industry are usually dominant as regards river flow regulation, such threats are very real.

Soviet specialists, like their American counterparts, are becoming more aware of the difficulty in accurately evaluating flood control benefits. Moreover, some are particularly critical of the wisdom of using flood control as a justification for constructing hydro-projects in the low-lying areas of European Russia, if huge, often fertile, areas must themselves be permanently flooded in the name of downstream flood control. Further, such shallow reservoirs are not particularly effective for flood control purposes; the extensive series of low-head reservoirs on the Volga and Kama Rivers, for instance, are capable of retaining less than one-half of the average volume of flood waters on the lower Volga.[12]

Still further land around some of the Volga reservoirs undergoes periodic flooding, moreover, due to the raising of the reservoirs to levels above their designed heads, in the interests of increasing power output during periods of peak flow. These additional periodically flooded lands can amount to as much as 20 per cent of the normal total surface area of the reservoirs themselves, or an additional 500,000 hectares.[13] In addition, the existence of the reservoirs causes a rise in the water table in the area surrounding them. This waterlogging of the land is often quite extensive, and in forest zones can kill off large numbers of trees for a considerable distance beyond the maximum limits of the reservoir.

In arid regions, a tendency for shorelines to collapse along some reservoirs, such as the Tsimlyansk and Mingechaur, has been noticed. This is doubly unfortunate, since rivers in these drier areas tend to carry large amounts of sediment, causing the silting of reservoirs to be a major concern. Even in a moist climate, poor logging practices within the watershed of a reservoir can greatly increase the rate of silting. Two reservoirs in the Soviet Union, the Ak-Su in Dagestan and the Shterovskoye in the Donets Basin, were reported to have been largely ruined by silting in less than six years.[14]

An increasing concern for the conservation of associated resources has led to the questioning or reconsideration of several proposed projects. For example, the final decision on the exact location for the Lower Kama dam has been long delayed while planners examine variants (including a 200 km levee) that will minimize its effect on the agricultural resources of the Tatar Autonomous Republic, already reduced by the expansive Kuybyshev reservoir. Losses attributable to the Lower Kama dam as planned would include 125,000 hectares of agricultural land and 44,800 hectares of forests. Early plans called for improving only 11,500 new hectares of arable land, less than half of what would have been flooded. Petroleum extraction in the area would have been adversely affected. Also, with reference to the earlier discussion of benefit–cost analyses of Soviet projects, Soviet writers have asserted that the costs of relocating the inhabitants from the Lower Kama reservoir area were understated by almost 50 per cent, that funds were not allocated for schools, hospitals, and other facilities in the resettlement towns, and that under-evaluation of the lost agricultural land led to an unrealistically low cost for the hydroelectricity produced by the dam.[15]

The wisdom of building a large new reservoir on the Kuban' River near Krasnodar has been questioned. Although it would irrigate about 110,000 hectares (a net gain of 75,000 arable hectares, subtracting the flooded farm land), it would also destroy fish runs on the Kuban' river, and the loss of the Kuban''s flow into the steadily salinizing Sea of Azov would greatly damage the remaining fisheries in that body.

Other major projects appear to have been permanently scrapped, or at least indefinitely consigned to a planning limbo. The development plans for the Volga, for example, have envisioned a series of nine dams, the last of which would be the Lower Volga dam, constructed in the braided floodplain between Volgograd and the Caspian Sea. A whole host of objections have arisen concerning this dam, however, including its disastrous effects on fish and wildfowl breeding areas, its inundation of the fertile floodplains, and the high rate of evaporation it would entail. To save the floodplain alone would have required a dike 300 km long costing 230 million rubles. Although this dam had the backing of Khrushchev while he was in office, it has apparently since been set aside.

However, problems still exist in this region. A water dividing dam is being constructed which will shunt 65 per cent of the river's flow through the eastern (Buzan) channel, and only 35 per cent through the former main channel on the west, the primary sturgeon migration route. It was estimated that 90 per cent of the sturgeon reaching the divider dam will be cut off from their upstream spawning grounds.[16] In addition, rice farms and a cellulose mill using reeds, not to mention the falling level of the Caspian Sea, have greatly reduced the extent of the natural breeding grounds in the lower Volga and Volga delta region.

Another project which has apparently been abandoned is the huge Lower Ob' dam, the original design of which would have created a reservoir over 2000 km long with a surface area of 11,340,000 hectares, a lake larger than the entire state of Ohio. Although a subsequent variant would have lowered the height of the dam from 42 to 37 meters, an enormous area still would have been flooded. Agriculture in West Siberia would have been dealt a sharp blow, since the best drained lands in that vast lowland plain are often found adjacent to the banks of the rivers. Drainage networks, not large reservoirs, are the main need of agriculture in this region. The reservoir also would have produced local climatic changes curtailing further the already short growing season. The area which the raised water table would have waterlogged was estimated in hundreds of thousands of square kilometers, an expanse greater than the entire area which has been laboriously drained for agricultural purposes in the whole of the Soviet Union.[17]

In addition, about 400 million m^3 of timber would have been flooded by the reservoir, but the removal of only about one-third of this was called for by the plans. Important deposits of peat, oil, metals and natural gas would be inundated by the reservoir, and it would also have covered parts of the Ivdel–Ob′ railroad and Igrim–Serov pipeline. In all probability, it was the vast extent of the newly discovered oil and gas resources in Western Siberia, rather than the loss of farm land, timber, and fisheries, which really sealed the fate of this project. However, smaller variants are still being studied.

Even larger reservoirs on the middle Ob′ and lower Irtysh and Yenisey Rivers have been designed in the past which would have connected all these rivers with a deep water transportation network and produced many millions of kilowatts of power (the 'Davydov plan'). However, the rich oil fields of the West Siberian lowland have probably negated these plans permanently.

Since this section has necessarily dealt only with problem areas that have arisen in Soviet water resources development planning, one might conclude that the Soviet Union has been extremely negligent in the planning of hydrotechnical projects. This would not be an entirely fair assessment. Soviet water development technology is quite high, and most projects have been well engineered and are, on balance, of great importance to the national economy. However, the seemingly inevitable procession of conflicts, foregone options, and unforeseen consequences have irrepressibly made their appearance in Soviet river development projects, even as they have in those of the United States.

Generally, the more obvious associated costs, such as relocations, timber removal, preparing new farm land and fish hatcheries, etc., have been included in the total cost estimates of large Soviet multiple-purpose developments. Nevertheless, externalities and the many other inherent uncertainties of feasibility studies, discussed earlier, cannot help but raise the questions of whether the true cost of hydroelectric power in the Soviet Union is not being greatly underestimated, and whether certain Soviet projects, particularly those involving very large reservoirs, may not represent far greater social and financial burdens than was thought to be the case at the time such dams were planned.[18] The degree of contemporary Soviet concern over the far-reaching effects of these large reservoirs is well illustrated by the sheer volume of material that has been published regarding them. Between 1958 and 1966 alone, for example, the Kuybyshev reservoir was studied in no less than 1138 publications, the Rybinsk reservoir in 994, and the Tsimlyansk in 666; twenty reservoirs in all were the subject of at least 100 different works in that nine-year period.[19]

This is a consideration being stressed by some of the foremost Soviet specialists on water resources utilization. S. L. Vendrov and G. P. Kalinin, among others, have urged that

> one of the main tasks in future hydroelectric projects is increasing the utilization of mountain streams and the reduction of any further flooding [i.e. inundation by reservoirs and waterlogging], as well as the rationalization of present flooding, of lowland areas with a relatively high population density.[20]

Many similar opinions have been expressed in the succeeding years, and the tabling of such controversial projects as the Lower Volga and Lower Ob′ dams would seem to indicate that those who speak for comprehensive resource development and conservation are beginning to gain a degree of influence more commensurate with that of the proponents of low cost hydroelectric power. As yet, however, the concept of 'total basin planning' is still very seldom encountered in Soviet literature on water resource utilization.

On the other hand, the concept of 'total river development' appears to remain an *a priori* assumption of Soviet planners. In the words of a leading economic geographer, '. . . by the start of the next century the rivers of the European part of the U.S.S.R., the Caucasus, and Central Asia will have been completely subjected to engineering controls'.[21] Although the word 'completely' certainly overstates the case, it nevertheless points out the prevalent 'beaver syndrome' which for years has characterized Soviet water resource planning, and the tendency to underestimate the value of free-flowing rivers for fish, wildlife, soil, and forest conservation, and for aesthetics and recreation as well. There is reason to believe, however, that the current re-examination of clearly questionable projects might indicate a change in this respect, and that the pleas of ecologists and others for a more carefully thought out and balanced form of water resources planning are beginning to take effect.

Irrigation and water conservation in the arid regions

River systems which have been developed primarily for irrigation purposes present an additional set of conservation problems. These arise from the fact that irrigation is a consumptive use of water, whereas hydroelectric power and navigation are not (except for reservoir evaporation and seepage). For this reason, care must be taken to avoid wasting the diverted water, and to foresee the consequences both to the areas which will be deprived of the normal river flow, and to the areas which will receive the new water.

About 10 million hectares of land were developed with irrigation networks by the end of 1965, plus another million hectares lying in estuaries

Table 7.2 *Distribution of irrigated land in the Soviet Union*

	Thousand hectares as of November 1, 1965		
Republic	Land accessible to irrigation water	Land with completed irrigation networks	Land actually irrigated in 1965
Uzbek	3179	2752	2620
R.S.F.S.R.	1832	1510	1458
Kazakh	1823	1255	1146
Azerbaydzhan	1599	1278	1116
Kirgiz	1095	861	839
Turkmen	904	514	514
Ukraine	514	503	496
Tadzhik	473	468	437
Georgia	355	348	344
Armenia	258	249	233
Moldavia	75	74	67
Total, U.S.S.R.	12,107	9812	9270

From *Narodnoye khozyaystvo S.S.S.R. v 1965 g.*, pp. 362–3.

and flood-plains. Of these 10 million hectares, about 60 per cent are to be found in Central Asia and Kazakhstan (Table 7.2).[22]

Problems which have arisen in the development of irrigation in the Soviet Union have centered around unfavorable consequences to the rivers and lakes from which the water was taken, losses and wastage of the diverted water, and the secondary salinization of soils in irrigated regions.

The withdrawal of water from a river for irrigation purposes always imposes measurable changes upon the hydrological regime of that river, the magnitude of which will vary with the coefficients of evaporation and return flow. Where the former is high and the latter low, a major decrease in the volume of river flow is possible. In the United States, examples of this are the San Joaquin and Colorado Rivers; in the Soviet Union, the Don and Amu-Darya.

Irrigation water is taken from the Don primarily via the Tsimlyansk Reservoir. Although the amount of water withdrawn is much less as a percentage of the total flow than is the case with the Colorado, a reduction in the average annual flow of the Don has nevertheless been observed. The flow is currently about 10 per cent less than it was a few decades ago, and by 1980 it may possibly be reduced by as much as 30 per cent.[23] Increased transpiration from agricultural crops is a major cause of reduced run-off into rivers in arid areas. The increasingly harmful effects that this reduction in the flow of the Don is inflicting on the fish resources of the Sea of Azov was mentioned in Chapter 5. In addition, rivers developed for hydroelectric power which flow through areas being increasingly used for agriculture may in the future have to operate at a lower level of power output than was originally envisioned.

Another irrigation project involving the problem of preserving a natural water body is that of central Armenia. By far the greatest source of water for the quarter-million irrigated hectares in Armenia is Lake Sevan. At a surface elevation of 1900 meters (6200 feet), it is the largest natural water body in the Transcaucasus. Lake Sevan is drained by the Razdan River, which is almost fully developed for its hydroelectric potential. These man-made developments, together with the high rate of natural evaporation from Lake Sevan (about 90 cm or over 1.2 billion m^3 annually), have resulted in a rate of depletion of water from the lake which has been in excess of its natural recharge capabilities. By 1960, the surface area of Lake Sevan had decreased from 1416 km^2 to about 1250 km^2.

The solution which was first proposed for this problem essentially entailed the sacrifice of the lake itself. It called for a planned drop of the surface level of the lake by 50 meters, which would have reduced the surface area of the lake by 80 per cent, and the amount of evaporation correspondingly. The duckpin-shaped lake would then have existed only in its smaller, deeper, northern section, the larger southern portion having been drained entirely.

As the result of strong local objections, this plan was set aside for one which will decrease the surface area by only 13 per cent, but which will also require extensive inter-basin diversions of water from other rivers into Lake Sevan. These diversions include a 48-km tunnel from the Arpa River to the south, now under construction, and the possibility of two or three other diversions.

Although these diversions are expensive, and may involve a considerable short term subsidy, they will hopefully permit a significant increase in the irrigated area in the Armenian lowlands, as well as the preservation of most of the diverse natural resources of Lake Sevan. Additional details of the Lake Sevan problem are available in the articles by Vermishev, Valesyan, Mikirtitchian, and Greenwood, and the article 'Work on the Arpa–Sevan tunnel' cited in the bibliography.

Similar situations have developed with regard to the surface levels of the Caspian and Aral Seas, resulting from developments on the Volga and Amu-Darya Rivers. These will be discussed in detail in the next section.

The conservation of irrigation water once it has been diverted from the rivers is a major problem in the Soviet Union, as it is anywhere else in the world. Based on rough estimates only about one-third to one-half of all diverted irrigation water is actually beneficially used by crops, both in the U.S.S.R. and the U.S.A.[24] The main causes of water loss are seepage from irrigation canals, evaporation from canals, aqueducts, reservoirs, and fields, and the growth of phreatophytic plants.

Seepage is perhaps the greatest source of lost irrigation water. The Kara-Kum Canal in the Turkmen Republic, one of the largest irrigation diversion schemes in the country, is a case in point. From this canal 'losses through filtration are especially large, almost equalling the annual discharge of the Murgab and Tedzhen rivers. These waters form a whole series of lakes and seepages along the canal that are overgrown with reeds known for their tremendous and useless evaporating capacity.'[25]

The Murgab and Tedzhen rivers are two of the three major rivers flowing into the Turkmen Republic from the mountain ranges to the south, both of whose flows have been incorporated into the Kara-Kum irrigation system. Soviet specialists are expending considerable efforts to reduce seepage losses, centering on ways to line canals and even whole reservoirs so as to prevent filtration out of them. Between 1963 and 1967 alone, the Soviet reclamation journal *Gidrotekhnika i melioratsiya* carried at least eleven articles on various aspects of the seepage problem.

The most effective method of preventing evaporation losses is through the use of pipelines, rather than open canals. Although more expensive, Soviet experts have recommended an increased use of such closed distribution systems. Experiments involving the use of monolayer films on reservoirs have been carried out with some reported success, and such a treatment may be employed in the effort to reduce the evaporation from Lake Sevan.[26]

Phreatophytes are water-consuming plants which grow along waterways in arid regions. Research on chemical and other means of eliminating them is being conducted in both the United States and the U.S.S.R. In the Soviet Union they are most troublesome in the lower reaches and deltas of such rivers as the Volga, Amu-Darya, and Syr-Darya, where concerted campaigns to eliminate them have been conducted. The growth of reeds became such a problem in the Kara-Kum Canal that various exotic fish species were introduced into the canal which hopefully would feed on the plants. One of the varieties, the white carp from the Amur River in the Soviet Far East, has proved effective. The placing of the carp in the canal is in one sense ironic, since one conservation problem confronting irrigation system planners is preventing fish in rivers and reservoirs from entering irrigation head-works.

Another area of concern to Soviet planners is the silting of reservoirs and canals. In addition to the phenomena of bank collapse and stream sediment, mentioned previously, blowing sands can be a major problem in the desert regions. This has been true in the case of the Kara-Kum Canal, parts of which extend through areas of moving sands in the Kara-Kum desert.

Finally, there is the problem of inefficient application of delivered water to the fields. This is an area in which proper research and training, and perhaps charging for the use and waste of water, can be of much corrective value. However, there is often a planning problem involved here as well. Complaints have been raised in the press about irrigation projects which allocated too little capital to the preparation of the fields, so that the delivered water was unusable. Much land in the U.S.S.R. now accessible to irrigation water lies unused every year, and the extent of this problem is indicated by the difference between the figures in the first and third columns of Table 7.2. One case of this nature which was cited was in Stavropol' Kray, where in 1964 1.2 billion m^3 of water was transported through the main irrigation canals, but 500 million m^3 of this was written off as unused, due mainly to lack of funds to prepare the fields for receiving irrigation water.[27]

Further problems in the use of irrigation water arise even after it has been applied to the crops. By far the most serious of these is secondary salinization – the creation of salt deposits on or near the surface in irrigated areas. It has been estimated that as much as a million hectares may be so affected in the Soviet Union, or about 10 per cent of the country's total irrigated area.[28]

Methods for averting secondary salinization include periodic 'washing' of the soil, and the construction of artificial drainage networks. Unfortunately, both have frequently been overlooked in Soviet planning. Such an area is the Golodnaya ('Hungry') Steppe near Tashkent, where

> under the administrative pressure of the 'anti-drainage lobby', no provision was made in the irrigation plans of the Golodnaya Steppe for the construction of the required drainage structures, and calculations of water requirements...failed to provide for the additional water necessary to wash salt out of the soil...Irrigation development in the area fell into a difficult situation...[29]

The same source notes that a similar problem has occurred along the Kara-Kum Canal, especially in the Murgab oasis area, where an insufficient drainage network resulted in the salinization and waterlogging of portions of this oasis. In the Transcaucasus, expansion of irrigation based on the Mingechaur project is being hampered because of soil salinization, compounded by the natural occurrence of heavily mineralized ground water. A 'substantial portion' of all of Azerbaydzhan's irrigated land has become saline.[30]

The controlling of secondary salinization is thus a significant problem in several of the major irrigated areas of the U.S.S.R. Reflecting on this, and on the fact that the irrigated area in the Soviet Union increased only from 9.3 to 9.8 million hectares between 1960 and 1965, I. P. Gerasimov acknowledged that 'statistics show that the total area of abandoned lands in irrigated oases continues to match the total area of newly irrigated land'.[31] It seems clear the Soviet planners will have to be more generous with funds for supplemental drainage facilities if the problem of secondary salinization is to be adequately handled.

The Soviet Union has been researching numerous alternative ways of supplying irrigation water to usable arid lands other than diverting rivers. One obvious source is underground fresh water lenses, which are of frequent occurrence in the drier parts of Central Asia. To date, however, Soviet irrigation planners have depended somewhat less on underground supplies than have their American counterparts, pending further study of these resources. This is perhaps a wise approach, considering the problems which have arisen out of the rapid rate of ground-water depletion in such regions as the Llano Estacado of Texas in the United States. However, in certain areas, such as the Kuban' and around Moscow, underground reserves have been widely developed, and a lowering of the water table has been recorded. Pollution of underground reserves has also been noted. Strict controls and measurements of underground water use will be required if a quantitative and qualitative deterioration of sub-surface water resources is to be prevented.

Other potential water sources being investigated include the use of urban and industrial sewage waters for irrigation, the use of weakly salinized water for both irrigation and livestock, the induced melting of glaciers in the Pamirs and Tyan-Shans, the distillation of desert waters by both natural freezing and solar evaporation, and of course, weather modification. Several articles concerning these and other methods of supplementing water supplies in arid regions may be found in the February 1963 and June 1968 issues of *Soviet Geography: Review and Translation*. Research of such breadth and scope reflects the determination with which Soviet planners are searching for ways to make their arid regions more productive. Hopefully, this research will also include the full range of effects of these frequently unproven approaches on other related elements of the environment.

The Caspian and Aral Seas problem

One of the most significant 'secondary costs' of developing the hydropower and irrigation resources of the Volga and Central Asian basins has been the consequent effect of these projects on the inland seas of the U.S.S.R. Both the Caspian Sea and the Aral Sea are threatened with extensive losses of water and decreases in size.

Perhaps the most outstanding Soviet example of the far-reaching effects of intensively developing a natural resource complex is the troublesome 'Caspian Sea problem'. Since about 1929, the level of the Caspian Sea has dropped by almost 3 meters (about 8 feet), with severe consequences for the economy of the entire Volga–Caspian basin.

The probable causes of the drop have been two. First, the Caspian basin has undergone a long-term cyclical increase in temperature and decrease in precipitation over the past century. This has resulted in a decline in the runoff of the Volga River, which contributes almost 80 per cent of the stream discharge into the Caspian Sea. Secondly, man's activities have abetted the decrease in stream flow, through withdrawals for agricultural and urban uses, evaporation from reservoirs, the initial filling of the reservoirs, and other related uses of river water. Of these two causes, S. N. Bobrov estimates that about 80–90 per cent of the recent inflow deficit can be attributed to man's activities, rather than to climatic changes.[32] The average annual deficit in the water balance of the Caspian Sea from 1929–62 was 30.7 km^3, or a total decrease in the volume of the sea of 1012 km^3. The largest single cause of this deficit is evaporation from the large reservoirs on the Volga and Kama rivers, which, when all dams are built, will amount to about 14 km^3 a year.[33]

The economic consequences of the drop in the level of the Caspian Sea have been great. The sea is of major importance in terms of navigation, fisheries, and mineral extraction, and all have suffered.

The ports along the Caspian Sea have been hard hit. The lowering surface level has caused the shoreline to retreat many kilometers seaward; in the area of the Volga delta this extension of the shore has amounted to 25–30 km. As a result, the shipping approaches to such major port cities as Baku, Astrakhan', Krasnovodsk, and Makhachkala are being kept operable only at the cost of constant dredging. Some smaller ports have been closed entirely.[34]

The damage to the Caspian–Volga fisheries has been illustrated previously (Table 5.5). Not only have migration routes been blocked by dams, but vast areas of river deltas and shallow bays which formerly were rich spawning grounds (and waterfowl habitats) have been dried up and lost to the economy; these areas used to account for 80 per cent of the total Caspian fish catch. In addition, fishing fleets can no longer navigate in some of the diminishing river channels, and some fishing collectives have lost all water access to the Volga River and Caspian Sea.[35]

Major wildlife preserves, such as Gasan-Kuli, Astrakhan′, and Kyzyl-Agach, have increased in area but have also suffered greatly in performing the functions for which they were established.

Mineral industries, particularly petroleum, are having difficulties at coastal ports and sea-side loading installations. Extraction of the extensive sodium sulphate deposits in the Kara-Bogaz-Gol has been complicated.

Agriculture in the delta areas, particularly of the Volga, has been affected as arms and channels have slowly dried up. Some orchards and other crop lands have been abandoned, and in other areas water must now be pumped in at added expense. The failure to develop about three million hectares of irrigable land in the trans-Volga and lower Volga regions, as envisioned in Stalin's 1950 plan for the transformation of the European south of the U.S.S.R., undoubtedly reflected in large part a concern over the effect of such a massive withdrawal of Volga water on the already falling level of the Caspian. In addition, the decrease in the surface area of the Caspian from over 400,000 km^2 in 1929 to about 370,000 in 1962 has to a small degree adversely affected microclimatic conditions in the area immediately surrounding the Sea.

The magnitude of the problem has, of course, evoked many suggestions for remedying the situation. Some have serious drawbacks and are not being actively considered. These include (*a*) diverting the huge Siberian rivers southward (the 'Davydov plan'), a scheme which would have all the defects, and more, that were discussed earlier in connection with the proposed Lower Ob′ project, and (*b*) building a relatively simple channel from the Sea of Azov through the Manych depression to the Caspian, which would prohibitively increase the salinity of the Caspian Sea.[36]

The solution agreed upon involves the diversion of about 40 km^3 of water a year from the Pechora and Vychegda rivers in northern European Russia into the Kama, the main affluent of the Volga. Such a diversion has been discussed for almost 200 years, and the specific need for it which would result from the intensive development of the Volga was recognized even in the earliest days of the Soviet state. Yet, no provision for such a water transfer was included in the final plans for the Volga cascade of dams.

The diversion plan under consideration at present, involving three main reservoirs of equal surface elevation and two connecting canals, was worked out in the early 1960s. In addition to supplying the needed water to the Caspian Sea, as well as water for irrigating much new land in the lower Volga region, this diversion would allow an addition 11 billion kWh of power to be generated by the existing dams on the Kama and Volga. This is an amount equal to the yearly output from the Kuybyshev power station, the largest hydroelectric station in the U.S.S.R. outside of eastern Siberia.

The three proposed reservoirs will be quite large, and have serious implications for other resources in Perm' Oblast and the Komi Republic, where they are to be located (Table 7.3).

Table 7.3 *Characteristics of the Pechora–Vychegda–Upper Kama Reservoir System*

Feature	Pechora Reservoir	Vychegda Reservoir	Upper Kama Reservoir	Total
Reservoir area (km²)	9,950	2,910	2,690	15,550
Volume (km³)	184.5	34.5	16.0	235.0
Mean annual flow at dam site (km³)	34.1	8.3	27.7	70.1
Population to be resettled	20,000	13,000	27,000	60,000
Arable land to be flooded (ha.)	1,400	4,300	7,800	13,500
Hay meadows to be flooded (ha.)	5,300	15,600	14,400	35,300
Pastures to be flooded (ha.)	1,000	1,000	2,400	4,400
Commercial timber to be flooded (million m³)	52	18.5	8.5	79

From S. L. Vendrov, 'Geograficheskiye aspekty problemy perebroski chasti stoka Pechory i Vychegdy v basseyn r. Volgi', *Izvestiya Akademii nauk S.S.S.R., seriya geograficheskaya* (1963) No. 2, p. 37.

It can be seen from Table 7.3 that the cost of this project will be high in terms of relocation, and in terms of forfeited agricultural and forest resources. But these figures should not be considered as revealing the full story:

In view of the extensive level lands in the Pechora zone, rising only 1 to 3 meters above the water table and made up of fluvio-glacial sands that will become progressively waterlogged, trees and other present plant associations will inevitably perish... The area of such condemned forests... is yet to be accurately determined, but it would now seem to be at least as large as the reservoir itself.[37]

In addition to the agricultural land lost by flooding, certain climatic and hydrologic changes brought about by the creation of the reservoirs would work to the disadvantage of not only agriculture, but navigation and fisheries as well. It should be noted, though, that the climatological consequences of the project have been thoroughly studied in advance.

A variation of the project has been suggested which would call for two small reservoirs and a pumping station on the Pechora, instead of the one large reservoir. This would help to minimize agricultural losses, which would become even greater if a planned additional regulating dam on the Lower Pechora were built.

Another aspect of the Caspian problem should be noted. Although there is general agreement that additional water should be diverted into the Volga system, there is not agreement on how much of it should be devoted to stabilizing the level of the Caspian Sea itself. One school of thought would raise it back to the 1929 level to revitalize the fisheries and ports; another would stabilize it at about the present level, which would allow much more of the diverted water to be used for irrigation. Another variant, the 'Apollov plan', would build a dike across the north end of the Sea, with the northern ports and fisheries benefiting from about a four meter higher level than the larger southern end.[38] A conference on the optimum level of the Caspian Sea was held in Moscow in October of 1969.

It should be noted that historically, long-term climatic variations have produced wide fluctuations in the level of the Caspian Sea. Prior to about A.D. 1200, it was much lower than at present, ranging between 30 and 36 meters below sea level. It then rose rapidly, and during the fourteenth century, it was at its highest, about 20 meters. Another low of around 28 meters was realized in the sixteenth century, from which it climbed to a high of around 23 in the early 1800s. For a century it fluctuated around 26 meters, prior to its recent drop to 29 meters which started about 1930.

This disagreement over the level at which the Caspian Sea should be stabilized, together with the difficulties that the proposed new reservoirs would present to the economy of the Komi A.S.S.R., help to account for the long delay in actually commencing the project. Although the plan was apparently approved in 1961, construction had not yet begun as the Sixties were drawing to a close. Although this may largely have been the result of a temporary natural stabilization of the sea's surface level, it might also be an indication that closer scrutiny is today being given to new large-scale projects, hopefully reflecting an increased awareness of the far-reaching complexities involved.

The status of the Aral Sea is not so acute for two reasons. First of all, it is of much less economic significance, and secondly, a dramatic decrease in its level has not yet taken place. Since the turn of the present century, its level has varied only between about 52 and 53.5 meters above sea level, and was near the latter level in 1960. Since 1960, there has been a drop of about one meter.

The most immediate cause of this recent drop has been the diversion of water from the Amu-Darya River into the Kara-Kum Canal for irrigating new fields in the Turkmen Republic. Approximately 4.7 km^3 per year are being diverted at present, but eventually 33 km^3, or about half of the flow of the Amu-Darya at the diversion point, will flow through this canal.[39] The significance of this is that the Kara-Kum Canal is the first large-scale irrigation project in Central Asia which diverts water entirely out of the basin of the Aral Sea.

The Kara-Kum Canal is the largest inter-basin diversion of river water in the Soviet Union. Other major inter-basin diversions include the North Crimean Canal, which annually transfers about 10 km^3 of Dnepr River water to the Crimean Peninsula, and the Irtysh–Karaganda Canal, which can divert 75 m^3 per second from the Irtysh to the steppes around the city of Karaganda.

Although the Syr-Darya has been tapped extensively for irrigation for over three decades, its flow into the Aral Sea has decreased very little. This is apparently due to a high rate of return of the diverted water into the Syr-Darya, plus a lower rate of transpiration losses from field crops as compared to the rate from the large areas of phreatophytic plants which they replaced in the lower reaches of the river.

However, the increasing diversion of water out of the Aral basin via the Kara-Kum Canal does seem to make inevitable a continued lowering of the surface level of the Aral Sea. There is still insufficient information on both the probable rate and ultimate limit of this lowering. One specialist suggests that the Sea would stabilize itself after falling 12 meters due to artesian pressure from underground waters, even if 90 per cent of the river flow into the Sea were diverted for irrigation.[40]

As with the Caspian, a disagreement exists over whether the Aral Sea should be stabilized at its ‘normal’ level, or be allowed to drop and be stabilized at a lower level, or, for that matter, be allowed to dry up altogether. Arguments in favor of maintaining its present level revolve around the greatly depleted fish and fur industries which depend on the flow of fresh water, and the unresolved question of the climatic effects of allowing the sea to dry up. The proponents of allowing it to continue to fall, who are many and highly qualified, stress the greater priority of irrigation, and the fact that the presumed adverse climatic consequences cannot be scientifically substantiated at this time.[41]

A ‘compromise’ position of allowing it to fall only to a certain point has strong arguments both pro and con. The decreased surface area would permit less evaporation, and the drying up of the present delta areas would eliminate vast areas of phreatophytes which transpire many cubic kilometers of river water annually. However, the increased salinity would all but eliminate fresh water fisheries. Even now, salt water species of fish are being experimentally introduced into the Sea.

The problem of the Aral Sea is far from resolved, and most Soviet writers on the subject agree only on the need for much more study of the probable consequences of the continued drying up of this unique inland sea.

Soviet investigation of ocean resources

A review of water resources management and conservation in the U.S.S.R. would not be complete without a *résumé* of Soviet oceanographic research relating to the natural resources of the seas. In the past decade, such research in the Soviet Union has gained considerable momentum.

There are four main categories of resources available from the oceans at the present level of technology: fish and other marine food resources, tidal electric power, fresh water through desalinization, and a vast reserve of mineral resources. The latter three will be reviewed here briefly, Soviet ocean fisheries having been discussed in Chapter 5.

The Soviet Union is presently constructing its first experimental tidal power station in Kislaya Bay, a short distance west of Murmansk on the Barents Sea. A small facility of only about 1000 kW, it was authorized in 1962 but the first 400-kW generating unit did not go into operation until late 1968. Assuming the successful operation of this experimental station, a much larger one is planned for Lumbovskaya Bay on the Kola Peninsula, about 300 km east of Murmansk. Utilizing tides with a maximum range of up to 7 meters, a 5 km-long dam enclosing a storage basin of 70 km^2 will be constructed with an installed generating capacity of 320,000 kW. A still larger station is being planned for Mezen Bay on the White Sea, where 9-meter maximum tides would permit a first-stage dam across the mouth of the Mezen River with a capacity of 1.3 million kW. Subsequent developments in this bay could possibly have as much as 14 million kW of installed capacity.[42]

Soviet technicians have made great strides in lowering the cost of desalinized water, and their existing plants are of considerably greater capacity than similar pilot plants constructed to date in the United States. The need for such purified water in the Soviet Union is greatest not along the oceans, but rather in the arid interior, and various types of pilot plants are being tried out, particularly on the east side of the Caspian Sea, in the Kazakh and Turkmen republics. The extent of the progress made through 1968 is summarized in Table 7.4.

Table 7.4 *Technical and economic characteristics of saline-water conversion installations in operation or planned in the U.S.S.R.*

Type of unit	Place of operation or design institution	Date of construction or status	Number of units	Capacity (m^3/day)	Cost of fresh water (rubles/m^3)	Unit investment (rubles/m^3)
Straight distillation	Gasan-Kuli (Turkmenia)	1907	12	16	14.60[a]	14.20
Three-chamber evaporators	Krasnovodsk No. 1 Heat-and-power station (Turkmenia)	1946–53	40	3,700	2.63[a]	
Pilot-plant long-tube vertical distillation	Shevchenko (Kazakh S.S.R.)	1963	1	4,320	1.43[a]	2.47
Industrial distillation plant	Shevchenko (Kazakh S.S.R.)	1967	1	13,500	0.50	0.95
Experimental flash distillation (5 stages)	Krasnovodsk No. 2 Heat-and-power station (Turkmenia)	under const.	1	1,200	0.72	2.73
Flash distillation plant (42 stages)	Bek-Dash (Turkmenia)	design	1	3,840	0.50	0.43
Vapor compression distillation	Krasnovodsk Fish cannery (Turkmenia)	design	1	1,000	0.49	1.31

Multichamber distillation unit working on mobile atomic reactor ARBUS						
70,000 thermal kW	Experimental models (Moscow)	1964	1	12,000	0.42	—
15,000 thermal kW		1964	1	3,000	0.68	—
Dual-purpose atomic water-conversion and power station with reactor of 1 million thermal kW	Shevchenko (Kazakh S.S.R.)	under const.	8	120,000	0.06	—
Dual-purpose atomic water-conversion and power station, including cost of 1.95 million thermal kW reactor and steam generator	Economic and project design (Moscow)	design	—	404,000	0.15	0.33
Experimental electrodialysis, Type EOU-NHPM-12, for ground water containing 4 grams of salt per liter	Vodyanskiy (Volgograd Oblast)	1963	4	12	1.15	2.42
Experimental electrodialysis (3 stages for Caspian Sea water (13.5 grams/liter))	Bek-Dash (Turkmenia) designed by Plastics Institute (Moscow)	1968	1	200	0.85	1.03
Electrodialysis unit Type EDM-300 (for 2.1 grams/liter)	Aktogay RR sta. (Kazakh S.S.R.)	1966	4	100	0.30	0.92

Table 7.4. (*continued*)

Type of unit	Place of operation or design institution	Date of construction or status	Number of units	Capacity (m^3/day)	Cost of fresh water (rubles/m^3)	Unit investment (rubles/m^3)
Electrodialysis unit Type EDU-300-11 (for 3.5 grams/liter)	Gyaurs RR sta. (Turkmenia)	1967	2	50	0.40	0.71
Mobile electrodialysis unit Type SEKhO-1 (for 5 grams/liter)	Experimental model (Moscow, Farm Electrif. Institute)	testing	1	7	0.60	3.77
Mobile distillation unit	Mounted on KRAZ-214 truck	serial manufacture	1	48	1.94	2.74
Natural freezing	Takhta Rayon (Turkmenia) (experimental site)	1963	$4 \times 45.2m^2$	0.5	16.60	105.00
Natural freezing	Project designs of Water Engineering Design Institute	design	$10 \times 440m^2$	20.0	1.85	35.50
Solar greenhouse still	Bekharden State Farm (Turkmenia) (experimental)	design	$4 \times 600m^2$	7.9	5.52	39.70

[a] Based on 1966 accounting data.

From Kolodin (1967) in translation cited, pp. 495–6. A subsequent article by the same author on Soviet desalinization efforts was translated in *Soviet Geography: Review and Translation*, **11**, No. 10 (December 1970), pp. 858–64.

Soviet investigation of ocean resources

Soviet oceanographic scientists have shown considerable interest in the extraction of mineral resources from the sea. The U.S.S.R. has been extracting oil from beneath the Caspian Sea for many years, but current interest is strong in devising economic ways of recovering minerals both from sea water itself and from the rich nodular concentrations of manganese and iron on the ocean floor. In 1970, the Soviet Union began mining titanomagnetite from the ocean bottom off Iturup Island in the Kuriles.[43] Ocean mining for tin in the Laptev Sea is to begin in 1971, and off-shore mining for gold will commence by 1972. In December of 1970 the U.S.S.R., joined only by the other communist nations, voted against a United Nations resolution to convene a conference in 1973 to form an international body to direct the harvesting of the ocean floor's resources.

To assist in its various oceanographic endeavors, the Soviet Union has a fleet of more than a dozen well-equipped research vessels, and numerous oceanographic research centers in various parts of the country. Early in 1968, the Supreme Soviet of the U.S.S.R. issued a decree claiming all the mineral wealth of the country's continental shelves as the exclusive state property of the Soviet Union, in conformity with the 1958 Geneva Conference on the Law of the Sea.[44]

Soviet planners have also been considering some very grandiose schemes for diverting ocean currents so as to influence favourably the climate of the U.S.S.R. Two projects in particular might be mentioned. The first would build a dike across the narrow (7-km) Tatar Straits between the Asiatic mainland and Sakhalin Island, so as to block the cold currents entering from the Sea of Okhotsk, and thus warm the coastal areas of Primorskiy and Khabarovsk Krays. The second, of immense potential consequences, would dam the Bering Strait between Siberia and Alaska. Needless to say, vast environmental, not to mention political, considerations would have to be resolved before either project, especially the latter, could receive serious consideration.

Projects such as these, however, do illustrate the scope of Soviet thinking in their resolve to master the natural resources of the sea, and to make them feasible substitutes for the less plentiful water, mineral, and protein resources of the land. It is quite probable that the pace of Soviet oceanographic research will greatly intensify in the 1970s. This exploration of the sea, however, will not reduce the need for more comprehensive planning of the U.S.S.R.'s internal water resources, if further 'unforeseen difficulties' stemming from their development are to be avoided in the future.

8 Environmental pollution and environmental quality in the Soviet Union

Victors are not judged...

Perhaps the most common misfortune resulting from a national emphasis on rapid economic expansion, as both the United States and the Soviet Union have done in their separate ways, is the use of the atmosphere and water bodies as depositories for large quantities of waste materials. With time, this inevitably leads to a significant degree of environmental deterioration, and may even pose serious public health hazards.

Unfortunately, it has been the 'normal' course of action in both countries to disregard these deteriorating conditions until such a biological hazard developed that it was no longer possible to ignore them. Both countries are now finding it mandatory to combat the pollution of their waterways and atmosphere, not basically out of aesthetic considerations (which ought to be sufficient reasons in themselves) but for the more urgent reasons of public health and the preservation of other resources dependent upon these polluted media.

In the U.S.S.R., the problem has been compounded by the political-philosophical assumption that improper natural resource exploitation is the result of the capitalist mode of production, and that a socialist economy necessarily pursues the wisest possible use of natural resources. While it would probably be an overstatement to suggest that communist theorists have believed that serious air and water pollution could not develop under a socialized economy, they no doubt have assumed that the process of centralized national planning would naturally take care of such matters quickly and efficiently.

Unfortunately, pollution control costs no less money under socialism than it does under capitalism, and Soviet planners have simply been more enthusiastic about putting the available investment capital into steel mills and oil wells than into industrial and municipal purification facilities, whose output, after all, does not often show up in gross national product statistics.

In the last few years, however, the necessity for such facilities has become increasingly apparent, and provisions for their construction are starting to be allocated on a somewhat broader scale.

Water pollution in the U.S.S.R.

Supervision over the prevention of water pollution in the Soviet Union, and responsibility for the establishment of water quality standards, is charged to the U.S.S.R. Ministry of Public Health, and to corresponding ministries in each of the Union republics. The U.S.S.R. is responsible for setting national norms, standards, and principles, for regulating the use of water bodies situated in more than one republic, and for overall supervision of republic water conservation and purification work. The fifteen republics are directly responsible for the day-to-day prevention of pollution and other misuse of water resources within their own territories, and for enacting the necessary legislation to carry this out (see Appendix 22, Articles 5 and 6). The various public health ministries operate a nation-wide State Sanitary Inspectorate, which has the responsibility of monitoring pollution levels and reporting and invoking fines for infractions, and also for supervising and approving the implementation of measures which would eliminate and prevent the causes of such pollution. Other concerned ministries, such as fisheries and agriculture, also operate inspection services.

Basic research on pollution control is carried out by several different research institutes, such as the All-Union Research Institute for Water Supply, Sewerage and Hydrological Engineering Equipment (acronym: VODGEO) and its branches which are located throughout the country, at all departments of public health connected with medical schools, and at various sanitary and epidemiological field stations.[1]

Although regulations regarding the discharge of waste water appeared in 1923 and 1929, the first meaningful anti-pollution legislation in the U.S.S.R. was not enacted until May 17, 1937, and that dealt only with protecting sources of water procurement.[2] Updated decrees on pollution control were enacted in 1943, 1954, and 1961.

On April 22, 1960, the U.S.S.R. Council of Ministers published a decree entitled 'On Measures for Regulating the Use and Strengthening the Conservation of the Water Resources of the U.S.S.R.', which defined basic policy and responsibility for water pollution control, and reconfirmed the duty of the Ministry of Public Health to establish water quality norms and regulations. The R.S.F.S.R. conservation law of October, 1960, restated these policies for the territory of the Russian Republic (Appendix 26). Article 223 of the penal code of the R.S.F.S.R. stipulates that 'the pollution of rivers, lakes, and other bodies of water or water sources by unpurified or harmful effluent waters. . . is punishable by corrective labor for a period of up to one year or a fine of up to three hundred rubles.'[3]

In 1970, a comprehensive set of national policy guidelines governing all aspects of water use in the U.S.S.R. was enacted (Appendix 22). Several articles relate to the prevention of water pollution, and call for all new factories and communities to have means available to prevent the pollution of water. More specifically, these principles state that

> the discharge of industrial, household, and other types of waste materials into bodies of water is prohibited (Article 38).
>
> The discharge of sewage is permitted only in cases in which it does not lead to an increase in the pollutant content of the particular body of water above established norms, and on the condition that the water users purify the sewage up to the limits established by the agencies for the regulation of the utilization and conservation of water.
>
> If the above requirements are violated, the discharge of sewage is to be limited, halted, or prohibited...In cases threatening the health of the population, the agencies...have the right to halt the discharge of sewage, up to and including discontinuation of the operation of industrial or other facilities...(Article 31).[4]

Similar but less detailed guidelines covering pollution prevention were also included in an act entitled 'Principles of Public Health Legislation', passed in December of 1969.

Despite the existence of the above regulations and penalties, the pollution of rivers and other water bodies of the Soviet Union has been and remains a very serious problem. From the time of the rebuilding of Russia's industrial base in the early 1920s until 1962, the quantity of waste waters dumped into the waterways of the Soviet Union increased by approximately twenty times.[5] By 1966, about 16 billion m^3 of industrial waste was being dumped annually into the rivers and lakes of the Russian Republic alone, of which only 4.5 billion had undergone purification.[6]

Nor is the situation with regard to municipal wastes any better. In 1960 it was stated that only fifty-one Soviet cities had sewerage treatment plants, and that over 60 per cent of all the U.S.S.R.'s sewage was being discharged into natural water bodies without preliminary treatment.[7] Two years earlier, a study revealed that domestic sewage was causing intensive bacteriological contamination for tens of kilometers below the large cities on the Volga, demonstrating that even a river of this size was not able to cope with the amount of municipal sewage being discharged into it. Bacterial pollution of the Kama River was cited in the same study in connection with the frequent occurrence of helminthosis in areas along the river.[8]

It would be fair to say by way of summary that most of the rivers in the industrialized parts of the U.S.S.R. suffer from some degree of pollution over at least a portion of their length. Most heavily polluted are the rivers on the west slopes of the Urals in the Kama River watershed, the rivers of the Donbass, and the rivers of the Central Industrial Region, particularly from Moscow eastward to the Volga. However, the portion of almost any river that lies below a major industrial city is generally adversely affected. The major sources of harmful wastes, in addition to municipal sewage, are enterprises of the pulp and paper, petroleum, chemical, and metallurgical branches of the economy (Table 8.1).

The Northern Donets River, in the heavily industrialized Donbass region of the eastern Ukraine, is an example of a major river greatly deteriorated by pollution. In 1960 it was receiving: rotting organic matter from two sugar refineries near Belgorod; aldehydes, fatty acids, and soap from a Shebekino factory; waste material from the major industrial city of Khar'kov via the Udy River; calcium chloride from the Slavyansk Soda Plant; pollution from the major metallurgical centers of the Donbass via the Kazennyy Torets River; and, further downstream, ammonia and other chemicals from the Rubezhansk and Lisichansk Chemical Plants, the Donets Soda Plant, and from ore concentration plants near Lugansk. Khar'kov, a city of over a million, was in 1960 drawing its municipal water from the Northern Donets, and its residents were advised to boil it before drinking it.[9] Despite considerable publicity in the press, the situation in the Northern Donets and the other rivers of the Donbass has improved only slightly in the years since 1960.

Even in rural and forested areas, rivers are polluted by fertilizers, pesticides and herbicides, logging operations, and floated logs which have sunk on their way to the mill. The Volga and Kama have suffered particularly from this latter phenomena, and local fisheries on the Volga have been destroyed by sunken and rotting logs.[10]

Soviet fisheries, in fact, have suffered extensively from river pollution, in an example which is all too common the world over of the incompatible uses of a vital natural resource. Numerous rivers have become totally unfit for even scavenger species of fish. In rivers where fish runs are still possible, mass deaths occasionally occur due to improper waste discharge. The Lysva Metallurgical Plant discharged effluents into the Lysva River on May 21, 1967, which resulted in the death of 1050 tons of spawning bream.[11] The total loss to the fishing industry caused by water pollution has been estimated at around 100,000,000 (new) rubles a year.[12]

At least one fish kill resulting from thermal effluents has been reported, on the upper Volga. In the spring of 1970 a large number of dead fingerlings were encountered below the thermal power station at Kostroma, with the cause attributed to the water being heated several degrees above normal. The problem has been alleviated by discharging the heated water into a smaller tributary first.[13]

Some instances of the pollution of underground waters have also been reported. This has become a problem in such areas as the Tatar and Bashkir Autonomous Republics, Gor'kiy and Tula Oblasts, and in towns around the city of Moscow.[14]

The large inland lakes and seas of the Soviet Union have not been

Table 8.1 *Major polluted rivers and lakes of the Soviet Union*[a]

River or lake	Points of pollution	Industries causing pollution	References[b]
Angara	Bratsk, Irkutsk, Angarsk	Wood processing, oil refining	1(108), 14, 16
Baykal, Lake	Various lumbering operations	Wood decay, pulp, sewage	5, 8, 22
Belaya	Ufa, Sterlitamak, Salavat, Ishimbay	Oil refining, petrochemical	1(110), 2, 6, 11
Berezniki	Bobruysk	Wood processing, chemical	13
Biryusa	Suyetikha	Wood hydrolysis	1(108), 16
Caspian Sea	Apsheron Peninsula	Oil drilling, processing	10, 17
Chusovaya	Pervoural'sk, Revda, Chusovoy, Lys'va	Non-ferrous metallurgy, chemical	3, 6, 12, 15, 21
Desna	Bryansk Oblast	Wood processing	18
Dnestr	Stryy, Dubossary, Kishinev	Oil drilling, sewage, food processing	25
Don	Rostov	Ships' wastes, food processing, sewage	12, 13
Irtysh	Omsk	Petrochemical	12, 16, 18
Kal'mius	Donetsk, Makeyevka	Metallurgical, chemical	3, 12
Kama	Perm', Krasnokamsk	Oil refining, wood processing, chemical, wood decay	1(107), 4, 6, 11, 15
Kama Reservoir	Solikamsk, Berezniki	Chemicals, pulp and paper, fertilizers	1(108), 2, 4
Kazennyy Torets	Kramatorsk, Dzerzhinsk, Konstantinovka	Metallurgical, chemical	3, 12
Klyazma	Orekhovo-Zuyevo, Vladimir	Chemicals, textiles	7, 9, 18
Kuban'	Krasnodar	Industrial fibers	26
Ladoga, Lake	Syas', Priozersk, Kirishi, Lyaskelya	Pulp and paper, oil refining	1(108), 22, 24
Miass	Chelyabinsk	Chemical, sewage, metallurgical	9, 12
Moskva	Lyubertsy, Voskresensk	Oil refining, chemicals	1(97, 116), 7, 18, 21
Northern Donets	Belgorod, Slavyansk, Lisichansk, Lugansk, Rubezhnoye	Metallurgical, chemicals, food processing	2, 3, 11, 18

River or lake	Points of pollution	Industries causing pollution	References[b]
Northern Dvina	Kotlas	Pulp and paper	2, 12, 14, 18
Ob′	Novosibirsk, Kuznetsk Basin	Metallurgical	2, 18
Oka	Dzerzhinsk, Kolomma, Ozery	Chemicals, textiles	2, 7, 11, 12, 21
Onega, Lake	Kondopoga	Pulp and paper	1(108), 22
Osuga	Kuvshinovo	Pulp and paper	22
Selenga	Ulan-Ude	Meat packing, sewage	5, 16, 19
Sukhona	Vologda, Sokol	Wood processing	18, 22
Tom′	Kemerovo, Tomsk, Novokuznetsk	Metallurgical, chemical	3, 16, 18
Udy	Khar′kov	Metallurgical, chemical	3
Upa	Tula	Metallurgical	21
Ural	Orenburg, Orsk, Magnitogorsk	Oil refining, metallurgical	6, 11, 20
Uvod′	Ivanovo, Kokhma	Chemicals, textiles	21
Volga	Kalinin, Konakovo, Kazan′, Gor′kiy, Kuybyshev, Syzran′, Saratov, Volgograd	Oil refining, wood processing, textiles, sewage, thermal	1(110), 2, 6, 11, 13, 23
Yenisey	Krasnoyarsk	Wood processing, chemicals	1(108), 2

Other large rivers mentioned in the literature as being heavily polluted, but for which supporting details were not given, include: Onega, Vaga, Kura, Terek, Kos′va, Neyva, Iset, Vychegda, Velikaya, Seym, Dnepr, and Tagil.

[a] Table 8.1 is not presented as necessarily being a complete list of polluted rivers in the Soviet Union; it represents only those for which the author was able to locate specific references concerning pollution. It is possible that other equally contaminated rivers exist, or that other points and sources of pollution on the cited rivers exist. It is also possible that in the interval since the cited references were published, some of the sources of pollution listed above may have been eliminated.

[b] Key to sources used in compiling Table 8.1 (as listed in the bibliography): 1 Blagosklonov *et al.*, 2 Danilov, 3 Demin and Bilenkin, 4 Gaydar and Bogatenkov, 5 Gribanov, 6 Gurvich and Kibal′chich, 7 Iosifov, 8 Ishkov, 9 Karmanov, 10 Kiselev and Luk′yanenko, 11 Kuznetsov and L′vovich, 12 Litvinov, 13 L′vovich, A. I., 14 Merkulov, 15 Mironov, 16 Nasimovich, 17 'O merakh. . .', 18 Razin and Gangardt, 19 Semenov, 20 'Trevoga o reke', 21 Vendrov *et al.*, 22 Volkov ('Uroki Baykala'), 23 'Zagryaznyayushchiye Volgu – nakazany', 24 Zakharov, 25 Bogatenkov, 26 Kir′yanov. Numbers within parentheses indicate page references within source cited.

immune from pollution, either. Such large bodies of water as Lakes Onega, Ladoga, Baykal, and Balkhash, and the Ivankovo and Kama reservoirs, are the recipients of large amounts of industrial wastes. Lake Ladoga and the Kama Reservoir have been especially hard hit by the effluent waters from pulp and paper mills and other industrial plants located along their shores and on the rivers which empty into them.

Large areas of the Caspian Sea are heavily affected by oil pollution, both from wastes from refineries along the shore, and from losses at off-shore drilling operations (Figure 8.1). Sewage and waste material and bilge water from ships are also contributing factors, and the total result has been severe losses in fisheries and wildlife habitats in the areas of oil development. So serious has the problem become that a special decree of the U.S.S.R. Council of Ministers was put forth on October 3, 1968, directing the several ministries concerned and the councils of ministers of the four republics bordering on the sea to take extensive, expressly specified measures during the period 1968–75 to remedy the situation.[15] Oil spills have also been reported along the Black Sea coast, especially near the oil trans-shipment ports of Tuapse and Novorossiysk.[16] A large oil fire was reported to have occurred on the Caspian Sea near Baku in the fall of 1970.[17] It is not known whether the extensive drilling and shipping on the Caspian and Black Seas has produced any spills of the magnitude of the *Torrey Canyon* or Santa Barbara disasters, but the potential constantly exists. A major oil spill did occur, however, during the winter of 1970–1, when the main oil pipeline from the new Mangyshlak oilfields in Kazakhstan to Kuybyshev broke, threatening serious pollution of the Ural and other smaller nearby rivers.[18]

Numerous articles have appeared in recent years which demonstrate a keen interest on the part of Soviet scientists in preventing the contamination of the world's oceans by petroleum spillages and radioactive wastes. In 1962 an agreement was reached between the U.S.S.R. and Rumania for the control of oil pollution on the Black Sea, and an international conference the same year forbade the dumping of crude oil in the Baltic Sea, which is still polluted by German nerve gas dumped there after World War II. Despite this academic interest, Soviet practices on a day-to-day level still leave much to be desired. Many Soviet seaports and even oceanside resorts still utilize the sea as a waste water depository, and Soviet ships are rather lax in their habits of dumping shipboard refuse and bilge water. Although virtually all ships dump their wastes in the oceans, Soviet ships often do not wait until they are far at sea to do so. Passengers have observed large Soviet vessels dumping their holding tanks and other refuse in such scenic areas as the inner harbor of Helsinki and the St Lawrence River. This trait is made all the more unfortunate by the very

Figure 8.1. Oil scum along shore of Caspian Sea at Baku.

large amount of shipping that takes place on the inland waters of the U.S.S.R.

Soviet efforts to correct the generally sad state of pollution in the U.S.S.R. have had certain positive results. A few factories, such as the Novo-Gor'kiy and Novo-Yaroslavl' oil refineries, have been made into models of waste water purification. The Saratov refinery has an advanced recycling system which has greatly reduced local pollution of the Volga, and refineries at Krasnovodsk no longer discharge oil-polluted water into the Caspian Sea. Factories along the Kama, in the Donbass, and elsewhere have begun to extract many compounds from their waste waters which, when recovered, not only reduce pollution but represent a valuable, re-usable by-product. Many cities, such as Moscow, Kuybyshev, Ryazan', Zaporozh'ye, and others either have or are building modern, efficient sewage treatment plants. As a result, certain rivers, such as the Belaya, Northern Donets, Dnepr, Ural, and local sections of the Volga and Moskva have undergone some (but far from total) improvement in the condition of their waters. The Moskva River through the city of Moscow is 'flushed' on occasion, when upstream storage levels permit it. One such 'flushing' in the spring of 1970 reduced the carbon dioxide content of the river to 6.6 mg/liter, compared to 25.1 mg/liter in 1965, and increased the transparency to 30 cm as compared to 4 cm in 1965.[19] Nevertheless, the number of factories with efficient purification facilities and cities with adequate sewage treatment plants remains small.

An examination of Soviet reports on the water pollution situation in that country reveals three major areas of weakness which have hindered a more rapid elimination of this problem: poorly defined norms, ineffective enforcement regulations, and delays or postponements in the construction of purification facilities.

In the mid-1950s, the State Sanitary Inspectorate drew up a comprehensive list of maximum permissible norms for the discharge of various types of waste products into rivers. However, these norms were essentially based on human health factors rather than on marine biological tolerances. As a result, they were attacked as being much too high as long ago as 1960.[20] Although it is probable that many alterations have been made in these norms since 1960, nevertheless similar attacks on their leniency appeared in 1963 and in 1967.[21]

The method of penalizing pollution violations was formerly to levy fines against the enterprise involved. This proved to be ineffective if not in some instances self-defeating:

> In Lugansk Province alone the coal industry pays pollution fines of millions of rubles...The fines have not sufficed as a brake. In fact, they have proved an insidious device. The fines are turned over to the local Soviets. With these funds

they pave the streets of settlements, build club houses, and lay water mains... The local Soviets begin to regard pollution indulgently, if not favorably.[22]

In some cases the sum of the fines levied would have been sufficient to have built the needed purification facilities. Fines levied directly on the managers responsible for the pollution were a possible penalty, but were too small. Today, such personal fines, larger in amount, and even dismissal and sentencing are being increasingly employed against those guilty of allowing intolerable pollution, and their names and transgressions are being more often publicized in the press.

The preceding chapter noted that water is disbursed as a free commodity in the Soviet Union. N. Mel'nikov, Chairman of the Academy of Sciences' Commission for the Study of Productive Forces and Natural Resources, has made the interesting proposal that enterprises should not only have to pay for the water they consume, but also for the water they pollute.[23] There can be little doubt that if a factory had to pay standard water rates for the entire flow of a major river below its waste water outlet pipes, the pollution would be halted rather quickly. It is intriguing to contemplate such a course of action as a remedy for river pollution even in the United States.

Perhaps the major problem, however, is that many purification facilities called for in the plans of new factories are completed only long after the plant goes into operation, if at all. In other cases, plants have been greatly expanded, leaving the original treatment facilities inadequate to handle the increased load. Of course, the cost is high. Expansion of the Novo-Yaroslavl' oil refinery will necessitate additional purification facilities costing five million rubles. The waste treatment plant at the Severodonets chemical factory cost 7 million rubles, and the purification installations at the Salavat and Ufa refineries cost 20 million rubles each. It is easy to imagine economic planners and factory managers attempting to side-step these costs.

A major portion of the problem originated after World War II, when it was felt that the need to reconstruct the country's war-damaged industrial capacity as quickly as possible justified the omission of costly purification installations at the new plants. However, the situation has not improved greatly even to the present. In 1965, for example, the situation existed where, out of 150 purification installations in the Western Urals Economic Council, only 25 per cent were working normally. About 30 per cent were operating at below planned capacity, and some 45 per cent were not in use at all or were not working satisfactorily.[24] Such conditions are rather commonplace. The problem appears to be a lack of sufficient concern on the part of plant managers plus an inefficient system for enforcing regulations from above, all of which is nourished by the familiar catalyst of economic expediency. As a consequence, the needed improvements are often slow in being realized.

The State Sanitary Inspectorate in theory has the right to close down enterprises which are in serious violation of pollution norms, and to 'veto' the opening of new plants which do not have adequate purification facilities. In practice, these steps are rarely taken. Pressure from the proper places can keep violating firms in business, and new plants frequently open despite the health inspectors' objections. As two examples of this, both the Perm′ and Solikamsk paper mills opened new units before the associated waste treatment plants were finished, even though the Sanitary Inspectorate representatives had failed to give their approval. All this, it should be noted, is in direct violation of the 1960 Russian Republic Conservation Law, Article 4 of which says in part: 'It is forbidden to put into operation factories, shops, or components thereof which discharge waste water without carrying out measures to ensure its purification' (Appendix 21).

In addition to these three major problem areas, the Soviet Union faces the same difficulty that exists in the United States, that of improving water quality in river basins which generally lie in more than one political unit. A successful approach to pollution abatement using basin-wide control agencies has not yet been developed in the U.S.S.R.

The seriousness of the water pollution problem has been recognized by Soviet scientists and conservationists for some time, even though, incongruously, they still generally classify water as an 'inexhaustible' resource (Chapters 1 and 7). What has been mainly lacking is a conviction on the part of economic and Party planners that the problem is sufficiently urgent so that the necessary measures, in terms of funds, manpower, and administrative follow-through, simply must be made.

An indication that such an understanding may be evolving was the emphasis on water pollution which appeared in many of the speeches of the 23rd Congress of the C.P.S.U., held from March 29 to April 8, 1966. For example, in the directives for the 1966–70 Five-Year Plan, Premier Kosygin stated that 'there must be an improvement in the protection of the fish population in rivers and reservoirs. Those guilty of polluting and clogging up rivers, lakes, and reservoirs with unpurified drainage waters from industrial and municipal plants should be called to account.'[25] Similar statements appeared in the speeches at the 24th Congress of the C.P.S.U., held in 1971.

In addition to better enforcement, writers on the pollution problem in the Soviet Union stress that in the future much more advanced systems of extracting chemicals from waste water, sewage irrigation, and industrial water recycling must be put into operation. However, the key question is whether the Party's call for 'improvement' in the situation carried with it provisions for the necessary monetary, supervisory, and enforcement measures as well. Lacking these, little real improvement in the condition of the Soviet Union's polluted streams and lakes can be expected in the near future.

The Lake Baykal controversy

An anti-pollution campaign which is deserving of special mention is the effort to preserve the quality of the waters of Lake Baykal. This controversy is significant in part because of the world-wide attention it attracted, but more so because of the unprecedented extent to which public opinion on this dispute was both voiced in, and given widespread publicity by, Soviet news media. There was even a movie being filmed in 1970 based on the threat of pollution to Lake Baykal.

The scientific uniqueness of Lake Baykal needs to be emphasized. Located in a deep intermontane trench near the Mongolian border in Eastern Siberia, it is both the most voluminous (23,000 km^3) and deepest (1620 meters) fresh water lake in the world (see Figure 4.6). Its waters, which are particularly pure and clear, are inhabited by about 600 species of plants and over 1200 types of animal life, more than three-quarters of which are found nowhere else in the world. Among these are the Baykal sponge, a species of fresh water seal (the *nerpa*), the omul (an important commercial fish), and unique species of marine life which are found only at very great depths. It could reasonably be stated by way of analogy that, from the scientific standpoint, the preservation of Lake Baykal represents the same magnitude of priority as does the preservation of the Grand Canyon in the United States.

Despite common awareness of these unusual natural characteristics of Lake Baykal, industrial activity has been developed along its shores and in its basin which has threatened a dangerous rise in the quantity of polluted water which would be discharged into the lake. Although small processing plants have existed around the lake for some time, the focal point of the recent controversy has been two new pulp mills which are being constructed on the southern part of the lake, and the increased tempo of logging which would have to take place in the region to supply them.

The acceleration of logging activity is of concern not only for its scenic disutility, but more so for the increase in soil erosion which it could produce in the lake's basin. Approximately 50,000 hectares of forested land will have to be cut annually to supply the needs of the two new mills, and it has been estimated that 400,000 tons of soil would be washed away from the cut area, as compared to a natural loss of 60,000 tons at present.[26] Much of this cutting would take place in semi-arid areas with sandy soils, under climatic conditions where natural reforestation is a very slow process. As a result, streams in these areas could be carrying up to seven times as much sediment into Lake Baykal as they do at present. Assurances that cutting would be shifted to a less potentially harmful location were called by Oleg Volkov, a leading opponent of the new lumber mills who had visited many logging areas around the country, 'the kind of good intentions with which the road to hell is paved'.[27] In addition to the sediment, considerable debris from log-floating operations on the lake and its tributaries would be left to rot in the lake's waters.

However, the problem posed by the waste materials from the two newly constructed pulp mills has received the bulk of the attention in the Soviet press. The larger one is now in operation at the town of Baykalsk near the southernmost point of the lake; the other is being built near the delta of the Selenga River. Both would be in a major seismic area.

The controversy which arose centered around whether or not the planned purification facilities at the new plants would be adequate to prevent even that small amount of pollution which would upset the lake's delicate ecological balance. The State Forestry Committee and the then existing Ministry of the Lumber, Pulp-and-Paper, and Wood-Processing Industry stood in defense of the new plants, stating that they were necessary to the economy, suitably located, and would result in no harmful consequences to Lake Baykal since the discharged water would be raised to drinking-water standards. And if pollution were found to be occurring, a 40-mile pipeline would be built to carry the Baykalsk plant's effluent to the Irkut River, which does not flow into Lake Baykal.

The attack on the plants originated with knowledgeable spokesmen from the natural sciences, but soon spread to persons with other than a scientific background, such as the writers Oleg Volkov and Mikhail Sholokhov. The protests, spearheaded by Volkov and others, and the rebuttals to them, appeared frequently during 1965 and 1966, primarily in the newspaper *Literaturnaya gazeta.*[28] They reached the highest possible level in an address by Sholokhov to the 23rd Congress of the Communist Party of the U.S.S.R. in April of 1966.

The dissent reached its climax in an outspoken letter to the newspaper *Komsomol'skaya pravda*, printed on May 11, 1966. The letter, which questioned the basic wisdom of the planned projects and called for the dismantling of the new mills, was signed by more than thirty distinguished scientists, artists, and writers, including several prominent members of the U.S.S.R. Academy of Sciences. It stands as one of the most dramatic and noteworthy public appeals on behalf of natural resource conservation which has ever been publicized by the Soviet news media.[29]

Despite the collective prestige of the signers of this letter, which would have amounted to an immutable mandate for action had it occurred in the United States, there was never any real possibility that the mills would be dismantled. However, the unprecedented public outcry on this issue did call for a review of the situation, and a special committee was set up within the State Planning Committee to study the problem. One positive change that did occur was the decision to ship the wood pulp to plants located elsewhere for further processing, rather than doing so at the Baykal mills.

In 1965 a proposal was brought forth for a comprehensive administration for the natural resources of Lake Baykal, to be termed a 'Lake Baykal National Park' (discussed in Chapter 4). It would have supervisory authority over the processing plants and all other economic activity within its borders, including tourism. However, little has been heard concerning this idea in the last few years, and its current status is unknown.

The problem of pollution from the Baykalsk mill was resolved instead, simply by authorizing several million additional rubles for expanding the plant's purification facilities. Also, a special inspection committee was established to watch over the quality of the water discharged from these facilities into the lake.

Did all this mean that the preservation of the lake's natural treasures had thereby been assured? Unfortunately, the question still could not be answered in the affirmative with any degree of certainty. A strangely ambivalent article by Volkov in the fall of 1967 asserted that 'it should now be frankly admitted that the builders have refuted those who considered it impossible to achieve the design indices for purifying the plant's wastes . . . Now one can safely say that what is observed in other places has not been and will not be tolerated at Lake Baykal.'[30]

However, in the same article, Volkov also acknowledged that violations of the effluent norms have occurred (in part because here, as so often elsewhere, initial production runs commenced before the purification plant was completed). Further, the plant has yet to be operated at anywhere near full capacity. Nor did Volkov fail to imply again that even standards of waste discharge far stricter than 'what has been observed in other places' might still do considerable harm to the fragile and unique biology of the lake. Indeed, he and others continued to stress that only time and much research will reveal whether the assigned discharge-water standards are in reality 'safe' for the lake's flora and fauna, by which time, of course, irreversible damage may have been done. In addition, the U.S.S.R.'s Minister of the Fishing Industry has charged that compliance with the norms for per cent of pollutants in the treated water was being achieved only by considerable dilution of the discharged waste water with pure water from Lake Baykal before it is discharged into the lake.[31] The situation was still sufficiently uncertain that in 1968 the director of the Lake Baykal Limnological Institute again called for the construction of the Irkut River diversion pipeline, for basic changes in logging practices around the lake, and for the use of the partly constructed Selenga plant for some purpose other than pulp processing.[32]

Other threats to the waters and biology of Lake Baykal remained unresolved, as well. Log rafting continued as the main form of timber transport to the mills around the lake, and the entire question of logging quotas, areas, and practices remained unanswered. Timber debris and sunken logs in the rivers flowing into the lake, and pollution of the lake's primary affluent, the Selenga River, have contributed to a sharp decline in the catch of the lake's most important commercial fish, the migratory omul. Finally, domestic sewage from the numerous small towns along the lakeshore does not undergo purification, nor from the city of Ulan-Ude.

These several lingering problems and uncertainties clearly called for some form of more comprehensive regulations. Such action was realized on February 8, 1969, when the U.S.S.R. Council of Ministers issued a special resolution on the use of Lake Baykal and the natural resources of its basin. The provisions of the resolution appear to be fairly strict. Those dealing with forestry include the exclusion of recreational and soil protective forests (including all stands on slopes of 25 per cent or more) from commercial exploitation, the clearing of sunken logs and wood debris from Lake Baykal and its affluent streams, and the prohibition of tractor skidding of logs in the Baykal basin. Other provisions, aimed at the prevention of effluent pollution, include the completion of purification facilities at both the Baykalsk and Selenga pulp mills, the prevention of other industrial communal sewage from entering the lake or its tributaries, and a ban on the construction of any new industrial enterprises whose operations would contribute to environmental pollution or other conservation infractions. The various ministries involved have been advised of their personal responsibility for carrying out these measures, with control over the prevention of pollution resting with the U.S.S.R. Ministry of Reclamation and Water Resources. The U.S.S.R. People's Control Committee is to assure fulfillment by the various ministries of their respective assignments in protecting the quality of Lake Baykal's waters.[33] Supplemental regulations to safeguard the lake were published in *Pravda*, September 24, 1971.

All this reads very well, and it is to be hoped that the above regulations will be rigidly enforced, and will in fact ensure the integrity of Lake Baykal's unique character. However, it seems clear, as Volkov and others emphasize, that only time will render the final verdict on the effort to preserve Lake Baykal. Whatever the outcome, the nation-wide campaign on behalf of its preservation stands out as a singular example of conservation 'lobbying' on behalf of the public and the scientific community in the Soviet Union. It remains to be seen to what extent public opinion on matters of natural resource conservation will continue to be heard in view of the increased internal control of writers and other intellectuals which began in the late 1960s.

Air pollution in the U.S.S.R.

Air pollution represents equally as urgent a problem in the Soviet Union as does water pollution. The supervisory organs are the same – the Ministries of Public Health, and their subordinate State Sanitary Inspectorates. Under the U.S.S.R. State Sanitary Inspectorate there is a Committee on the Sanitary Protection of the Atmosphere, formed in 1958, whose responsibilities include the determination of maximum allowable limits for the various categories of atmospheric pollutants. A list of these allowable limits is presented in Appendix 27. The series, *U.S.S.R. Literature on Air Pollution and Related Occupational Diseases*, has translated numerous articles on air pollution control research in the Soviet Union from such journals as *Gigiyena i sanitariya*, *Gigiyena truda i professional'nyye zabolevaniya*, and many others, as well as longer technical works by such leading Soviet authorities as V. A. Ryazanov.

The first specific air pollution control legislation for the Soviet Union was passed in 1949 in a decree entitled 'On Measures for Combating Air Pollution and Improving the Sanitary-Hygienic Conditions of Populated Areas'. The section of the 1960 Russian Republic conservation law pertaining to air pollution is included as part of Appendix 26.

An important part of air pollution control in the Soviet Union has been the establishment of sanitary clearance zones around industrial plants; these zones delineate the distance a factory must be from any residential area. There are five categories of buffer zones, varying from 2000 meters in diameter for Class I industries to 100 meters for Class V. The other zones are: Class II, 1000 meters; Class III, 500 meters; and Class IV, 300 meters.[34] The widest zone (Class I) takes in those industries which universally tend to be the worst air polluters: non-ferrous metal smelters, petroleum refineries, pulp-and-paper mills, cement and fertilizer mills, and a wide variety of chemical processes. Whether or not these zones take into account such factors as prevailing wind is unclear, but certainly a point downwind would be in greater pollution danger than a point equidistant upwind. A chemical plant built near Tolstoy's country estate at Yasnaya Polyana (near Tula) is apparently inflicting at least aesthetic if not biological injury to the area of the estate, which presumedly is located outside the control zone. Thermal electric stations, which often burn soft coal, sulphurous oil, or even shale or peat, are not included in the list as they are not manufacturing enterprises; but they are also frequently serious contaminators of the atmosphere and are usually located well on the outskirts of cities. Furthermore, in the drier regions of the Soviet Union, severe dust storms often constitute a source of severe air pollution even in the major cities (Chapter 3).

The worst air pollution in the U.S.S.R. is found in the major industrial regions: the Donbass, the Urals, and the Kuznetsk Basin; and, within these, in such cities as Magnitogorsk and Novokuznetsk. In some industrial cities in the Donbass, particulate matter in the air is so dense that it is possible to travel through the area in a closed passenger train compartment and in so doing accumulate a solid coating of dust and soot particles upon all exposed surfaces in the closed compartment. Forests and other vegetation have been killed off for distances of 10 km or more from industrial plants in such cities as Novaya Gubakha, Karabash, and Krasnoural'sk in the Urals.[35]

Soviet writers acknowledge the considerable cost to the economy which this pollution represents. In addition to increased expenses for public health, electricity, and air travel, a rather appreciable loss of actual natural resources takes place. For example, it has been estimated that the Soviet cement industry loses 16–20 per cent of its total production into the atmosphere.[36] Losses of natural gas associated with petroleum drilling and refining were noted in Chapter 6. One copper smelting plant alone annually discharges into the atmosphere 1200 tons of lead, 1800 tons of zinc, and 300 tons of copper.[37] As perhaps the extreme example, the amount of sulphur discarded as a component of industrial waste gases has been cited as being several times the entire industrial demand for this mineral.[38]

Part of the explanation of this latter phenomenon, and of the air pollution problem around Soviet cities in general, lies in the relative scarcity of low-sulphur oil for use in thermal power plants. The Soviet Union defines low-sulphur oils as those containing less than 0.5 per cent sulphur. The main oil fields in the U.S.S.R., in the Volga–Ural region, today provide about 70 per cent of the country's total supply, and these oils contain significant amounts of sulphur. As a result, the percentage of all oil extracted in the Soviet Union having greater than 0.5 per cent sulphur has increased from 6 per cent in 1940 to 63 per cent in 1956 to over 80 per cent today.[39] It seems likely that many, perhaps most, of the oil-burning thermal power plants in the U.S.S.R. must rely on the use of relatively high-sulphur fuels.

The measures which have been adopted to combat air pollution in the Soviet Union have had mixed results. Contaminants originating from automotive exhausts are much less significant a factor in urban air pollution than is the case in the United States. Serious air pollution resulting from industrial sources, on the other hand, is at least as much of a problem in the U.S.S.R.

Automotive emissions are less of a problem, first of all, because the Soviet Union at present has only a fraction of the number of cars on its roads as are on American highways. However, in a large city such as Moscow, with its many trucks, buses, and taxis (not to mention industries), the potential for serious air pollution is still significant. To help reduce the problem in Moscow, the use of ethyl gasolines has been banned in that city, and an extremely low standard for atmospheric lead levels exists. Some cities, such as Khar'kov, use only electrified systems of public transportation. In addition, Soviet engineers are reported to be designing a type of catalytic muffler which will become standard equipment on all cars manufactured in the U.S.S.R.[40] Other research on controlling automotive emissions has been reported which also points to reductions in carbon monoxide and hydrocarbons, but which possibly would leave unresolved the problem of nitrogen oxides.[41] Research on electric cars has also been carried out, and led to the road testing of such a vehicle in late 1970.[42] With Soviet automobile production scheduled to increase several fold in the near future, it will be of more than passing interest to observe whether that country's urban planners will be able to avert the automotive-emission smog problems that currently plague almost every major American city.

A large gap appears to exist between the theoretical controls over industrial pollutants and the amount of control which actually takes place. The State Sanitary Inspectorate is authorized to order the closure, either permanently or temporarily, of any enterprise which is violating emission standards, and to forbid the opening of any new plant lacking the proper emission control devices. In practice, many plants have been shut down, but they have generally been either very small or ones having an outmoded technology. A serious polluter whose importance to the economy is beyond question, such as the Magnitogorsk steel mill, will certainly not be shut down out of pollution considerations.

It is equally difficult for the health inspectors to prevent the opening of new mills which will violate pollution norms. The reason is the same as it was for water pollution. Although the inspectors may issue restraining orders, it is frequently possible to find officials concerned with the fulfillment of production quotas in sufficiently high places to have the plant opened anyway. This happened, for example, both at new units of the Magnitogorsk works and at the Kachkanar ore concentrating plant.[43] The managers of plants which are persistent air pollution violators are rarely brought to trial on criminal charges, but even when they are, the fines are too small to be effective. A chief sanitary inspector can fine an individual 50 rubles, an enterprise 500; the chief sanitary physician of a city or a district may fine an individual 10 or an enterprise 100 rubles. Article 223 of the R.S.F.S.R. criminal code provides for a 300 ruble fine or one year of corrective labor. The net result to date has been that Article 12 of the Russian Republic Conservation Law (Appendix 26) is simply not complied with. Other republics, such as the Ukraine, Uzbekistan, Georgia, Azerbaydzhan, Lithuania, Kirgizia, and Estonia as of 1967 had no criminal liability for air pollution at all.

Unlike the case of water pollution, considerable funds are allocated for air pollution control measures, but these are often not fully used. The construction of pollution control facilities, given a low priority by everyone except the health officials, often is greatly prolonged. In the five years from 1960 to 1964, more than 25 per cent of the funds allocated for air pollution control purposes in the R.S.F.S.R. were not put to use.[44] A major reason for this appears to be that the All-Union Gas Purification and Dust Removal Association, which is responsible for the research, design, and construction of air pollution control equipment, is understaffed and has only one inadequate production facility. It also suffers from being only a subordinate agency of the Ministry of Petroleum Refining and Petrochemicals, and from that position it is difficult to effect the necessary research and coordination among the various other industrial ministries.[45]

As was the case with water pollution, the need for improvement was noted in speeches at the 23rd Congress of the Communist Party of the Soviet Union, held in 1966, and at the 24th in 1971. Of course, it had also been noted at the 22nd Congress in 1961, although somewhat less strenuously.

For the future, Soviet specialists have called for many familiar long-range solutions, such as electric cars and more efficient technological processes, as well as for certain practical short-run goals. One of the most feasible of these is to replace coal as a fuel with natural gas, electricity, and other 'clean' energy sources. Vast new gas fields have been discovered in Western Siberia, Central Asia, the Volga region, and elsewhere, and these may permit a reduction in the amount of peat, shale, and sulphurous coal which is now being burned in cities and in thermal electric stations.

The Soviet Union has made considerable progress in the development of nuclear power. In 1971, there were atomic power plants in operation at about six locations, with at least three more under construction. Soviet planners recognize nuclear power as an energy source free of the conventional contributions to air pollution, and feel they can cope successfully with the other problems associated with nuclear plants:

> Atomic power plants do not pollute the air. The problem of rendering the radioactive waste harmless and burying it has been solved. The facilities devised for this purpose and well tested in practice exclude hazards completely. This is why we may expect atomic energy to be used soon not only to generate electricity but to supply hot water for cities and steam for industry. Atomic power plants will be followed by atomic heating plants.[46]

More specific information on the disposal of spent fuel, thermal pollution, site selection, radiation safeguards, etc., is unfortunately lacking, although a few comments may be found in the report of a 1963 Atomic Energy Commission study team.[47] A news report in February of 1970 stated that an explosion at the nuclear submarine works at Gor'kiy had resulted in significant radioactive contamination along that portion of the Volga.

In a few cases where the quality of the urban atmosphere has been made a specific planning objective, the results have been quite good. Two commendable examples in this regard are the capital cities of Moscow and Alma-Ata. In general, however, air pollution remains a very serious problem in most industrial cities of the Soviet Union. The situation is not improved by the fact that air, like water, remains categorized as an 'inexhaustible' natural resource (Chapter 1). Needless to say, the full use of the allotted, and additional, investment funds for purification equipment is mandatory, especially for devices which will recapture valuable chemicals now going to waste. And, as with water pollution, more stringent enforcement of existing laws and more effective penalties for violations are essential.

The unfavorable conditions with regard to both air and water pollution in the Soviet Union stem from a combination of insufficient concern at the highest levels, and the usual amount of predictable (though not excusable) economic expediency at the managerial level. At the heart of the problem is the implicit understanding that the fulfillment of the economic plan is the paramount consideration, and that aesthetic and ecological considerations are clearly secondary. An editorial in the government newspaper *Izvestiya* very succinctly summarized the problem with the phrase 'victors are not judged'.[48] Until such time as the victors, as well as the underachievers, are made fully accountable for all aspects of their performance, it is unlikely that any real solution to the Soviet pollution problem will be found.

Other aspects of the quality of the Soviet environment

Air and water pollution are two integral parts, possibly the most important two, in an increasingly important aspect of modern life that has become known as the quality of the environment. In addition to air and water pollution, it would be remiss not to mention briefly certain other aspects of environmental quality in the Soviet Union.

Environmental quality

An extensive discussion of the quality of the environment in the U.S.S.R. is well beyond the scope of this study, as many components of environmental quality become further and further removed from the direct utilization of natural resources. Necessarily, they become intimately entwined with economic, social, cultural, and even psychological considerations.

In addition, any attempt to assess the quality of the Soviet environment runs the risk of arbitrarily superimposing an American value system on a much older and in many ways quite different Russian (and other) cultural history. Nevertheless, a brief review of a few other aspects of Soviet environmental quality might be useful, if only by way of comparison with experiences in the United States.

Some have already been mentioned: surface mining in Chapter 6, and the impact of tourists on protected environments in Chapter 4. Those to be discussed here deal primarily with the quality of the urban environment. This is a very broad subject deserving of a separate investigation, and only a few key features will be noted here.

Among the most important aspects of quality urban environment is the creation of parks and green belts within and around cities. The majority of Soviet cities are reasonably well endowed in this respect. Several cities which required extensive rebuilding after the Second World War, such as Kiev, now enjoy unusually large areas of parks and woodlands within the city limits. It should be noted that here is one instance in which the free availability of land works to a distinct advantage. In fact, given this consideration, it is perhaps surprising that more 'mini-parks' of a block or two in area have not been created in the central districts of major cities such as Moscow.

In addition to city parks, almost all large Soviet cities have extensive regions surrounding them designated as green belts (Table 8.2). Whatever wooded areas exist within the designated regions are classified as protected forests, to be preserved for aesthetic, health, and recreational considerations (Appendix 26, Article 7, and Appendix 5, Article 34). Inasmuch as almost every Soviet city is pushing new housing developments further and further into its outlying areas, it seems probable that occasional conflicts between green belts and urban and industrial expansion may arise, and whether these 'inviolate' suburban forests will remain inviolate in every such case is open to question.

With regard to urban and suburban planning in the U.S.S.R., it should be noted that the sprawling networks of independent suburbs and overlapping administrative regions which surround most American cities generally do not exist in the Soviet Union. City limits are expanded so that the new 'bedroom communities' on their extremities stay within their boundaries. A major expansion of the Moscow city limits, which incorporated five major suburban areas into Moscow proper, took place in 1961.[49] The resort city of Sochi, as another example, extends for about 100 km along the Black Sea coast, from the district of Lazerevskoye to the district of Adler. Volgograd (formerly Stalingrad) extends for 70 km along the Volga, and Novosibirsk for 40 km along the Ob′.

The preservation of historical and cultural monuments in cities has

Table 8.2 *Green belt zones around selected Soviet cities*

City	Population (1970)	Radius of green belt zone[a] (km)	Forested area (ha.)
Moscow	7,061,000	50	463,000
Leningrad	3,950,000	30	149,000
Khar′kov	1,223,000	?	103,000
Riga	733,000	30	100,000
Gor′kiy	1,170,000	25	77,000
Kiev	1,632,000	30	65,000
Perm′	850,000	25	65,000
Vil′nyus	372,000	30	62,000
Tbilisi	889,000	30	59,000
Sverdlovsk	1,026,000	25	53,000
Kazan′	869,000	25	52,000
Chelyabinsk	874,000	30	35,000
Kemerovo	385,000	30	31,000
Ordzhonikidze	236,000	20	31,000
Kishinev	357,000	30	26,000
Minsk	916,000	30	25,000
Ufa	773,000	30	22,000
Cheboksary	216,000	15	15,000
Tallin	363,000	30	10,000
Tashkent	1,385,000	30	6,000

[a] The green belt need not necessarily be in the form of a circle surrounding the city; the ‘radius’ figure given may in some cases only indicate the average outer limit or even the farthest single point of the green belt from the city center.

From V. P. Tseplyayev, *Lesa S.S.S.R.*, Moskva: 1961, p. 27.

had a rather checkered history, particularly with regard to religious artifacts and structures. In recent years, a somewhat greater concern for the cultural heritage of the nation has been discernible, as evidenced by the creation in 1966 of the ‘All-Russian Society for Safeguarding Historical and Cultural Monuments’.[50] The 1967 appropriations for renovation and restoration work almost doubled the 1966 amount, and numerous plans for restoring objects of historical significance are now being worked out.

Of course, the determination of what is of historical significance, and the type of interpretation it is to receive, remains basically a political (i.e. Party) decision.

Two further forms of sensory pollution which are found in most large cities are noise pollution and visual pollution (junkyards, billboards, etc.). In addition to the normal noises of large cities, Soviet cities and transportation conveyances have been plagued by a further noise nuisance – government loudspeakers. Although these are now starting to be removed from many urban centers, they are still to be found in numerous communities and on trains, boats, and other transportation media. In addition, it should be recalled that Soviet cities are characterized by a much higher frequency of trolley cars and heavy truck transportation than are American cities. Whereas cheap, reliable mass public transportation is one strong point of Soviet cities, the street cars and commuter trains do constitute a considerable noise irritant. Anti-noise provisions were included in the 1969 Public Health Law Principles, defining general responsibility:

> *Art. 24. The Prevention and Elimination of Noise.* – The executive committees of the local Soviets and other state agencies, enterprises, institutions, and organizations are obliged (upon recommendations made by agencies and institutions of the Sanitary-Epidemiological Service or by agreement with it) to carry out measures to prevent, reduce the intensity of, or eliminate noise in production premises, housing and public buildings, in courtyards, in city streets and squares, and in other populated areas.
>
> It is the duty of all citizens to observe the rules for preventing and eliminating noise in everyday living conditions.[51]

No discussion of sonic-boom or side noise problems associated with the Soviet supersonic transport (SST) has been observed as yet.

Many forms of visual pollution which blight American cities are absent from Soviet cities. One significant contributing factor to date has been the much smaller number of automobiles, which has prevented the appearance of junkyards, unkempt gas stations, abandoned cars, advertising billboards along highways, and so forth. Secondly, government ownership of all the land facilitates zoning, screening, and cleaning up other eyesores that inevitably develop. Unfortunately, the same centralized government that has banned private enterprise advertising gives itself unrestricted license to strew propaganda posters and billboards all along streets, on buildings, and in even the most beautifully cultivated parks and gardens. It has lately gotten into the neon sign business as well, with some urban planners recommending more neon signs as a cure for Russia's monotonously repetitious style of urban construction. Considerable dialog on the subject of urban planning in the Soviet Union, particularly as regards architectural design, appeared in the Soviet press around 1960, and at a lesser pace ever since. Most of these articles have been translated in the weekly *Current Digest of the Soviet Press*.

The problem of automobiles with respect to air pollution was noted above. However, other ramifications of their use need to be studied by Soviet planners, as the U.S.S.R., with the aid of French and Italian firms, takes steps to increase automobile production from 185,000 units a year in 1964 to over 700,000 in the early 1970s. First, residential parking areas and downtown garages are virtually nonexistent. Secondly, inter-city roads in the Soviet Union, with few exception, are woefully inadequate to handle large volumes of traffic. Both of these situations seem to foreordain large capital investments, concerning which too little study has been devoted to date. Thirdly, it can be assumed that increasingly large numbers of people will have greatly enhanced mobility for pursuing recreational and leisure time endeavors. Even at present, the planning of mass recreation in the U.S.S.R. is rather disorganized, and the provision of resorts, inns, garages, campsites, beaches, and numerous other such facilities is far short of the demand. It appears that thus far the tremendous impact of the coming 'automobile explosion' on the beaches, mountains, nature preserves, transportation infrastructure, and urban environment of the U.S.S.R. in general has been disastrously underestimated by Soviet planners.

This is not to say that it has been ignored in the press. Many articles have specifically pointed out the problems that the mass use of cars will bring about, stressing in particular the need for city-center bypass highways and parking garages.[52] Unfortunately, such warnings are no guarantee at all that such action will be taken. As a result, critically needed paved highway construction is to grow by 20 per cent a year during the 1970s, but automobile production is scheduled to increase by almost double that amount annually even in the pre-1970 period. While it is possible that the latter (or both) target figures may not be met, a worsening situation would seem to be in the making. The tardiness of Soviet planning for the automobile age is well illustrated by the fact that not until 1968 was there created a state agency to build and operate a nationwide network of gas stations, garages, and parts stores. Here is a very definite area where the American experience could serve as an invaluable guide, if Soviet planners were so inclined.

Much more could obviously be said about the Soviet urban environment, but a more detailed treatment than the above, intended as an overview of the Soviet situation only, must await a separate study. In summary it can be said that the centralized planning of Soviet cities, and the forced rebuilding of many of them after the war, has imparted to the urban environment of the U.S.S.R. many desirable qualities, but also in many respects that it is still far from satisfactory. Chief among these are air and water pollution, which for too many years were virtually ignored. The situation having now been forced, just as it has been in the United States, changes are slowly beginning to appear. Nevertheless, so many years of neglect will require many additional years of remedial construction and investment in order to restore adequately the quality of the Soviet Union's internal waters and urban atmospheres.

9 Conservation in the Soviet Union: a summary of attitudes, problems and trends

...the Soviet experience provides no panaceas...

The preceding chapters have examined some of the specific features and problems in each of the major areas usually associated with the concept of natural resource conservation. It is now necessary to attempt to bring all the preceding material together and to formulate for the whole of Soviet society whatever valid generalizations and conclusions can be drawn from the foregoing case studies.

Before doing this, it will be useful to review the dominant features of the political system with which we are dealing, and the extent and manner of the control it exercises over the use and conservation of its territorial natural resources. As elsewhere, the form of government involved is of fundamental importance, but in the case of the U.S.S.R., the political ideology which lies behind it is equally important. This can be demonstrated by the influence of that ideology on one particular key component of both environmental quality and of the ideology itself: population control. This subject is deserving of special consideration, and a separate section is devoted to it. The chapter concludes with a summary compilation of the more significant official attitudes, administrative problems, and current trends which characterize Soviet natural resource management and conservation today.

Political aspects of Soviet natural resource development and conservation

Politics and conservation are inseparable. Politics represents the implementing force behind natural resource conservation, and in many cases the motivating force as well. Any study of conservation within the context of a particular society must take into consideration the political structure of that society, and nowhere is this more true than for the Soviet Union. The nature of the Soviet political system – socialist in its economics and communist in its political theory – has very basic and important implications for the use and conservation of that country's natural resources.

Attitudes, problems and trends

At the national level, it may be noted that there are certain similar elements running through the Soviet and American conservation efforts. Among these are a high degree of federal encouragement and support for conservation activities, federally funded research organizations, large territories withdrawn from economic use by the government for conservation purposes, a high level of academic interest and research on natural resource conservation, and a lesser but significant amount of private interest and concern. It goes without saying, however, that there are substantial differences too.

Certainly dominant among these is the totally centralized organization of the economy. The utilization of all natural resources is planned by the various levels of government, and their extraction, fabrication, and all aspects of their conservation is carried out by agencies and enterprises directly under governmental supervision. The question of environmental perception in the Soviet Union, therefore, becomes essentially a question of governmental perception – of the way in which the Soviet government views the U.S.S.R.'s environment and the complexes of natural resources which comprise it. The government bears full responsibility for the state of natural resource management and conservation which exists at any given time.

The role of the Communist Party in all of the above needs to be emphasized here, for it is the Party that runs the government, and not vice versa. The Party formulates the basic directions of the national economy, and supervises the fulfillment of these plans. Its representatives exert their influence vertically through the economy from the highest levels of the Council of Ministers to the smallest local machine shop. Without any loss of precision, the phrase 'the Party' could be substituted everywhere for the word 'government' in the preceding paragraph.

It would seem logical to assume that the centralization of economic planning and the absence of the intricate parliamentary formalities which characterize the American congressional system would at least have the advantage of making it possible to expedite the passing of conservation legislation and to execute the letter and spirit of such laws once they are enacted. Such, however, is not necessarily the case. Although legislation can in fact be passed fairly swiftly in the U.S.S.R. when needed, there are several serious drawbacks also inherent in the system.

First, there is only one effective lobby in the Soviet Union, and that is the fully understood and immutable emphasis on industrial expansion. The voices of conservationists, while present, are weak by comparison, and certainly hold no threat of voting an unreceptive Central Committee member out of office.

This represents an important distinction between the United States and the U.S.S.R. If, in the United States, private enterprise displays poor conservation practices, there are still two avenues of recourse open for correcting the situation – public opinion and government regulation. But if the Soviet central planning mechanism is lax in these regards, there is no effective avenue of recourse. The Party–government supervision of both resource exploitation and environmental conservation has strong built-in conflicts of interest, and brings to mind the analogy directed by some towards our own Atomic Energy Commission of 'foxes guarding henhouses'.

Secondly, because of the nature of the centralized Soviet planning apparatus, there is generally a longer 'lead time' for new techniques in the U.S.S.R. than in the United States. That is, the innovations of research organizations, even if their need is appreciated, may have to wait years before they can be worked into the already formulated long-range plans for development and implementation in the economy. Many sources cited in this study pointed to these delays as a significant factor in retarding the improved management and conservation of natural resources in the Soviet Union.

Thirdly, as has been noted throughout this study, the laws that are passed are often much too generalized, and the procedures set up for enforcing them are, in many cases, almost totally ineffective. Despite centralized planning, the sheer size and bureaucratic complexity of the country, together with a variety of conflicts over economic priority, have seriously hindered the implementation of many desirable conservation practices in the Soviet Union. Delays in the construction of pollution control facilities represent a good example of this. Inter-ministry coordination and cooperation are often less than desirable, and represent one of the less unique aspects of the U.S.S.R.'s problems. They are akin to the squabbles between, for example, the Departments of Agriculture and Interior, or the Corps of Engineers and the Bureau of Reclamation, which are familiar to American conservationists. Also, as Chapter 4 noted, positive conservation measures that can be quickly enacted, such as the establishment of *zapovedniki*, can be just as quickly abolished.

The second main differentiating feature of the Soviet system is the total nationalization and free availability of the country's lands and waters and their component resources. Only the state can exploit natural resources; only the state can conserve them. As noted in the previous chapter, this works both ways. Whereas Soviet cities are generally well planned and landscaped, their streets and parks abound with loudspeakers and propaganda posters, and municipal sewage treatment is given too low a priority. Nor has the government proven that it will necessarily be more thoughtful of the natural environment in constructing housing, industries, and resorts than a private developer would be. The worst examples of commercial Philistinism are missing, to be sure, but the general implementation of rural and recreational landscape planning is still in its infancy in the Soviet Union.

A definite advantage of the Soviet economic system, however, – one that is partially intended and partially not – is that the Soviet Union has not yet developed into a 'waste-maker' society. The unrestrained competition to seduce the consumer into buying what he doesn't need by the use of gaudy and unnecessary packaging, misleading advertising, constant design changes, psychological pressures, 'beautifying' chemical additives of uncertain side-effects, etc., is largely absent from the Soviet economy. Although some improvement in the quality and attractiveness of Soviet consumer goods is definitely needed, it is at least theoretically possible for them to achieve this without the conspicuous consumption and manufacturer arrogance that accompanies product marketing in the United States.

Another strong point in the Soviet system is the amount of government sponsored research that is done in the biological and social sciences pertinent to natural resource conservation. This is carried out primarily through the vast network of institutes of the U.S.S.R. (and republic) Academy of Sciences, and in total scope and volume far surpasses that carried out in the United States by non-academic conservation research organizations. This Soviet research is particularly strong in the area of the physical and biological characteristics of natural complexes and ecological interrelationships.

In the social sphere, there is less awareness on the part of individuals of conservation conflicts than in the United States. The phenomena of periodically intensified governmental censorship, a low level of individual mobility, and a supply of natural resources generally exceeding the demand have combined to make the average Soviet citizen rather unaware of conservation conflicts in the exploitation of his country's natural wealth. This is a situation that shows some signs of changing at present, however, as the government press begins to give more coverage to conservation issues.

In addition to the above characteristics imparted to Soviet conservation by its economic and governmental setting, the ideological foundations of the system lend to it still other unique features.

Most important among these from the standpoint of natural resource conservation has been the continuing goal of rapid industrialization. Slogans such as Lenin's famous statement that 'Communism equals Soviet power plus the electrification of the entire country' have remained the watchword of planners in the U.S.S.R. ever since he coined the adage in 1920. The development of such leading industries as ferrous metallurgy, petroleum, machine construction, power and chemicals is understood to have unquestioned priority, and persons throughout the countryside who become disturbed over the inevitable result of such a policy on the local environment find it very hard to fight city hall, when city hall extends all the way up to the Council of Ministers and the canonized precepts of Lenin. This fortifies the entrenched position of the planners who, knowing where the priorities lie, logically feel little incentive to familiarize themselves with the pleas of ecologists and conservationists. Only if and when pressure along these lines comes from above, rather than from below or from outside, is remedial action apt to be taken in amounts commensurate with the intensity of such pressure.

A second major drawback of Marxist ideology is the belief that only under capitalism is the rapacious, wasteful, or improperly planned use of natural resources possible. This, it is believed, is the inevitable result of the competition for resources by private entrepreneurs who are guided only by their individual urge for profits (Marxist theory usually assumes that a capitalistic economy is necessarily a *laissez-faire* one, and ignores the possibility of any governmental or public feedback or control whatsoever).

On the other hand, it is believed that a socialist economy, faced with the obligation to plan centrally the use of all its resource wealth, will necessarily do so in the wisest possible manner. This nineteenth-century Marxist view appears to reflect an understanding of conservation that sees waste and premature depletion as the only unwise forms of natural resource utilization. It is an approach, especially when coupled with the rapid industrialization syndrome, that is ill-equipped to select wisely from among alternate or incompatible ways of developing a given resource complex, or to foresee and avert the more subtle or delayed forms of environmental deterioration that accompany new industrial techniques and products.

Another ideological foundation of Soviet economic planning, one shared by most Americans, is that economic growth is necessarily good. Soviet planners, of course, believe not only that growth is good, but that their own growth rate must and will exceed American growth. No one in the U.S.S.R. has yet called for a zero economic growth rate in the interests of the environment, but one economist has suggested that it might be wise at least to slow down a little:

> In calculating national wealth, GNP and growth of well-being, economics continues to employ indices that compare direct costs and direct results, without regard for ecological consequences...This view gives an incomplete and frequently falsified picture of reality...We must place serious limitations on burn-

ing fuel, cutting forests and producing chemical pesticides and a number of other products, and seek new technological solutions that will not harm the biosphere. We must be willing to slow our economic growth by probably 7% to 10% for several years to divert funds to rescuing the balance of nature.[1]

It can be assumed, however, that this represents a minority opinion among Soviet economists.

The propagandizing of the infallibility of socialist planning has naturally abetted the general censorship-inspired hesitancy of individuals to speak out against obvious planning errors which have impaired environmental quality or judicious use of natural resources. Even so, public participation in pointing out such planning errors increased greatly during the 1960s, as noted in the section of Lake Baykal. However, increased internal control could be reinstated at any time, and in fact, such appeared to be happening, at least to overly outspoken intellectuals, at the end of the 1960s. Whether this will be extended to dissuade public criticism of the planning and administrative establishment, with all that this would imply for conservation in the Soviet Union, remains to be seen.

Another important facet of Marxist ideology, one that has great importance for the future environmental quality of the entire world, is the Soviet position on birth control and population limitation. Both the number of people currently living under communist governments and the extent of Soviet (and Chinese) influence in the developing countries call for a close examination of the Marxist position on this complex subject.

Soviet theory and practice regarding population control

The common denominator of all natural resource management and conservation problems is the human population which is dependent upon these resources – its size, its growth rate, and the composition of its demands. This is an equation that must be examined not only for the world as a whole, but also for each country individually. The conclusions which will be drawn for any given country will vary greatly according to its size, resource endowment, and level of economic development. Even within similarly endowed and developed countries, views as to the present or potential threat of overpopulation will vary tremendously depending upon the manner in which the viewer perceives his society, his environment, and his national goals. This has most definitely been true for the manner in which this problem has been seen through American and Soviet eyes, respectively, in recent years.

In the Soviet Union, the official position on the subject of population growth and birth control is well rooted in Marxist philosophy, as would be expected. Like many other positions based on classical Marxism, however, it has been possible to lend differing interpretations to the subject of birth control in accordance with changing times and conditions, and the Soviet position on this subject has in fact undergone some re-examination in the last few years.

Until the fall of Khrushchev from power in 1964, the official Soviet attitude was what might be termed the 'hard line'. This stated flatly that there was not and could not be any such thing as overpopulation in a socialist country, and that those in the West who worried about lowering the birth rate did so because of what Marxists assume is the inability of capitalism to provide for the agricultural and material needs of its people. This alleged situation was termed by Soviet writers 'relative overpopulation', to suggest that overpopulation was a relative condition unique to capitalist societies, caused by what Marx foresaw as inevitable unemployment. As Khrushchev put it in 1955:

> The more people we have, the stronger our country will be. Bourgeois ideologists have invented many cannibalistic theories, including the theory of overpopulation. They think about how to reduce the birth rate and the growth of population. Matters are different among us, comrades. If we were to add 100,000,000 to our 200,000,000, it would be too few.[2]

In actuality, Soviet population theory was somewhat ambivalent in the late 1950s, as contraceptives and abortions were legalized in the U.S.S.R. by a decree which was also enacted in 1955. While Khrushchev remained in power, however, the hard line continued to prevail. In 1963, the government press attacked Communist Chinese approval and encouragement of sterilization, calling it 'not Marxist'.[3] In 1964 there did not exist a single demographic research institute in the Soviet Union; two which had existed earlier in Kiev and Leningrad had since been shut down.[4] Currently, demographic research is carried on in a number of separate departments belonging to several different institutions.

Soviet 'hard line' demographers direct abuse at few men with the enthusiasm that they do at Thomas Malthus, the late eighteenth- and early nineteenth-century British cleric and economist. Malthus' concern for overpopulation and poverty led him to the conclusion that wars, famine and disease might be the inevitable result of a failure to check a geometrically increasing world population by means of popular moral restraint. Marxists enjoy coupling Malthus' pessimistic conclusions to their belief that capitalism cannot provide for a growing population, in order to 'prove' that capitalist countries must engage in wars to perpetuate themselves. Contemporary Western spokesmen for birth control as the basis of economic and social improvement, such as Paul Ehrlich and Julian Huxley, are labeled as merely 'neo-Malthusian'. Thus, for Soviet leaders, the subject of population control has ramifications extending even into foreign policy, and this, together with their belief in the unlimited productive capabilities of socialism, has made it all the more difficult for them to accept the majority viewpoint that birth control is urgently needed now in many countries of the world.

Nevertheless, a certain change in the Soviet outlook did begin to take place around 1965. Whether this represented the emergence of viewpoints suppressed under Khrushchev, or whether this view developed independently after his fall is uncertain; the former is more likely. The first public indication of change appeared in late 1965 and early 1966, when a prolonged discussion took place in the Soviet press. A leading Soviet demographer, Boris Urlanis, noted that in many of the lesser developed countries of the world an increase in urbanization had not brought about an automatic decrease in the birth rate, as it had in many industrialized countries, and that in some of these countries immediate efforts to stabilize the population were essential. He also indicated that the eventual answer to controlling world population would lie in what he termed 'intentional parenthood'.[5]

However, the old 'hard line' also dominated a few of these articles. The hard-liners almost always limit the discussion of demographic problems to the food supply question, and to the context of other countries (since in socialist countries, in theory, there can be no food supply problem and hence no overpopulation). In their view, birth control is not only unnecessary in the Soviet Union, but some even suggest that a high rate of births should be encouraged.[6] These writers are very optimistic about technology's ability to increase the world's food supply: 'A rational and highly industrialized agriculture and food industry of future society should be able, by using the potentialities of nature for increasing biological productivity, to satisfy completely the needs of mankind even if the world population rises to several tens of billions.'[7]

Both the hard-liners and the more liberal participants in the discussion stress that 'radical social-economic transformations' (i.e. socialism) are the basic cure for poverty and hunger in the developing nations, and that birth control measures alone cannot solve these problems. This is a basic Marxist position to which any Soviet demographer must adhere. Western demographers who call for permanent government support of birth control and who are not advocates of 'radical social transformations' are promptly termed 'neo-Malthusians' by their Soviet counterparts, even though they generally do recognize the need for social-economic reforms in capitalist countries.

With regard to the lesser developed countries, the hard-liners state that over the long term, a decreased birth rate (if needed) would be brought about by the natural tendency for birth rates to fall in conjunction with rising urbanization and industrialization; whereas moderates such as Urlanis can see the need for birth control measures in many countries even at present. It is unfortunate that the old ideological constraints prevent many Soviet demographers from realizing the extent to which a high birth rate hinders or even prevents economic advancement in many countries. Still, an important step had been reached – the public support of at least a few Soviet demographers for some form of birth control efforts in the poorer countries of the world.

Some of the more objective writers also spoke approvingly of birth control efforts in economically advanced countries, even in the United States, and noted that 'radical social transformations' have not automatically brought about population stability in all socialist countries (e.g. China).[8]

One of the articles at last made the fundamental point which had been completely lacking in the Khrushchevian hard line view of population control. That was the seemingly obvious point that there are many determinants of population saturation other than agricultural output. Unfortunately, the idea was not developed. Rather, the author, falling back on the food supply arguments, suggested that even today the summed per capita need for supporting agricultural land was approximately equal to the world supply of such land, so that the threat of overpopulation, for the earth as a single biologic unit, exists at the present time.[9] This world view, as opposed to an 'our society *vs* your society' approach, marks another encouraging improvement in Soviet thinking on the subject. Equally significant is the fact that in 1966 the Soviet Union began to support United Nations resolutions on population control and family planning, the first time in the history of the organization that they had done so.[10]

In the Soviet Union itself, it might be that actions speak louder than words. Despite the still officially expressed view that there can be no overpopulation under socialism (a view not shared by Engels, among others), the Soviet Union is more and more encouraging 'family planning'. The reason given, which represents the current position of the government, is the right of marriage partners to plan their own future (regardless of the effect of this on the birth rate), and not that there is a need to limit population growth. This is consistent with the apparent position of Lenin who, while attacking Malthus and 'neo-Malthusians', nevertheless said that the use of contraceptive methods should be the decision of the individuals concerned. Lenin approved the legalization of abortion in 1920, and except for the 1936–55 period, that enactment has remained state policy.

However, although Soviet families can theoretically choose the size of their families (and are opting for smaller ones at present), the government frequently utilizes the media to propagandize for larger families, especially in the Slavic republics. The probable reason for this is the need for more workers for the economy, necessitated in part by the low average productivity of the Soviet labor force. Nevertheless several research institutes dealing with birth control methods have now been created, and the production of intrauterine devices was to reach half a million by the end of 1968.[11]

The above notwithstanding, it would seem desirable for the Soviet Union to develop a more specifically stated national policy on its own internal population growth, one based on other than purely agricultural considerations. It is clearly irresponsible for Soviet demographers to continue jousting with foreign 'neo-Malthusians', while simply dismissing even the possibility of there ever being any overpopulation in their own country. Pollution levels and resource depletion rates are unquestionably a function of population size.

It would also be instructive to see a discussion of what might constitute an optimum population range for the Soviet Union. It is logically clear that a maximum limit exists, based on the carrying capacity of the land, however high the most optimistic theorist might wish to set it. However, virtually nothing has as yet been attempted by way of working out either a maximum limit or an optimum range, or even to suggest criteria by which such approximations could be made. Granted that this is a very difficult topic rife with many subjective pitfalls (and one receiving little attention even in the United States), it would nevertheless seem clear that sooner or later such discussions will become necessary, and that the present would not be too soon for initiating some serious thoughts on them.

While the Soviet Union has made a start towards approaching population problems from a less dogmatic position, there still remain some basic questions that need to be faced squarely. It is probably a mistake for Soviet ideologists to insist on treating the population question as basically one of political-economic theory, with food supply as the only determinant. While there are no doubt elements of this in it, in the case of economically and agriculturally advanced countries such as the United States and the U.S.S.R., it is more a problem of maintaining environmental quality. Many would cite population control, in fact, as the single most important element in ensuring the future environmental quality not only of the developed countries, but of the lesser developed ones as well. It would undoubtedly be to the advantage of the Soviet Union to begin viewing population control, in its own internal context at least, more in this light.

Synopsis of current problems and future prospects in Soviet conservation

Although the theory and procedures of proper natural resource utilization and conservation are generally well known to Soviet specialists in universities, research institutes, and public conservation organizations, it has been seen that these are frequently not reflected in the management of the resources in the field. Although some progress is being made, the continued existence of several major problem areas has been noted throughout this study: much marginal agricultural land has been drawn into production, and dust storms continue to plague the dry steppes; forest resources are very inefficiently used; large reservoirs drown out vast timber reserves and fish runs; the poaching of fish and wildlife is rampant; and air and water pollution have become at least as bad as in the United States.

Why do these conditions exist within the purview of the world's most inclusively planned national economy? The source materials and case histories examined during the course of this study suggest a number of fundamental causes, all interacting with, and many reinforcing, one another. Only a few of them could be termed unique to the Soviet system.

One basic set of problems is inherently associated with the expansive Soviet bureaucracy. Problems of communication and coordination have been noted above, as has the tendency for delays in constructing purification facilities and introducing new conservation techniques. However, due to the low level of effectiveness of the public conservation lobby in the Soviet Union, the consequences of this lack of inter-ministerial coordination are more significant for the U.S.S.R. than are corresponding situations in the United States. Previous chapters have noted that ministries whose product is vital to the industrial growth of the U.S.S.R., such as electric power, are able to act with relatively little concern for the problems of other ministries enjoying a lower priority, such as fisheries. In the United States, pressure on Congress from sportsmen and the fishing industry can often effect a reasonable 'balance of power' in such cases, but to conduct this type of conservation lobbying in the Soviet Union is considerably more difficult. Under the present Soviet economic system, only directives from the highest Party levels can effectively bring about improvements in inter-ministry cooperation in the interests of natural resource conservation. But this, of course, is also where the directives for continuing the most rapid possible pace of industrial growth originate. With more effort, a better compromise between these occasionally conflicting goals might be realized.

The problem of communication between governmental and non-governmental personnel appears to be greater in the Soviet Union than is the case even in the United States, perhaps due in part to the ascendant position and implied infallibility of the socialist planning mechanism. In general, it can be said that Soviet natural resource specialists in universities and institutes are considerably more attuned to conservation problems than are governmental planners and administrators, whose primary interests involve quantitative, rather than qualitative, considerations.

Some increase in the influence of the former, however, can be noted in recent years. One of the more significant developments of the 1960s, involving a basic change in the former Soviet approach to natural resource exploitation, was the decreasing emphasis given to large hydroelectric projects, particularly those which would have altered natural conditions over huge areas. The cancellation of the Lower Volga and Lower Ob′ dams, and spirited debates concerning the wisdom of other projects, were noted in Chapter 7. Although economic considerations played an important role in most of these cases, the opinions of knowledgeable persons in the natural sciences were not without their effect. Even so, some method of still further improving communication between the planners and the natural scientists, and increasing the influence of the latter in natural resource decisions, would be desirable.

The researchers, however, sometimes lag behind the planners in recognizing one important aspect of natural resource conservation: it frequently costs a great deal of money. Since all funding in the U.S.S.R. is ultimately state funding, it is easy to request expensive cures for pollution, reclamation, and other conservation problems, but actually to obtain funding priority for these projects in the naturally tight state budget is as difficult in the Soviet Union as it is in the United States. It should be noted that Soviet economists are ultra-conservative by Keynesian standards, and the national economy is always operated on the principle of a balanced annual budget. *Gosplan*, of course, is well aware of these costs, and the fact that the allocated capital for such items as air pollution control is sometimes not fully utilized may make it more difficult for conservationists to elicit funds for other projects. This is particularly true in the light of the inherent resistance of Soviet planners to devote funds to uses such as water pollution control, which do little to raise industrial production indices, but which might involve a major sacrifice in the rate of economic growth for the cause of environmental quality.

Another similarity between the Soviet Union and the United States is that neither country's economic planning mechanism seems capable of incorporating the full range of social costs into its natural resource development planning. Intangible and delayed secondary costs are as inadequately reflected in *Gosplan*'s proposals as they often are in those of both private and governmental developers in the United States, and the long-range deleterious effects of certain types of natural resource exploitation are equally apt to be overlooked. In the U.S.S.R., as to an extent in the United States, comprehensive planning is further hindered by the lack of rational planning units (e.g. drainage basins for water resources planning), and the lack of any type of monetary evaluation of the resource being developed.

An effort by both Soviet and American planners to perfect means for identifying all of the ultimate costs to society of developing a natural resource complex in a given way would be a race between the two countries well worth running.

One of the most significant aspects of the gap between the planners and conservationists is the continuing predilection of Soviet planners, concerned about their country's agricultural limitations and emboldened by their faith in the technical forces of socialism, to think of themselves as in a contest with nature, and to view 'the transformation of nature' as the most direct and desirable cure for a variety of economic problems. Such ideas, though acknowledged to need much more careful planning than they received under Stalin, are still very much promoted in the Soviet Union today. Proposals for diverting the large Siberian rivers, for damming ocean straits, for eliminating deserts, and for melting glaciers have been cited throughout this study. Even larger, more questionable projects involving many other countries are in the minds of some Soviet planners and propagandists, who seem quite unconcerned about either secondary effects to the environment or national sovereignty.[12] While similar plans are not unknown in the United States (N.A.W.A.P.A., for example), the harsh light of informed public dialog soon places them in a more cautious perspective. N.A.W.A.P.A. (North American Water and Power Alliance), as envisioned by an American engineering firm, would divert water from arctic rivers southward through the Rocky Mountain trench to Canada, western United States and northern Mexico, at a cost of around 100 billion dollars.

The difference between the prevailing Soviet and American attitudes is well illustrated by the fact that, in a review of the book *Future Environments of North America*, the reviewers, both prominent Soviet geographers, express surprise on only one point:

> The reader is thus impressed by the fact that these American authors, in discussing the basic problems of regional development, display great restraint on the subject of large-scale transformation projects designed to raise the level of resource use. We are thinking, for example, of programs providing for the diversion of water resources. . .the realization of grandiose irrigation, drainage, and amelioration (including warming) projects; large scale efforts to change the climate, etc.[13]

This Soviet reaction is quite understandable when it is considered that one of the reviewers, I. P. Gerasimov, who is chairman of the U.S.S.R. Scientific Council for the Study of Natural Resource Problems and perhaps the most prestigious geographer in the U.S.S.R. today, wrote in 1966 that 'the creation of the material and technical foundations of a communist society involves the planned transformation of the natural environment and of the national economy, in order to fully overcome the unfavorable influences of the spontaneous forces of nature [and for other economic and social goals].'[14] He goes on to suggest that the first of the three major problem areas of the geographic discipline in the future will be 'the development of theory and the elaboration of scientific programs for a planned transformation of nature, which is necessary for effective natural resource use. . . '.[15]

These excerpts indicate the official importance attached to the transformation of nature in the Soviet Union, with its post-Stalin qualification of careful advance planning. So widespread and accepted is this idea of transforming nature that Gerasimov has even defined the concept of the conservation of nature in terms of its transformation, as noted in Chapter 1.

The inclusion of nature transformation studies as a basic component of future geographic research illustrates the strength and pervasiveness in the U.S.S.R. of the idea that nature must undergo widespread human reorganization in order to be useful to man. In concert with the lack of any true statutory wilderness areas in the Soviet Union, it tends to suggest that the old master-slave attitude of man towards nature is still to be found in official Soviet thinking, and that a more sophisticated appreciation of man's subtle interrelationships with nature, and of the intrinsic worth of even the most extreme natural landscapes, is still largely missing from the orthodox Soviet planner's value system. In the final analysis, man's ability to transform nature to serve certain of his needs can never mean that he has conquered nature, only that he has, at best, learned how to achieve a more complex understanding of its laws.

The intent of the foregoing is not to suggest that large-scale hydrotechnical and other such man-made engineering works are not often desirable and useful, only that there is a grave danger in institutionalizing them as the very foundation of all social progress, and in the process casting the natural environment into some sort of implied villain's role. Soviet planning philosophy needs to appreciate that the concept of environmental quality hardly needs to be limited to man-made environments. This is especially true in a society that believes it cannot use natural resources unwisely. Nevertheless, Marxist ideology very definitely suggests that 'technological solutions' exist to all environmental problems, and that Soviet science is best equipped to find them.

The above considerations lead into the subject of conservation education. During the Stalin era, natural resource use education consisted primarily of propagandizing just such 'large scale transformation of nature' projects, while at the same time the biological sciences were stagnating under the influence of Lysenko. Trofim Lysenko rejected the standard genetic teachings of Mendel and Morgan and claimed, with Stalin's backing, that environmentally acquired traits could be inherited and passed on genetically. Today, the views of the natural scientists are given much greater consideration, and a genuine effort to teach school children the principles of conservation is being made in the Soviet Union. Indeed, if what Soviet writers claim is being done is actually the case, it may surpass present efforts in the United States.[16]

However, note might again be made that the textbook classification of air and water as inexhaustible resources remains as a curious anachronism. It is not so much the physical accuracy of this term which is open to question as it is the psychological implications which can arise out of its use. In very recent years, an attack on this concept appears to be gaining some momentum. I. P. Gerasimov, in a recent article, notes that the answer to the question of why the Soviet Union still retains poor natural resource utilization policies 'should begin with the admission that in our country the false idea of the inexhaustibility of natural resources, and the dangerous tradition of the gratuitous and uncontrolled use of the blessings of nature, has taken firm root'.[17] As Gerasimov suggests, much more work needs to be done in conservation education among the Soviet people in general, if such widespread practices as poaching, indiscriminate timber cutting, and pollution are to be brought under control. Even existing conservation education programs in the Soviet Union tend to be counterbalanced by the Party and government press, which continue implicitly to romanticize grandiose transformation schemes and to teach that natural resources cannot be used unwisely under socialism.

It is probably this latter article of faith which accounts for another interesting characteristic frequently encountered in Soviet books and monographs on conservation. This is the tendency in many scholarly writings to discuss conservation problems primarily in the context of other countries, and to pass over or understate obvious domestic examples. It would be considered rather unscholarly for an American author to open an article on water pollution with a discussion of the Northern Donets and Kama rivers, with only passing reference to Lake Erie and polluted American streams, yet the reverse is still encountered in many Soviet writings. Newspaper articles, surely more widely read than natural resource monographs, can be very outspoken about conservation problems in the U.S.S.R., but the authors of more formal works still seem compelled to prove that, no matter how poorly an issue may be handled at home, the situation is definitely worse in the West. A greater degree of candor concerning Soviet problems would be a healthy development in the effort to advance public conservation education. This lack of candor can take on humorous proportions. For example, visitors to Baku have been known to be assured by their Intourist guide that there is no oil pollution of the Caspian Sea, even though the view in Figure 8.1 is encountered only a few hundred feet from the main Intourist hotel.

In part because of the very nature of a bureaucracy, and perhaps in part because no one wants to be held responsible for natural resources being unwisely used in a socialist economy, another characteristic appears in Soviet resource management: buck-passing is endemic. Enterprise managers blame designers, designers blame planners, planners blame administrators; and any official at any of these levels who is worth his salt has a ready list of friends in higher places on whom he can call for support in an emergency. No one seems to have a sign saying 'the buck stops here'. In the Soviet system, it is clear where the buck should stop: at the local (or regional) Party headquarters. But all too often, this is precisely where the 'friends in higher places' are to be found, who, like the managers, have a prime interest in seeing that production targets are met.

Here is perhaps the key weakness in the Soviet system as regards natural resource conservation – the supremacy of plan fulfillment. Much can be overlooked so long as an enterprise has carried out its primary obligation and met its quotas. Examples of pesticides used 'in the interests of the harvest' which killed wildlife and plan fulfillment bonuses which exceeded pollution fines have been given already. The seriousness of the problem has been pointedly underscored in the government press:

> Local Soviets and the various inspection teams vested with considerable powers in the struggle against the destroyers of nature display timidity and excessive delicacy 'in the showdowns.' Indulgence and protection from consequences cover up the affliction.
>
> Its chief symptom is the still prevalent belief that increased output of goods compensates for and glosses over the moral and material losses. Victors are not judged, as the saying goes.[18]

This unwillingness to enforce regulations on successful managers will no doubt remain as a ponderous brake on Soviet conservation efforts so long as fulfilling the plan and meeting production goals remain the *sine qua non* of the Soviet economic system.

The future of conservation in the Soviet Union would seem to depend on the nature of the Soviet response to the more critical of the causal shortcomings and problems noted above. Specifically, there are several courses of action which the Soviet government needs to consider seriously, all of which would appear to be basic to achieving its goals of improved natural resource utilization and environmental quality. These include:

(1) instilling in Party personnel and high government planners an understanding of the need to carry out in practice, not just on paper, strict conservation measures in all branches of the economy, an awareness that these measures are equal in importance to fulfilling output or profit targets,

and a willingness to allocate the necessary funds and make the necessary sacrifices to do so;

(2) providing for effective personal penalties for all violations of conservation laws, and ensuring that no exceptions to such enforcement are permitted for any reason;

(3) putting aside certain outmoded and retrogressive concepts which continue to undermine the cause of conservation in the Soviet Union. The most pernicious of these are: (*a*) the continued propagation of the idea that there can be no unwise use of natural resources under socialism; (*b*) the still lingering notion that certain natural resources are inexhaustible; (*c*) the idea that nature must necessarily be extensively transformed by man in order to guarantee its usefulness (again it should be emphasized that it is not the actual need for such engineering works, but rather the dubious attitudes engendered by such a sweeping, romanticized, and misleading manner of presenting this concept, which is being called to question here);

(4) developing a more comprehensive system for planning the economic utilization of natural resources, particularly water resources, which would ensure a more thorough consideration of all the secondary effects and hidden social costs of such development on other branches of the national economy and on the environment in general;

(5) pricing (or other means of monetary accountability) for the use, and for the waste or pollution, of land and water resources;

(6) more public education, particularly for adults, concerning the need for natural resource conservation, and even more encouragement of public discussion and reporting of conservation problems and violations.

Although the above list is by no means exhaustive, it is along lines such as these that the Soviet government must move, and as rapidly as possible, if it is to fulfill its stated commitment to correct the current widespread deficiencies in the use of its natural resource endowment, and to provide for the greatest possible rate of improvement in the quality of its citizens' environment.

The record of the Soviet Union's first fifty years would seem to suggest that the centralized planning of an economy *per se* provides no necessary guarantee that natural resources will be judiciously used, or that pollution of the environment will not occur. As in the United States, numerous agencies to control these phenomena exist, but extravagances, wastes, and environmental contamination still occur. The problems of economic expediency, bureaucratic inefficiency, pressures from 'above', overly lenient enforcement, neglect of intangibles and externalities, fallacious concepts of inexhaustibility, and public indifference remain unsolved in the Soviet Union, or at best only partially solved, even as they do in the United States.

To date, then, the Soviet experience provides no panaceas for the universally encountered problems relating to natural resource development and conservation. This is not to suggest that the Soviet system lacks the means to resolve these problems, any more than the conservation shortcomings of the United States imply insoluble inadequacies in our system. In both countries it is a matter of will, not of means. The fact that the United States excels the U.S.S.R. in some areas of conservation is no grounds for complacency; it is also worse in some, and outstanding in very few. In the final analysis, there is little to be gained from an exercise in intercontinental polemics over who is 'least worst'. There is still much to be done in both countries and, as yet, still time to do it. The livability of our spaceship Earth rests in the balance.

Appendices

Appendix 1. Annotated glossary of common Soviet conservation terms

1. 'Complex use' (*kompleksnoye ispol'zovaniye*). This term is receiving increasing use in articles on natural resource development, and implies not only the use of a resource in as many ways as possible and desirable (i.e. 'multiple use'), but also the study of the subsequent integrated and interrelated effects which the employment of such a resource will have on other resources. For example, it asks how new irrigation projects will affect ground water and salinity, how large-scale logging activities may affect soils, local climate, and water quality, how new dams will affect fish runs and wildlife habitats, and so forth.

2. 'Conservation of nature' (*okhrana prirody*). See text.

3. 'Large-scale reproduction' (*rasshirennoye vosproizvodstvo*, also translatable as 'expanded regeneration'). As noted in the passage from Bannikov cited in the text, this term is somewhat similar to the concept known in the United States as 'sustained yield'. However, it also carries the notion of restoring and increasing useful species of flora and fauna (and even varieties of non-living resources, such as water) which have been depleted in the past, or which would be useful to man if available on a larger scale.

4. 'Rational use' (*ratsional'noye ispol'zovaniye*). The implications of this phrase have been defined as follows: 'A rational approach assumes that all so-called renewable resources – such as timber, soil fertility, grazing lands, and fauna – will be used only in such amounts and by such techniques as to allow complete renewal and sometimes even further growth. Non-renewable resources, such as minerals, should be used without waste and, if possible, with a growth of reserves through exploration for additional resources.'* 'Rational use' thus embraces the preceding idea of large-scale reproduction as one of its basic components. The phrase, however, is quite general, and as a guide to conservation activities it would seem to have the same limitations as the earlier-coined phrase 'wise use'.

Appendix 2. Metric conversion table

Metric	*English*
1 centimeter (cm)	= 0.394 inch
1 meter (m)	= 3.281 feet
1 kilometer (km)	= 0.621 mile
1 hectare (ha)	= 2.471 acres
1 square kilometer (km^2)	= 0.386 square mile
1 kilogram (kg)	= 2.205 pounds

* V. V. Pokshishevskiy, 'Na poroge tret'yego tysyacheletiya', *Nauka i zhizn'* (1968) No. 2, p. 71.

1 metric ton (m.t.)	= 1.102 short tons (2204.6 pounds)
1 cubic meter (m^3)	= 1.308 cubic yards

English	*Metric*
1 inch	= 2.540 centimeters
1 foot	= 0.3048 meter
1 mile	= 1.609 kilometers
1 acre	= 0.4047 hectare
1 square mile	= 2.590 square kilometers
1 pound	= 0.4536 kilogram
1 ton (short ton).	= 0.9072 metric ton
1 cubic foot	= 0.0283 cubic meter

Appendix 3. Excerpt from the 1960 Russian Republic law 'On the Conservation of Nature in the R.S.F.S.R.'*

'Nature and its resources in the Soviet state represent the natural basis for the development of the national economy, serve as the source of uninterrupted growth of material and cultural assets, and provide optimum conditions for the work and relaxation of the people.

The Soviet social system and the planned management of the economy create the possibility of rationally utilizing the natural wealth of the Russian Federation.

During the years of Soviet power in the R.S.F.S.R., considerable work has been carried out on the organization of nature conservation and the rational use of natural resources. However, in the work on the conservation of nature substantial shortcomings still exist.

During the period of the intensive construction of communism, the intensity with which the rich natural resources of our country have been drawn into economic utilization has increased, and the distribution of productive forces on its territory has greatly improved. This calls forth a need to establish a system of measures for directing the protection, rational use, and broad scale reproduction of natural resources.

The conservation of nature is a most important state problem and a concern of all the people.

For resolving problems of the national economy which concern the development of new regions, the reconstruction of existing ones, the transformation of river systems, the application of irrigation over vast areas, and the utilization of individual natural resources, ministries and departments should take into account the interests of other branches and of the national economy as a whole, and also the needs of the people.

With the goal of strengthening the conservation of nature and ensuring the rational use and replenishment of natural resources, the Supreme Soviet of the Russian Soviet Federated Socialist Republic decrees:

Art. 1. Natural objects, subject to protection. On the territory of the R.S.F.S.R. all natural resources, both those which are being drawn into economic production and those which are not, are subject to state protection and regulated use:

* From *Pravda* (October 28, 1960) p. 2.

(*a*) land;
(*b*) mineral resources;
(*c*) water (surface, underground, and soil moisture);
(*b*) forests and other natural vegetation, and green belts in populated areas;
(*e*) typical landscapes, and rare and noteworthy natural objects;
(*f*) health spa locales, protective forest-park belts, and suburban green belts;
(*g*) animal life (useful wild fauna);
(*h*) the atmosphere.

(Articles 2 through 12 deal with specific natural resources and are presented in subsequent appendices.)

Art. 13. Determination of the quantity and quality of natural resources. Ministries, departments, and *sovnarkhozi** which are engaged in the utilization and replenishment of natural resources are responsible for organizing and carrying out a quantitative and qualitative evaluation of them, by means of compiling cadastres, qualitative surveys, specialized maps, etc.

The Central Statistical Administration of the R.S.F.S.R. Council of Ministers is entrusted with organizing and maintaining the evaluations of natural resources which are carried out by the ministries, departments, *sovnarkhozi*, and executive committees of local Soviets of Workers' Deputies of the R.S.F.S.R.

Art. 14. Planning the utilization of natural wealth (*resources*). In drawing up plans for the development of the national economy, planning and economic organizations are responsible:

(*a*) for taking into account the interrelationships among the resources listed under Article 1, so that the exploitation of one resource will not inflict damage on others;
(*b*) for providing, in the exploitation of useful natural resources, not only for the full satisfaction of the current needs of the country, but also for the preservation and renewal of these resources by means of their wide-scale reproduction;
(*c*) for regularly and systematically forecasting and procuring appropriations and other material needs for work on the conservation and replenishment of natural resources;
(*d*) for not permitting reductions in the size of useful natural areas (forests, meadows, bodies of water), if on such areas there is not to be created more valuable natural environments, or economic enterprises, transportation routes, or settlements;
(*e*) for maintaining the maximum protection for valuable natural objects during the planning and carrying out of industrial, transportation, communal, and other types of construction.

The planning of the rational and integrated utilization of natural resources, and also of measures to ensure their replenishment on the basis of large-scale reproduction over the whole territory of the R.S.F.S.R. is charged to the State Planning Commission of the R.S.F.S.R. Council of Ministers and to the All-Russian Council of the National Economy.

Art. 15. Supervision over the conservation of nature. The Council of Ministers of the R.S.F.S.R., the Councils of Ministers of the A.S.S.R.s, the executive com-

* An abolished form of regional economic council.

mittees of kray, oblast, district, city, settlement, and village Councils of Workers' Deputies, ministries, departments, and *sovnarkhozi* are to ensure control over the observance of conservation laws that are in force and over the fulfillment of measures for preserving and replenishing natural resources by institutions, enterprises, organizations, state and collective farms, and private citizens.

Art. 16. Participation in the conservation of nature by public organizations. The conservation of nature is the concern of all the people. Public organizations (of trade unions, young people, scientists, etc.) and voluntary societies participate in it, attracting a wide variety of workers, farmers, and intelligentsia.

Guidance over all public work in the field of nature conservation is carried out by the All-Russian Society for Assisting the Conservation of Nature and the Beautification (*ozeleneniyu*: lit., 'planting of greenery') of Populated Areas.*

The R.S.F.S.R. state planning agency (*Gosplan R.S.F.S.R.*), the All-Russian Council of the National Economy, ministries, and departments may call upon the All-Russian Society for Assisting the Conservation of Nature and the Beautification of Populated Areas to participate in the consideration of plans involving complex measures for the utilization and transformation of nature and of proposals for large scale projects affecting the preservation and reproduction of natural resources.

For assisting state organs, the local units of the All-Russian Society for Assisting the Conservation of Nature and the Beautification of Populated Areas may establish public conservation inspectorates, whose activities are coordinated with other public inspectorates (hunting, fishing, etc.).

The duties of the public conservation inspectors are honorary.

Art. 17. Scientific-research work on problems on the conservation of nature. Scientific-research institutes and higher educational institutions are to include in their plans for scientific work subjects relating to the conservation of nature, and to carry out a systematic study of permissible norms for the utilization of natural resources and of possible methods for their replenishment.

Art. 18. Teaching the fundamentals of nature conservation in educational institutions. With a view to instilling in youth a feeling for a considerate attitude towards natural wealth and habits of the correct use of natural resources, training in the principles of conservation is to be included in school programs and in the corresponding sections of textbooks in natural science, geography and chemistry; and compulsory courses in conservation and the replenishment of its resources are to be introduced into higher and specialized secondary educational institutions, corresponding to their curricula.

Art. 19. Popularizing the problem of the conservation of nature. Cultural-educational institutions and organizations, publishing houses, museums, theatres, radio, television, editorial staffs of newspapers and magazines, and voluntary societies shall widely publicize the goals of the conservation of nature and the reproduction of natural wealth.

Art. 20. Accountability of the directors of departments and enterprises. Institu-

* This is apparently the same organization which is elsewhere referred to in this study as the 'All-Russian Society for the Conservation of Nature'.

tions, enterprises, and organizations which are granted the use or exploitation of sections of land or other natural wealth are obliged to provide for the conservation, rational exploitation, and replenishment of such natural resources.

For the unlawful destruction or spoilage of natural wealth, the directors of institutions, enterprises, and organizations, as well as other persons directly guilty in causing the cited damage, shall be held accountable in accordance with established legal procedures.

Art. 21. Accountability of citizens guilty of unlawfully using or damaging the riches of nature. Citizens who are guilty of unlawfully using or damaging natural wealth shall be held in administrative or criminal liability, accountable for the losses caused by them, in accordance with established legal procedures.

Art. 22. Planning and carrying out measures for the conservation of nature. The Council of Ministers of the R.S.F.S.R. is charged with planning and carrying out the necessary measures for the conservation of nature in the Russian Soviet Federated Socialist Republic which result from the present law.

N. Organov, President of the Presidium of the Supreme Soviet of the R.S.F.S.R.
S. Orlov, Secretary of the Presidium of the Supreme Soviet of the R.S.F.S.R.
Moscow, 27 October, 1960.'

Appendix 4. Recommended outline for courses in conservation in higher educational institutions of the Soviet Union*

1. Introduction.

2. The effect of human activities on nature:

(*a*) growing utilization of natural resources as a result of the development of human society;
(*b*) relative extent of contemporary utilization of natural resources in the world and in individual countries;
(*c*) alterations to the natural scene undertaken in the interests of man;
(*d*) unfavourable changes in nature and the consequences of irrational utilization of natural resources;
(*e*) biogeochemical activities of man.

3. History of nature conservation.

4. Nature conservation in our time:

(*a*) natural resources of the U.S.S.R.;
(*b*) conservation legislation;
(*c*) regulations concerning the utilization and conservation of natural resources;
(*d*) measures designed for the conservation, restoration and enrichment of natural resources;
(*e*) conservation and enrichment of natural environments around cities and industrial centers;

* From L. K. Shaposhnikov (1965) p. 46.

(*f*) regions enjoying special protection;
(*g*) administration of regulations designed for the conservation of nature;
(*h*) role of educational establishments in popularizing the concept of nature conservation;
(*i*) participation of the general public in nature conservation;
(*j*) the role of science in nature conservation.
5. Nature conservation in foreign countries.
6. International aspects of nature conservation.

Appendix 5. Text of the 1968 U.S.S.R. Land Legislation Act

'Principles of land legislation of the U.S.S.R. and the Union republics.* (*Pravda* and *Izvestiya*, Dec. 14, pp. 2–3; *Vedomosti Verkhovnogo Soveta S.S.S.R.*, No. 51, Dec. 18, Item 485, pp. 846–67. Complete text.) The Great October Socialist Revolution destroyed Tsarist Russia's land system of semi-serfdom, which had doomed the peasantry to poverty and had retarded the development of the country's production forces. The Oct. 26 (Nov. 8), 1917, Decree 'On Land' of the Second All-Russian Congress of Soviets abolished private ownership of the land forever; all land was made national property and was turned over to the working people without charge for its use.

State ownership of the land, which arose as the result of nationalization, constitutes the basis of land relations in the U.S.S.R. Land, which in the conditions of private ownership served as a tool for the exploitation of man by man, is used in the U.S.S.R. for development of the country's production forces in the interests of all the people.

State ownership of the land played an immense role in ensuring the victory of socialism in the U.S.S.R. It created the possibility for the most expedient location of all branches of the national economy and was one of the most important conditions for the transition of land use to socialist forms.

With the creation, during socialist construction, of conditions for mass collectivization of separate, individual farms, the peasantry, under the guidance of the Communist Party and with comprehensive aid and support from the working class, embarked on the path of socialism. As a result of the implementation of Lenin's cooperative plan and the victory of the collective-farm system, the peasant question found its own genuine solution.

State ownership of the land has promoted the creation of the material and technical base of communism in our country, the gradual transition to communist social relations and the liquidation of the difference between city and countryside.

Land in the U.S.S.R. – one of the most important resources of Soviet society – is the chief means of production in agriculture and is the spatial basis for the location and development of all branches of the national economy. Scientifically substantiated, rational use of all land, the protection of it and the all-out increase of soil fertility are nationwide tasks.

* The draft Principles of U.S.S.R. and Union-Republic Land Legislation were published in the *Current Digest of the Soviet Press*, **20**, No. 30; representative coverage of the subsequent discussion of the draft was carried in Nos. 32, 43, 44, 48 and 49; and Surganov's final report to the Supreme Soviet on the new law appears above. Changes in and additions to the draft version are included. Passages dropped from the draft are not indicated in this text.

1968 Land Legislation Act

I. GENERAL PROVISIONS

Art. 1. Tasks of Soviet land legislation. The tasks of Soviet land legislation are: the regulation of land relations for the purpose of ensuring rational use of the land and creating the conditions for increasing its effectiveness, the protection of the rights of socialist organizations and citizens and the strengthening of legality in the sphere of land relations.

Art. 2. Land legislation of the U.S.S.R. and the Union Republics. Land relations in the U.S.S.R. are regulated by these Principles and other acts of U.S.S.R. land legislation promulgated in connection with them and by the land codes and other acts of land legislation of the Union republics.

Mining, forest and water relations are regulated by special U.S.S.R. and Union-republic legislation.

Art. 3. State ownership of land in the U.S.S.R. In conformity with the U.S.S.R. Constitution, land in the Union of Soviet Socialist Republics is state property, i.e. the property of all the people.

Land in the U.S.S.R. is the exclusive property of the state and is granted only in usufruct. Open or concealed actions that violate the right of state ownership of the land are forbidden.

Art. 4. The single state land fund. All land in the U.S.S.R. constitutes a single state land fund, which consists, according to its primary designated purpose, of:

(1) agricultural land granted in usufruct to collective and state farms and other land users for agricultural purposes;

(2) land occupied by populated areas (cities, urban-type settlements and rural communities);

(3) land occupied by industry, transportation, resorts and preserves and other nonagricultural users;

(4) state forest land;

(5) state water-resources land;

(6) state reserve land.

The procedure for assigning land to these categories and transferring land from one category to another is to be determined by U.S.S.R. and Union-republic legislation.

Art. 5. Competence of the U.S.S.R. in the sphere of regulating land relations. The following are subject to U.S.S.R. jurisdiction in the sphere of regulating land relations:

(1) disposal of the single state land fund within the limits necessary to exercise the powers granted to the U.S.S.R. by the U.S.S.R. Constitution;

(2) establishment of the basic provisions for land use and land management;

(3) establishment of long-range plans for rational use of the country's land resources to meet the needs of agricultural production and of other branches of the national economy;

(4) establishment of plans for all-Union land-reclamation measures and other measures for increasing the fertility of the soil, as well as establishment of the

basic provisions for protecting the soil from erosion, salinization and other processes that cause the condition of the soil to deteriorate;

(5) establishment of state control over the use of land;

(6) establishment of a single U.S.S.R. system of state land inventory, state registration of land uses and the procedure for conducting a land cadastre;

(7) establishment of the procedure for compilation of the annual U.S.S.R. land balance.

Art. 6. Competence of the Union Republics in the sphere of regulating land relations. Subject to Union–republic jurisdiction in the sphere of regulating land relations are disposal of the single state land fund within the republic and establishment of long-range plans for its use, establishment of plans for reclaiming land, combating erosion and increasing soil fertility, as well as regulation of land relations on other questions that are not within the competence of the U.S.S.R.

Art. 7. Land users. Usufruct of the land in the U.S.S.R. is granted to:

collective and state farms and other state, cooperative and public agricultural enterprises, organizations and institutions;

industrial, transport and other nonagricultural state, cooperative and public enterprises, organizations and institutions;

U.S.S.R. citizens.

In cases stipulated by U.S.S.R. legislation, usufruct of the land can also be granted to other organizations and individuals.

Art. 8. Use of the land free of charge. Use of the land is granted free of charge to collective and state farms and other state, cooperative and public enterprises, organizations and institutions and U.S.S.R. citizens.

Art. 9. Time limits on land use. Land is granted in permanent or temporary usufruct.

Land use without a time limit set in advance is deemed permanent.

Land occupied by collective farms is assigned to them in unlimited usufruct, i.e., in perpetuity.

Temporary usufruct of land can be for a short term – up to three years – or for a long term – from three to ten years. In case of production necessity, these terms can be extended for a period not exceeding, respectively, the time limits for short-term or long-term temporary usufruct.

Art. 10. Procedure for granting the usufruct of land. The usufruct of land sections is granted by assignment.

The usufruct of land sections is assigned on the basis of a resolution of a Union-republic or autonomous-republic Council of Ministers or on the basis of a decision by the executive committee of the appropriate Soviet in accordance with procedure to be established by U.S.S.R. and Union-republic legislation. Resolutions or decisions on the granting of land sections are to state the purpose for which these sections are assigned and the basic conditions for use of the land.

A land section already in use is granted to another land user only after expropriation of the given section in accordance with the procedure stipulated by Art. 16 of these Principles.

Land acknowledged, in accordance with established procedure, as suitable for the needs of agriculture must be granted first of all to agricultural enterprises, organizations and institutions.

Nonagricultural land, land unsuitable for agriculture or agricultural land of inferior quality is granted for the construction of industrial enterprises, housing facilities, rail-roads, highways, power transmission lines and trunk pipelines, as well as for other nonagricultural needs. Land sections for the above-mentioned purposes are granted from state forest land primarily from areas not covered by forest or from areas occupied by scrub and growth of very little value. Land sections for construction on areas of mineral deposits are granted upon agreement with agencies for state supervision of mines. Power and communications lines and other utility lines are installed chiefly at the side of highways, existing routes, etc.

It is forbidden to begin using a granted land section before the respective land-management agencies have established (at the locality) the physical boundaries of this section and before a document certifying the right to use the land has been issued.

The right to land use of collective and state farms and other land users is certified by state deeds granting the right to use land. The forms of the deeds are established by the U.S.S.R. Council of Ministers.

The procedure for legalizing temporary usufruct of land is established by Union-republic legislation.

Art. 11. Rights and obligations of land users. Land users have the right and the obligation to use land sections for the purposes for which they were granted to them.

Depending on the purpose of each land section granted in usufruct, land users have the right in accordance with established procedure:

to erect dwellings and production, cultural-everyday and other structures and installations;

to sow agricultural crops and to plant forests and fruit, ornamental and other trees and plants;

to make use of hayfields, pastures and other types of land.

Losses caused by land users are subject to compensation.

Violated rights of land users are subject to restoration in accordance with procedure to be stipulated by U.S.S.R. and Union-republic legislation.

The rights of land users can be restricted by law in the state's interests and also in the interests of other land users.

Use of land to derive nonlabor income is forbidden.

Land users are obligated to make rational use of land sections granted to them and to refrain from committing actions on their sections that violate the interests of adjacent land users.

Enterprises, organizations and institutions that mine minerals by the open-face or underground method or that conduct geological-prospecting, construction or other operations on agricultural and forest lands granted to them in temporary usufruct are obligated, at their own expense, to put these land sections in a condition suitable for use in agriculture, forestry or fishing, and when the above-mentioned operations are performed on other types of land, in a condition suitable for their use as designated. Land sections are to be put in suitable condition during operations, but when this is impossible, no later than a year after the completion of operations.

Enterprises, organizations and institutions that carry out industrial or other construction, mine mineral deposits by the open-face method or conduct other operations involving breakage of the soil cover are obligated to remove and store the fertile layer of soil so that it may be used to recultivate the land and increase the fertility of unproductive types of land.

Art. 12. Secondary use of land. In cases established by law, collective and state farms and other enterprises, organizations and institutions may grant, from the land allotted to them, land plots for use by secondary parties.

The procedure for and conditions of secondary use of the land are determined by Arts. 25, 27, 28, 41, 42 and 43 of these Principles and also by other U.S.S.R. and Union-republic legislation.

Art. 13. Protection of the land and increasing soil fertility. Land users are obligated to take effective measures for increasing the fertility of the soil, to implement a complex of organizational-economic, agrotechnical, forest-reclamation and hydrotechnical measures for averting wind and water erosion of the soil and to prevent salinization, swamp formation, contamination of the land and its overgrowth by weeds, as well as other processes that cause the condition of the soil to deteriorate.

Measures for land reclamation and protection, for cultivation of forest-shelter belts and for combating erosion of the soil, as well as other measures aimed at fundamentally improving the land, are stipulated in the state plans for development of the national economy and are implemented by the relevant ministries, departments and land users.

Agricultural land, especially irrigated or drained land, is subject to special protection. Collective and state farms and other enterprises, organizations and institutions that use agricultural land are obligated to safeguard, restore and increase soil fertility.

Industrial and construction enterprises, organizations and institutions are obligated to prevent contamination of agricultural and other land by production or other wastes, as well as by sewage.

U.S.S.R. and Union-republic legislation may establish material incentive measures for land users to encourage steps for safeguarding the land, increasing soil fertility and drawing unused land into agricultural use.

Art. 14. Grounds for terminating the right of enterprises, organizations and institutions to use land. The right of enterprises, organizations and institutions to use the land granted to them is subject to termination in full or in part, respectively if:

(1) there is no longer a need for the land section or for part of it;

(2) the term for which the land section was granted has expired;

(3) the enterprise, organization or institution has been liquidated;

(4) the need arises to expropriate the land section for other state or public needs;

(5) the land section has not been put to use for two years in succession.

The right to use land can also be terminated if the land section is used for a purpose other than that for which it was granted.

The Union-republic land codes can stipulate other grounds for terminating the right of enterprises, organizations and institutions to use land.

Art. 15. Grounds for terminating citizens' rights to use land. The right of citizens to use the land granted them is subject to termination in full or in part, respectively, if:

(1) use of the land section is refused voluntarily;

(2) the term for which the land section was granted has expired;

(3) all members of the household or of the family have moved to a different permanent place of residence;

(4) labor relations that involve the grant of a land allotment with the job are terminated, if not stipulated otherwise by U.S.S.R. and Union-republic legislation;

(5) all members of the household or of the family have died;

(6) the need arises to expropriate the land section for state or public needs.

The right to use a land section can be terminated in cases where a citizen commits the actions specified in Art. 50 of these Principles and also if the land section is unused for two years in succession or if it is used for a purpose other than that for which it was granted.

The Union-republic land codes can specify other grounds for terminating the right of citizens to use land.

Art. 16. Procedure for expropriation of land for state or public needs. A land section or part of it is expropriated for state or public needs on the basis of a resolution of a Union-republic or autonomous-republic Council of Ministers or on the basis of a decision by the executive committee of the appropriate Soviet in accordance with procedure to be established by U.S.S.R. and Union-republic legislation.

Expropriation of land sections being used by collective and state farms and other agricultural enterprises, organizations and institutions and land sections of cultural or scientific importance is permitted only in cases of special necessity.

In exceptional cases and only by a resolution of the Union-republic Council of Ministers, irrigated and drained land, arable land and land sections occupied by perennial fruit-bearing trees and plants and by vineyards can be expropriated for nonagricultural needs and land sections occupied by water-protection, protective and other first-category forests can be expropriated to be used for purposes unrelated to forestry.

Enterprises, organizations and institutions interested in expropriating land sections for nonagricultural needs are obligated, before the initiation of design operations, to reach agreement first with the land users and the agencies that exercise state control over land use concerning the location of the facility and the approximate size of the area to be expropriated.

Land sections being used by collective farms can be expropriated only with the consent of the general meetings of the collective farm members or meetings of authorized representatives, and land sections being used by state farms and other state, cooperative or public enterprises, organizations and institutions under all-Union or republic jurisdiction can be expropriated only by agreement with the land users and the respective U.S.S.R. or Union-republic ministries and departments.

Art. 17. Procedure for using land sections for prospecting. Enterprises, organizations and institutions that perform geological surveying, exploratory, geodesic or other prospecting operations can conduct these operations on all land, in accordance with procedure to be established by U.S.S.R. and Union-republic legislation, without expropriating the land sections from the land users. The periods for starting and the site of these operations are agreed on with the land users but if agreement is not reached, they are set by the executive committees of the appropriate district or city Soviets.

Enterprises, organizations and institutions conducting the operations mentioned in the first paragraph of this article are obligated, at their own expense, to put the land sections they occupy in a condition suitable for their designated use. Land sections are put in suitable condition during the operations or, if this is impossible, no later than one month after the operations are completed, excluding the period during which the soil is frozen.

Art. 18. Compensation of land users for losses caused by expropriation or temporary occupation of land sections. Losses caused land users by expropriation of land sections for state or public needs or by temporary occupation of land sections are subject to compensation.

The losses are compensated by the enterprises, organizations and institutions to which the land sections have been assigned and in accordance with regulations to be approved by the U.S.S.R. Council of Ministers.

Art. 19. Compensation for agricultural production losses connected with expropriation of land for nonagricultural needs. Enterprises, organizations and institutions to which land sections occupied by agricultural land are assigned for construction or other nonagricultural needs compensate agricultural production losses connected with expropriation of these sections (in addition to the compensation of land users' losses under Art. 18 of these Principles).

The amounts of agricultural production losses to be compensated and the procedure for determining these losses, as well as the procedure for using the funds involved, are established by the U.S.S.R. Council of Ministers.

Art. 20. State control over the use of land. State control over the use of all land has the task of ensuring that ministries, departments and state, cooperative and public enterprises, organizations and institutions, as well as citizens, observe land legislation, the procedure for using land and correct conduct of the land cadastre and land management so as to utilize land rationally and safeguard it.

State control over the use of all land is exercised by the Soviets and their executive and administrative agencies, as well as by state agencies especially authorized to do this, in accordance with procedure to be established by U.S.S.R. legislation.

II. AGRICULTURAL LAND

Art. 21. Agricultural land. All land granted for the needs of agriculture or designated for such purposes is deemed agricultural land.

Agricultural land is used by socialist agricultural enterprises, organizations and institutions in accordance with plans for the development of agriculture so as to meet the national economy's growing needs for agricultural output.

Reduction of the areas of irrigated and drained land, plow-land, valuable perennial fruit plantings and vineyards, as well as of other highly productive land, including conversion of them into less productive land, is forbidden, except in cases of special necessity to be stipulated by Union-republic legislation.

Art. 22. The granting of agricultural land in usufruct. Permanent usufruct of agricultural land is granted to:

collective and state farms and other state, cooperative and public agricultural enterprises and organizations – for the conduct of agriculture;

research, educational and other agricultural institutions – for the conduct of field research and for practical application and dissemination of scientific achievements and advanced experience in agricultural, as well as for production purposes;

nonagricultural enterprises, organizations and institutions – for the conduct of auxiliary farming;

citizens – for operating a personal farm without the use of hired labor.

Enterprises, organizations and institutions can be granted land sections for collective orchard cultivation and gardening in accordance with procedure and conditions to be established by Union-republic legislation.

In addition to land granted in permanent usufruct, the land users specified in this article can also be granted land in temporary usufruct.

When farms are enlarged or subdivided the boundaries and dimensions of land uses by collective and state farms and other state agricultural enterprises and organizations, as well as by research, training-and-experimental and other agricultural institutions, may be changed on the basis of scientifically substantiated land-management designs approved in accordance with established procedure.

Art. 23. Obligations of users of agricultural land. On the basis of scientific achievements and advanced experience and taking local conditions into account, collective and state farms and other enterprises, organizations and institutions using agricultural land are obligated:

(1) to stipulate in the plans for organizational-economic structure and in production-financial plans concrete measures for raising soil fertility and using the land rationally;

(2) to adopt, in conformity with zonal conditions and given farm specializations, the most effective system of farming and an economically advantageous combination of the given farm's branches; to introduce and master crop rotation; and to draw unused land into agricultural production;

(3) to develop irrigation, drainage and watering of the land, to improve meadows and pastureland, to treat the soil with lime and gypsum;

(4) to take steps against soil erosion, swamp formation and salinization of the land; to plant forest-shelter belts; to afforest and anchor sands, ravines and steep slopes; and to prevent contamination of the soil;

(5) to clear agricultural land of stones, scrub and undergrowth and to combat weeds and agricultural plant pests and diseases.

Art. 24. Land for collective farms' communal and auxiliary farming use. Land granted to a collective farm by state deed in permanent usufruct (in perpetuity) consists of land for communal use and auxiliary farmland. Auxiliary farmland is to be physically demarcated from land for communal use.

If there is insufficient auxiliary land to give collective farm households land plots according to the norms specified by the collective farm's charter, the area of the auxiliary farmland may be increased at the expense of land for communal use by decision of a general meeting of the collective farm members or their authorized representatives; this decision is to be approved by the province (or territory) Soviet executive committee, the autonomous-republic Council of Ministers or, in republics without province division, the Union-republic Council of Ministers.

Land assigned to collective farms is used within the farms on the basis of the collective farm's charter in conformity with these Principles and also with other U.S.S.R. and Union-republic legislation.

Art. 25. Right of a collective farm household to an auxiliary land plot. Every collective farm household has the right to an auxiliary land plot granted in accordance with procedure and within the limits of the norms to be stipulated by the collective farm's charter.

The right to the auxiliary plot granted is retained by collective farm households in which the sole able-bodied member has been inducted into the U.S.S.R. Armed Forces for a tour of active duty or occupies an elected post, has enrolled in study or has temporarily transferred to other work with the collective farm's consent or in accordance with the procedure for organized recruitment, and also when only minors remain in the collective farm household.

The right to use an auxiliary plot is also retained by collective farm households in which all members have become disabled because of old age or invalidism.

In conformity with the collective farm's charter, the collective farm households are granted pasture for their livestock.

Art. 26. Auxiliary land of state farms and other state agricultural enterprises, organizations and institutions. In conformity with the approved plan for disposition of land within the farm, the auxiliary land to be granted as auxiliary land plots to workers and employees within the norms established by Union-republic legislation is set aside and physically demarcated from the land granted in usufruct to the state farm or other state agricultural enterprise, organization or institution for agricultural needs.

If there is insufficient auxiliary land to provide workers and employees with land plots, the area of this land, on petition of the farm's officials, can be increased with the permission of the province (or territory) Soviet, the autonomous-republic Council of Ministers or, in republics without province division, the Union-republic Council of Ministers.

Art. 27. Granting of auxiliary land plots to workers, employees and other citizens living in the rural locality. State farms and other state agricultural enterprises, organizations and institutions grant auxiliary land plots or gardens, from land designated for this purpose, to permanent workers and employees of these organizations and also to teachers, physicians and other specialists working and living in the rural locality.

By decision of a general meeting of the collective farm members or their authorized representatives, the collective farm grants auxiliary land plots to teachers, physicians and other specialists working and living in the rural locality.

If unused auxiliary land is available on collective and state farms and other state agricultural enterprises, organizations and institutions, personal plots can be granted accordingly to workers, employees, pensioners and invalids living in the rural locality by decision of a general meeting of the collective farm members or their authorized representatives or by the administration of the state farm, enterprise, organization or institution; this decision is to be approved by the district Soviet executive committee.

When the workers and employees mentioned in this article go on pension for old age or invalidism, they retain the auxiliary plots previously granted to them in their former dimensions, as do the families of workers and employees inducted into the U.S.S.R. Armed Forces for a tour of active duty or enrolled in study – for the entire period spent in military service or the educational institution.

The categories of citizens listed in this article who own personal livestock are allocated land sections for livestock grazing from state reserve land, state forest land and land of cities and urban-type settlements, as well as from nonagricultural land. If there is no such land, sections for livestock grazing can be allocated, in accordance with established procedure, from the land of collective and state farms and other agricultural enterprises, organizations and institutions; the livestock owners compensate the land users for the expenses involved in maintaining and improving these sections.

Land sections for haying are granted to the above-mentioned categories of citizens from state reserve land, state forest land, land belonging to the rights-of-way of railroads and highways and other nonagricultural land.

Land plots are granted according to the procedure and within the limits of norms to be established by Union-republic legislation.

Art. 28. Procedure and conditions under which permanent land users grant agricultural land in temporary usufruct to other land users. Collective and state farms and other state agricultural enterprises and organizations that temporarily are not using part of the agricultural land assigned to them can turn these lands over in temporary usufruct to collective and state farms and other economic units that need them, by decision of the district Soviet executive committee. An economic unit that has received a section of land for a specified period compensates the land user for investment that the latter has made in the land but on which the latter receives no return by virtue of having granted it in temporary usufruct; such compensation shall be proportionate to the period of time the land is temporarily used.

Part of the land area of one economic unit is transferred to another economic unit in permanent usufruct according to the procedure stipulated in Art. 10 of these Principles.

Art. 29. Use of the land by one-family peasant farms. One-family peasant farms, which exist in some areas, use the sections of fields and auxiliary land granted to them for agriculture in accordance with procedure and with the limits of norms to be established by Union-republic legislation.

III. LAND OCCUPIED BY POPULATED AREAS (CITIES, URBAN-TYPE SETTLEMENTS AND RURAL COMMUNITIES)

Art. 30. Composition of urban land. All land within city limits belongs to the city's land.

This land includes:

(1) the land on which the city is built;

(2) common-use land;

(3) land for agricultural use and other types of land;

(4) land occupied by city forests;

(5) land occupied by rail, water, air and pipeline transport, the mining industry and others.

Art. 31. Procedure for use of city land. All land within city limits is under the jurisdiction of the city Soviets.

The procedure for establishing and changing city limits and for establishing and changing the economic disposition of the land in the territory of cities and for granting and expropriating land sections and the conditions for using them are determined by Union-republic legislation; the procedure for using the land as specified in Clauses 3–5 of Art. 30 of these Principles is determined by U.S.S.R. and Union-republic legislation.

Art. 32. Use of land by collective and state farms within city limits. The communal land of collective farms and the land of state farms and other state agricultural enterprises, organizations and institutions, when such lands are located within city limits and are not subject to urban construction or improvement on the basis of the city development plan, are assigned to them in permanent usufruct.

Housing, cultural-everyday and production construction and installations are located on this land by agreement with the city Soviet executive committees.

Art. 33. Transfer of the usufruct of land sections with transfer of property rights to structures in cities. When property rights to structures on city land are transferred, the usufruct of the land section or part of it is also transferred in accordance with procedure to be established by Union-republic legislation.

Art. 34. Suburban and green zones. The land outside city limits, which serves as a reserve for expansion of the city's territory and as space for the location and construction of the requisite installations connected with the improvement and normal functioning of the urban economy, as well as the land occupied by forests, forest parkland and other green plantings that perform protective and health and hygiene functions and are a place of recreation for the population, is set aside as the city's suburban and green zones respectively.

The procedure for setting aside suburban and green zones and the use of the land in them is established by Union-republic legislation.

Art. 35. The land of urban-type settlements. The provisions of Arts. 30, 31, 32, 33 and 37 of these Principles apply to the land occupied by communities classified in accordance with Union-republic legislation as belonging to the category of urban-type settlements.

All land within the limits of settlement boundaries is under the jurisdiction of settlement Soviets.

Art. 36. Land of rural communities. All land within the limits of the boundaries set for rural communities belongs to the land of these communities. Land of rural communities that are designated as having prospects of further development is demarcated from other land by establishing the boundaries of communities in conformity with the plans for their development. The land of rural communities that are not designated as having prospects of further development is demarcated from other land by way of the intrafarm disposition of the land.

Within the boundaries of rural communities the rural Soviet exercises control over the granting of all land sections and adopts decisions on granting land sections, except from the land of collective and state farms or other agricultural enterprises.

Land sections within the boundaries of the rural community that are assigned to collective and state farms and other agricultural enterprises are used by them for construction of housing cultural-everyday and production structures and installations, as well as for auxiliary land use in accordance with Arts. 24, 25, 26 and 27 of these Principles and other U.S.S.R. and Union-republic legislation. Housing, cultural-everyday and production structures and installations are sited on this land by agreement with the rural Soviet executive committees.

The procedure for establishing and changing the boundaries of rural communities, their designation as having prospects for further development and the procedure for using the land occupied by rural communities are determined by Union-republic legislation.

Art. 37. Granting of the usufruct of land sections to housing-construction and dacha-construction cooperatives and also to citizens for individual housing construction. Land sections are granted to housing-construction and dacha-construction cooperatives and also to citizens for individual housing construction from the land of communities, state reserve land and state forest land located outside urban green zones in accordance with procedure and conditions to be established by U.S.S.R. and Union-republic legislation.

IV. LAND OF INDUSTRY, TRANSPORT, RESORTS AND PRESERVES AND OTHER NONAGRICULTURAL LAND

Art. 38. Land of industry, transport, resorts and preserves and other nonagricultural land. The land occupied by industry, transport, resorts and preserves and other nonagricultural land is land granted in usufruct to enterprises, organizations and institutions for performing the special tasks entrusted to them (industrial production, transportation, the organization of resorts, preserves, etc.).

The size of the land sections to be granted for these purposes is determined in accordance with norms approved according to established procedure or with design and technical documentation; and the sections are assigned with regard for land development priorities.

The procedure for the use of land occupied by industry, transport, resorts and preserves and other nonagricultural land and the establishment of zones with special land use conditions (health protection zone, etc.) are determined by regulations concerning this land to be approved by the U.S.S.R. Council of Ministers and the Union-republic Councils of Ministers.

Art. 39. Resort land. Land sections that are of therapeutic importance, have favorable conditions for the organization of health measures and have been granted, in accordance with established procedure, in usufruct to health-resort institutions are categorized as resort land. Resort land is subject to special protection.

In the interests of ensuring the necessary conditions for the population's treatment and recreation, as well as for the purpose of protecting the natural treatment factors, health protection zones are established at all resorts. Within these zones it is forbidden to grant land sections in usufruct to enterprises, organizations and institutions whose activity is incompatible with protection of the natural therapeutic properties and favorable conditions for the population's recreation.

Art. 40. Preserve land. Land sections that have been allocated in accordance with established procedure and within whose boundaries are natural objects of special scientific or cultural value (representative or rare landscapes and groupings of plant and animal life; rare geological formations or rare types of plants and animals, etc.) are deemed preserve land.

Any activity that violates the natural complexes of preserves or threatens the continuation of natural objects of special scientific or cultural value is forbidden both within the preserves and within the boundaries of the protected zones established around the preserves.

Art. 41. Granting of land by industrial, transport and other nonagricultural enterprises, organizations and institutions for agricultural purposes. Industrial, transport and other nonagricultural enterprises, organizations and institutions, by decisions of district or city Soviet executive committees, grant temporary usufruct of land they are not using to collective and state farms, other enterprises, organizations and institutions and citizens for agricultural purposes under procedure and conditions to be established by U.S.S.R. and Union-republic legislation.

Art. 42. Land allotments connected with jobs. Land allotments connected with jobs are granted to individual categories of workers in transport, forestry, the lumber industry, communications, water resources, fishing and hunting, as well as in certain other branches of the national economy.

Land allotments connected with jobs are allocated from land in the usufruct of enterprises, organizations and institutions of the respective ministries and departments, and if there is a shortage of such land, from state reserve land and state forest land.

The list of categories of workers that have the right to land allotments connected with jobs, the size of the allotments, the conditions for granting them and the procedure for using them are determined by Union-republic legislation.

V. STATE FOREST LAND

Art. 43. State forest land. Afforested land, as well as land that is not afforested but is designated for forestry, is deemed state forest land.

By decisions of district or city Soviet executive committees, state forestry enterprises, organizations and institutions grant, from the state forest land in their usufruct, agricultural land not being used for the needs of forestry and the lumber industry to collective and state farms and other enterprises, organizations and institutions and citizens in temporary usufruct for agricultural purposes, if such usufruct is compatible with the interests of forestry.

The procedure for the use of state forest land is determined by U.S.S.R. and Union-republic legislation.

VI. STATE WATER-RESOURCES LAND

Art. 44. State water-resources land. Land covered by bodies of water (rivers, lakes, reservoirs, canals, inland seas, territorial waters, etc.), by glaciers and by hydrotechnical and other water-resources installations, and also land set aside as rights-of-way along the shores of bodies of water, as safety zones, etc., are recognized as state water-resources land.

The procedure for use of state water-resources land is determined by U.S.S.R. and Union-republic legislation.

VII. STATE RESERVE LAND

Art. 45. State reserve land. All land not granted to land users for permanent or long-term use is state reserve land.

State reserve land is granted in permanent or temporary usufruct to collective and state farms and other state, cooperative and public enterprises, organizations and institutions, as well as to citizens, in accordance with the procedure stipulated in Art. 10 of these Principles.

VIII. THE STATE LAND CADASTRE

Art. 46. The state land cadastre, its contents and purpose. The state land cadastre, containing an aggregate of reliable and essential information on the natural, economic and legal condition of the land, is conducted to ensure rational use of land resources.

The state land cadastre includes data on the registration of land uses, the inventory of the quantity and quality of the land, classification of the soils and economic appraisal of the land.

State land cadastre data serve the purposes of organizing effective use and protection of the land and planning the national economy, the location and specialization of agricultural production, land reclamation and chemicalization of agriculture, as well as the implementation of other national-economic measures connected with use of the land.

The land cadastre is conducted at state expense according to a uniform system throughout the U.S.S.R. The procedure for conducting the state land cadastre, the forms of cadastre documentation and the frequency with which cadastre data are checked and updated are established by the U.S.S.R. Council of Ministers.

IX. STATE LAND MANAGEMENT

Art. 47. Land management and land-management actions. Land management includes a system of state measures aimed at implementing decisions of state agencies in the sphere of land use.

The tasks of state land management are to organize the fullest, most rational and effective use of the land, to raise the standards of farming and to safeguard the land.

Land management includes the following land-management actions:

(1) formation of new land uses, as well as regulation of existing land uses to eliminate the fragmentation of land sections and other drawbacks in the disposal of land; clarification and revision of land-use boundaries on the basis of district planning maps;

(2) intrafarm organization of the territory of collective and state farms and other agricultural enterprises, organizations and institutions with introduction of

economically substantiated crop rotation and disposition of all other agricultural land (hayfields, pastures, orchards, etc.), as well as the development of measures for combating soil erosion;

(3) disclosure of new land for agricultural and other national-economic development;

(4) assignment and expropriation of land sections;

(5) establishment and revision of the boundaries of cities, settlements and rural communities;

(6) the conduct of topographical and geodesic, soil, geobotanical and other investigations and research.

Land management, including planning, prospecting, photographic surveying and inspection, operates at state expense. Land management is carried out by state land-management agencies.

Art. 48. Land-management documentation. Land-management plans and the documents granting the right to use land, as stipulated in Art. 10 of these Principles, are drawn up in the process of land management.

Land-management plans are drawn up with the participation of interested land users and after approval are marked off physically (at the locality) with a designation of land-use boundaries by standard boundary markers.

The intrafarm organization of territory, established in accordance with land-management procedure, is binding on collective and state farms and other agricultural enterprises.

X. SETTLEMENT OF LAND DISPUTES

Art. 49. Procedure for settling land disputes. Land disputes between collective and state farms and other state, cooperative and public enterprises, organizations and institutions and citizens are settled by Union-republic or autonomous-republic Councils of Ministers or by territory, province, regional, district, city, rural and settlement Soviet executive committees in accordance with procedure to be established by Union-republic legislation.

Disputes involving the collective and state farms and other state, cooperative and public enterprises, organizations and institutions of one Union republic on questions of the use of land located on another Union republic's territory are heard by a commission to be formed on the parity principle from among representatives of the interested republics. If a commission does not arrive at a concerted decision, the disputes on these questions are subject to review by the U.S.S.R. Council of Ministers.

Disputes between joint owners of individual structures on the land of cities or urban-type settlements and on land sections assignable by the rural Soviet executive committees in rural communities concerning the procedure for using a common land section are reviewed by the courts.

Property disputes involving land relations are reviewed in accordance with procedure established by U.S.S.R. and Union-republic legislation.

XI. LIABILITY FOR VIOLATION OF LAND LEGISLATION

Art. 50. Liability for violation of land legislation. Buying and selling, mortgaging, bequeathing, deed of gift, leasing, unauthorized exchanging of land sections and other transactions that in open or concealed form violate the right of state ownership of land are invalid. Individuals who are guilty of committing the above-mentioned transactions as well as of:

unauthorized occupation of land sections;

wasteful use of land and use of it for the purpose of deriving income not earned by labor;

spoilage of agricultural and other land or contamination of it with production or other wastes and sewage;

failure to carry out obligatory measures for improving the land and protecting the soil against wind and water erosion and other processes that cause the condition of the soil to deteriorate;

failure to return temporarily occupied land promptly or failure to fulfill obligations for putting it in condition suitable for its designated use;

bear criminal or administrative liability in accordance with procedure to be established by U.S.S.R. and Union-republic legislation.

U.S.S.R. and Union-republic legislation can also establish liability for other types of violations of land legislation.

Land sections occupied without authorization are returned to their proper quarter without compensation for expenses incurred during illegal use.

In cases to be established by U.S.S.R. and Union-republic legislation, incorrectly used land sections can be expropriated from land users that permit systematic violation of regulations for the use of land.

Enterprises, organizations, institutions and citizens are obligated to compensate the damage they cause as a result of violating land legislation.'

Translation from *Current Digest of the Soviet Press*, published weekly at Columbia University by the Joint Committee on Slavic and East European Studies appointed by the American Council of Learned Societies and the Social Science Research Council.

Appendix 6. Excerpt from the 1960 Russian Republic law 'On the Conservation of Nature in the R.S.F.S.R.' (selected articles dealing with land and agricultural resources)*

'*Art. 2. Conservation of the land.* All land, particularly arable land allotted to its various users as the basis for agricultural production, is subject to protection.

All land users are required to systematically carry out, in accordance with local conditions, a complex of agrotechnical, reclamation, and erosion-control measures for the purposes of protecting the soil cover and maintaining the most advantageous soil moisture and fertility regime.

Every state and collective farm or other organization is required to maintain data on the character and features of the soil on the lands which are allotted to it, so as to rationally organize the application of fertilizer, and to properly control the processes which regulate the life of the soil itself and influence the harvest of the crops grown on it.

An evaluation of agricultural lands on the basis of their economic worth and

* From *Pravda* (October 28, 1960) p. 2.

soil quality, and the compilation of a land cadastre, is the responsibility of the Ministry of Agriculture of the R.S.F.S.R.

The agricultural use of soils and other natural resources connected with soils (vegetation, water) must not lead to a reduction in the area of agricultural lands or to a deterioration in the quality of fertile lands.

On lands subject to water and wind erosion, land users must carry out a mandatory set of erosion control measures, determined in consideration of local conditions.

Enterprises and organizations which carry out construction and prospecting work and which extract mineral resources (including construction materials and peat) are required to carry out measures for re-establishing soil fertility on lands affected by their work which are suitable for agricultural use.

In carrying out agrotechnical and lumbering work, and also in road, hydrotechnical, and other types of construction, it is forbidden to use devices or methods that contribute to the development of wind or water erosion of soils (the washing away, blowing away, or water-logging of soils and ground, the growth of gullies, the blowing of sands, the formation of flash floods and landslides), that lead to the salinization or bogging of soils, or to other forms of loss of soil fertility.

The agricultural use of lands, the employment of which could lead to the development of the above cited harmful processes, is also forbidden.

Art. 6. Conservation of other natural vegetation. Natural (wild) vegetation other than forests is subject to protection and regulated use as a fodder base for domestic and useful wild animals, as a source of food products, medicines, and technical raw materials, and of seeds of wild plants for sowing, as a reserve of species which might be brought into cultivation, as a medium for securing the soil, and as an essential part of the geographic environment which influences climate and water regimes, and enriches the soil. Individual valuable, rare, or threatened species of vegetation are also subject to protection.

For purposes of maintaining, increasing, and also improving the productivity of natural vegetation, the pasturing of cattle should be regulated and carried out so as to avoid the overgrazing of pastures, with consideration for the time required for the growth of the vegetation and the condition of the soil, with an even use of an entire pasture area, and with limited pasturing following hay harvesting. State and collective farms and other organizations are required to take measures to improve natural feeding areas which are allotted to them.

Government procurers of raw materials from wild vegetation are forbidden to employ rapacious methods which would obstruct the reproduction of useful plants or cause the destruction of the vegetation cover.'

Appendix 7. *Zapovedniki* of the Soviet Union (1966)

Preserve	Number on Fig. 4.1	Republic or oblast	Year established[a]	Area in 1964 (ha.)	Per cent forested	Natural characteristics and main functions
Part 1: existing *zapovedniki* (1966)						
Adzhameti	1	Georgia	19??,[b] 1957	4,753	93.8	Preserves residual stands of Imeritian oak and Kolkhidian zelkova.
Aksu-Dzhabaglinskiy	2	Kazakhstan	1926	75,000	27.2	Preserves flora and fauna of the western Tyan-Shan mountains.
Alma-Ata	3	Kazakhstan	1960	71,700	88.0	Preserves flora, fauna, and watershed area on north slopes of the Tyan-Shans.
Aral-Paygambar	4	Uzbek	1965	4,000	?	Preserves an island-oasis in the Amu-Darya river.
Askaniya-Nova	5	Ukraine	1874, 1921	10,500[c]	0.0	Virgin steppe area, open-air zoological park with exotic species, gardens.
Astrakhan′	6	Astrakhan′	1919	72,500	5.7	In 3 sections; protects migratory birds and fish spawning grounds; studies ecological consequences of changes in the level of the Caspian Sea.
Babanauri	7	Georgia	1960	800	87.5	Preserves stands of Kolkhidian zelkova.
Badkhyz	8	Turkmen	1941	85,900[d]	37.8	Preserves Central Asian desert and plateau–steppe ecology, has saksaul and pistachio groves; breeds kulan.
Barguzin	9	Buryat-Mongol A.S.S.R.	1915, 1926	248,180	49.9	Typical Eastern Siberian deciduous tayga, dark coniferous tayga, and mountain tundra; studies tayga ecology and breeds Siberian sable.

Appendix 7. (*continued*)

Preserve	Number on Fig. 4.1	Republic or oblast	Year established[a]	Area in 1964 (ha.)	Per cent forested	Natural characteristics and main functions
Barsa-Kel'mes	10	Kazakhstan	1939	19,800	0.0	On island in Aral Sea; protects water fowl, research on desert ungulates.
Bashkir	11	Bashkir A.S.S.R.	1958[b]	72,100	85.6	In two sections; typical natural area of south Urals; large Kapov Cave.
Batsar	12	Georgia	1957[b]	3,052	96.7	Mountain ecology, residual yew groves.
Berezina	13	Belorussia	1925	73,600	74.2	Mixed forest ecology, fish spawning studies and breeding of beavers.
Borzhomi	14	Georgia	1959	13,600	82.3	Mineral springs, protects red deer.
Chatkal	15	Uzbek	1947	34,800	13.2	Natural zones in dry mountainous area, rich fauna and flora.
Chernomorskiy	16	Ukraine	1927	9,500	3.2	Black Sea littoral; protects nesting and migratory water fowl.
Darvin	17	Vologda, Yaroslavl'	1945	112,600	36.7	Studies ecological changes resulting from creation of Rybinsk reservoir.
Dilizhan	18	Armenia	1958	28,939	91.6	Protects flora and fauna of all zones on north slopes of the Lesser Caucasus.
Engure	19	Latvia	1960	1,340	0.0	Protects aquatic birds on Lake Engure.
Garni	20	Armenia	1958	27,100	55.0	Preserves rich flora and fauna of the Lesser Caucasus; scenic waterfalls.
Gasan-Kuli	21	Turkmen	1932	69,700	0.0	Marshes along Caspian Sea; preserves and studies wintering waterfowl.
Grini	22	Latvia	1957	700	100.0	Preserves and studies forest ecology.
Il'men	23	Chelyabinsk	1920	32,100	91.9	An exceedingly rich mineralogical preserve used for geological research.

Issyk-Kul′	24	Kirgiz	1958	781,600	1.3	Includes all of Lake Issyk-Kul′ and a small belt around it; research on economically useful birds and animals.
Kandalaksha	25	Murmansk	17th cent., 1932	180,400	39.6	Includes large Lapland preserve on Kola peninsula and islands in the Barents and White Seas; protects and studies arctic birds, studies reindeer.
Kavkaz	26	Krasnodar Kray	1888, 1924	266,300	61.4	Preserves and studies rich flora and fauna of western Caucasus, breeds European bison; includes Khosta grove.
Kedrovaya Pad′	27	Primorskiy Kray	1916	17,900	73.1	Protects unique area of Far Eastern mixed vegetation; biological research.
Khekhtsir	28	Khabarovsk Kray	1963	47,900	?	Contains and preserves flora and fauna of four Far Eastern vegetation zones.
Khingansk	29	Amur	1963	58,300	?	Preserves flora and fauna of the Amur region: biological research.
Kintrish	30	Georgia	1959	6,000	85.0	Preserves unique Colchis flora.
Kivach	31	Karelian A.S.S.R.	1931	10,315	77.7	Typical middle tayga flora and fauna; Kivach waterfall.
Kolkhida	32	Georgia	1959	500	80.0	Protects Colchis flora and fauna.
Komsomol′sk	33	Khabarovsk Kray	1963	32,200	?	Preservation and reproduction of local Far Eastern flora and fauna.
Kyzyl-Agach	34	Azerbaydzhan	1929	91,100	3.3	Two sections; major wintering area for aquatic birds, fish spawning; Girkan section has relict vegetation.
Lagodekhi	35	Georgia	1912	13,300	63.9	Preserves flora and fauna of the southern slopes of the east-central Caucasus.
Mariamdzhvar	36	Georgia	1959	1,100	81.8	Preserves unique pine groves.

Appendix 7. (*continued*)

Preserve	Number on Fig. 4.1	Republic or oblast	Year established[a]	Area in 1964 (ha.)	Per cent forested	Natural characteristics and main functions
Matsalu	37	Estonia	1957	11,000	9.1	Waterfowl nesting area on Baltic coast.
Mordov	38	Mordov A.S.S.R.	1935	32,100	96.0	Preserves section of Russian broadleaf forest; studies on the desman.
Moritssala	39	Latvia	1957	835	77.8	On Lake Usmas; ecological studies.
Nigula	40	Estonia	1957	2,730	22.0	Studies of peat marsh ecology.
Oka	41	Ryazan	1935	22,900	75.1	Ecological studies in mixed forest zone with riverine floodplains.
Pechoro-Ilych	42	Komi A.S.S.R.	1930	721,300	83.3	Typical middle coniferous tayga and tayga fauna; domestication of elk.
Prioksko-Terrasnyy	43	Moscow	1945	4,800	93.7	Mixed forests with some relict species; breeding of the European bison.
Ramit	44	Tadzhik	1959	16,100	4.3	Preserves native flora and fauna; Bukharian deer have been introduced.
Repetek	45	Turkmen	1928	34,600	?	Preserves a section of the Kara-Kum desert; desert ecology.
Ritsa	46	Georgia	1957[b]	15,928	91.6	Embraces Lake Ritsa and surrounding broadleaf forests; includes the Pitsunda pine grove on the coast.
Saguram	47	Georgia	1957[b]	5,500	87.3	Residential tertiary flora, local fauna.
Sary-Chelek	48	Kirgiz	1960	20,700	32.6	Preserves western Kirgiz flora and fauna, including many wild fruit trees.
Satapliyskiy	49	Georgia	1957[b]	300	100.0	Limestone caves with dinosaur fossils.

Sikhote-Alin	50	Primorskiy Kray	1935	426,600	95.1	Two sections – includes the former Sudzukhe preserve; preserves and studies typical flora and fauna of the Far Eastern and Manchurian zones.
Slitere	51	Latvia	1957	7,800	80.8	Preserves virgin forest and its fauna.
Stolby	52	Krasnoyarsk Kray	1925	47,200	96.1	Studies Siberian pine and larch forests; contains tall, wind-eroded granite columns; tourism.
Suputinka	53	Primorskiy Kray	1911, 1932	17,000	97.1	Preserves and studies typical southern Ussuri flora and fauna.
Teberda	54	Stavropol′ Kray	1935	90,300	38.0	Preserves north slope Caucasus flora and fauna; high alpine meadows, glaciers; tourism; has two sections.
Tigrovaya Balka	55	Tadzhik	1938	41,100	45.3	Flood-plain forests, desert and marsh wildlife; breeding of nutria.
Tsentral′no-Chernozemny	56	Kursk	1930	4,200	26.2	Preserves three sections of virgin steppe with oak groves.
Tsentral′no-Lesnoy	57	Kalinin	1960	21,400	87.4	Located at source of both the Volga and Western Dvina; mixed forests and moss swamp.
Turianchay	58	Azerbaydzhan	1958	12,700	40.2	Two sections; many unusual tree species, desert fauna; El′darskiy exclave protects only stand of Eldar Pine in the world.
Ukrainian Steppe	59	Ukraine	1961	2,100	0.0	Preserves four sections of the Ukrainian steppe, one of which is also of geologic interest.
Vashlovani	60	Georgia	1957[b]	6,600	36.4	Juniper and pistachio groves, steppe and semi-desert fauna.
Vayka	61	Estonia	1957	35	0.0	Nesting colonies of sea birds.

Appendix 7. (*continued*)

Preserve	Number on Fig. 4.1	Republic or oblast	Year established[a]	Area in 1964 (ha.)	Per cent forested	Natural characteristics and main functions
Viydumyae	62	Estonia	1957	593	100.0	On Saaremaa Island; protects unusual relict vegetation.
Volga–Kama	63	Tatar A.S.S.R.	1960	7,600	80.3	At confluence of Volga and Kama; forest and forest–steppe biota; two sections.
Voronezh	64	Voronezh, Lipetsk	1927	47,000	87.9	Two sections; mixed forest with much wildlife, studies of river beaver; Khopyor section has rich flood-plain ecology, breeds European bison.
Zaamin	65	Uzbek	1960	10,500	35.2	Preserves flora and fauna of north slopes of Turkestan Range.
Zakataly	66	Azerbaydzhan	1930	25,300	61.2	Encompasses the natural zones of the south slopes of the Main Caucasus.
Zeya	67	Amur	1963	82,300	?	East Siberian pine and larch forests and associated wildlife.
Zhuvintas	68	Lithuania	1937	3,167	0.0	Surrounds Zhuvintas Lake; preserves aquatic birds, especially the mute swan.
Part 2: *zapovedniki* undergoing a change of status in 1961						
Altay		Altay Kray	1932, 1957	914,780[e]	[f]	Siberian tayga, alpine tundra, rich wildlife; on Lake Teletskoye; abolished. Reactivated in 1967.

Amu-Darya	Uzbek	?	60,000	Delta marshes of Amu-Darya; abolished.
Azov-Sivash	Ukraine	1927	12,000	Since 1961 a state hunting reserve; included in 1966 data in Table 4.1.
Bartasskiy	Armenia	?	10,000	Flood-plain and forests along Tsov River; abolished.
Belovezhskaya Pushcha	Belorussia	13th cent.	74,200	Since 1961 a state hunting reserve; included in 1966 data in Table 4.1.
Childukhtaron	Tadzhik	?	15,000	Relict nut and maple groves; abolished.
Denezhkin Kamen	Sverdlovsk	1946	146,700	Urals tayga with belts of scrub, tundra, and bare rock; abolished.
Gazimaylik	Tadzhik	?	15,000	Natural zones of Kok-Tau Mts.; abolished.
Gyok-Gyol	Azerbaydzhan	1925	7,000	Area around Lake Gyok-Gyol, formed by a landslide in 1139; abolished, may now be a *zakaznik*.
Gumista	Georgia	1957	2,744	Black Sea coastal forest; abolished.
Iskanderkul	Tadzhik	?	30,000	Central Asian mountain zones, scenic lake at 2176 meters; abolished.
Kamennyye Mogily	Ukraine	1927	356	Granitic hills above stony steppe; now part of Ukrainian Steppe preserve.
Kemeri	Latvia	1957	28,900	Mixed forests, rare flora; abolished.
Khomutov Steppe	Ukraine	1926	1,028	Made part of Ukrainian Steppe preserve.
Khopyor	Voronezh	1935	16,200	Made part of Voronezh preserve.
Khosta	Krasnodar Kray	1930	238	Now part of the Kavkaz preserve; relict box and yew tree grove.
Kronotskiy	Kamchatka	1934	964,000	Volcanoes, one of world's four great geyser areas; abolished. Reactivated in 1967.

Appendix 7. (*continued*)

Preserve	Number on Fig. 4.1	Republic or oblast	Year established[a]	Area in 1964 (ha.)	Per cent forested	Natural characteristics and main functions
Krym (Crimean)		Krym (Crimea)	1920	30,200		Since 1961 a state hunting reserve; included in 1966 data in Table 4.1.
Kurgal′dzhino		Kazakhstan	1959	23,000 15,000 (1964)		Since 1961 a state hunting reserve; included in 1966 data in Table 4.1.
Lapland		Murmansk	1930	158,300		Made part of Kandalaksha preserve.
Lenkoran		Azerbaydzhan	1936	15,000 3,100 (1964)		Relict vegetation; made part of the Kyzyl-Agach preserve.
Mari		Mari A.S.S.R.	?	29,500		Central Russian mixed forest; abolished. May be activated again.
Mikhaylov Steppe		Ukraine	1928	202		Now part of Ukrainian Steppe preserve.
Myussera		Georgia	1934	2,545		Abkhazian flora and fauna; abolished. Reactivated in 1966.
Naurzum		Kazakhstan	1930	180,000		From 1961 to 1967 a state hunting reserve; then remade a *zapovednik*.
Pitsunda		Georgia	1957	1,111		Made part of Ritsa preserve; pine groves; now part of Myussera Preserve.
Pontic Oak		Georgia	1957	1,400		Stands of pontic oak; abolished.
Streletskaya Steppe		Ukraine	1948	525		Virgin grassy steppe; now part of Ukrainian Steppe preserve.
Sudzukhe		Primorskiy Kray	1935, 1957	140,000 116,500 (1954)		Far Eastern and Manchurian vegetation; made part of Sikhote-Alin preserve.

Tsiskarskiy		Georgia	1959	3,928	Virgin oak forest; abolished.
Zhigulevsk		Kuybyshev	1927, 1959	17,500	Oak and pine groves, picturesque cliffs and crags; abolished. It was reactivated in 1966.
Part 3: *zapovedniki* added, or under consideration, 1966–9					
Added:					
Algetskiy	1[g]	Georgia	1965	6,400	Mixed forests with relict vegetation.
Altay	2	Altay Kray	1932, 1967	864,000	Mountainous region of cedar and fir forests surrounding Lake Teletskoye. Had been abolished in 1961.
Girkan	3	Azerbaydzhan	1936	3,100	Relict vegetation; for periods administered by the Kyzy-Agach Preserve. Also known as Lenkoran Preserve.
Khopyor	4	Voronezh	1935	16,200	Previously part of Voronezh Preserve.
Kronotskiy	5	Kamchatka	1934, 1967	964,000[h]	Volcanoes, geysers and other thermal phenomena. Had been closed in 1961.
Lapland	6	Murmansk	1930	158,000[h]	Pine forests and mountain tundra around Lake Imandra. Previously part of Kandalaksha Preserve.
Myussera	7	Georgia	1934, 1966	3,600[i]	Abkhazian flora and fauna; now combined with Pitsunda Preserve.
Naurzum	8	Kazakhstan	1930	180,000	Pine forests, lakes and feather-grass steppe. 1961–7 a hunting preserve.
Servero-Ossetinsk (North Ossetian)	9	North Ossetian A.S.S.R.	1967	26,100	Various vertical zones on the north slopes of the Caucasus Mountains.
Sudzukhe	10	Primorskiy Kray	1935, 1957	116,500	Far Eastern and Manchurian vegetation; formerly part of Sikhote-Alin Preserve.

Appendix 7. (*continued*)

Preserve	Number on Fig. 4.1	Republic or oblast	Year established[a]	Area in 1964 (ha.)	Per cent forested	Natural characteristics and main functions
Zhigulevsk	11	Kuybyshev	1927, 1966	19,400		Pine and oak forests; plateau dissected by many gorges and cliffs.
Under consideration:						
Gumista	12	Georgia	1941, 1957	13,000(?)		Georgian broadleaf forests. Had been deactivated in 1961.
Gyok-Gyol	13	Azerbaydzhan	1925, 1967(?)	7,500(?)		Area around impounded Lake Gyok-Gyol. Wintering area for birds.
Kanev	14	Ukraine	1969(?)	?		In Ukrainian forested steppe region.
Karpatskiy (Carpathian)	15	Ukraine	1969(?)	?		Located in the Carpathian Mountains in western Ukraine.
Kursk	16	Kaliningrad	1969(?)	10,000(?)		A sand-dune spit extending into Baltic Sea.
Lugansk	17	Ukraine	1969(?)	?		In steppe area of eastern Ukraine.
Mari	18	Mari A.S.S.R.	1969(?)	?		Mixed-forest region north of Volga River.
Polesskiy	19	Ukraine	1969(?)	?		Boggy mixed forest region in North Ukraine.
Pontic Oak	20	Georgia	1957, 1969(?)	1,400		Stands of Pontic oak; had been abolished in 1961.
Tsiskarskiy	21	Georgia	1959, 1969(?)	3,928		Beech and oak forests; had been deactivated in 1961.

Now classified as State Hunting Preserves:

Azov-Sivash	22	Ukraine	(See Appendix 7, Part 2, for description of these areas)
Belovezhskaya Pushcha	23	Belorussia	
Crimean (Krym)	24	Ukraine	
Kurgal'dzhino	25	Kazakhstan	

[a] Where two years are given, the first is the year a preserve was first created on this location, the second is the year it was established or restored as a *zapovednik*.

[b] These preserves existed before 1951, but were abolished in that year and subsequently re-established.

[c] Soviet sources differ on the area of this preserve; some list it as only 500 hectares.

[d] Some sources give the area of the Badkhyz preserve as 75,100 hectares.

[e] For Part 2, the area given is as of January 1, 1961.

[f] Data on amount of forested area are not available for Parts 2 and 3.

[g] Numbers in Part 3 of Appendix 7 refer to Figure 4.2.

[h] Bannikov (1969) gives an area for the Kronotskiy Preserve of 977,000 hectares; for the Lapland Preserve he gives 180,400 hectares.

[i] Bannikov (1969) lists 2260 hectares as the area of the Myussera Preserve.

Sources used in compiling Appendix 7: *Zapovedniki S.S.S.R.* (1964); Bannikov, *Po zapovednikam Sovetskogo Soyuza* (1966); *List of National Parks and Equivalent Preserves* (1962); *Bol'shaya Sovetskaya Entsiklopediya* (1933); Bannikov (1969).

Appendix 8. Activities prohibited in Soviet *zapovedniki**

A. List as posted in the Khosta exclave of the Kavkaz *zapovednik*:

'On the territory of state *zapovedniki* the following are absolutely prohibited:

1. any kind of hunting, catching, or destruction by any means of wild animals;
2. destruction of nests and burrows;
3. fishing;
4. collection of eggs and down;
5. cutting or damaging trees and shrubs;
6. collecting fruit, berries, and mushrooms;
7. extracting fossils or minerals;
8. pasturing cattle;
9. trespassing on the territory of a *zapovednik* without permission.'

B. List as posted in the Ritsa *zapovednik*:

'In the *zapovednik* the following are forbidden: any hunting, destruction, or catching of animals, birds or fish; destruction of nests or burrows; collecting fruit, seeds or eggs; cutting or damaging trees and shrubs; working mineral deposits; pasturing cattle; harvesting hay; making campfires.'

Appendix 9. Excerpt from the 1960 Russian Republic law 'On the Conservation of Nature in the R.S.F.S.R.'†

'Art. 8. *Preservation of typical landscapes, and of rare and outstanding natural objects.* Typical landscapes, and rare or outstanding objects of organic and inorganic nature, are subject to preservation as characteristic or unique examples of natural conditions in particular zones and physical-geographic regions, valuable in terms of scientific, cultural-educational, or hygienic considerations.

The executive committees of local Councils of Workers' Deputies, in the interest of the present and future generations, are required to secure the preservation of examples of unspoiled nature and picturesque locales; of natural objects, valuable in historical-memorial respects; of locales for tourism and excursions, and places of relaxation and recuperation for workers; of outdoor laboratories for the study of naturally occurring processes; of centers for the reproduction and resettlement of valuable animals for the purpose of enriching hunting areas; and of individual types of rare and endangered plants and animals.

Art. 9. *State natural preserves (zapovedniki) and restricted areas (zakazniki).* The preservation of areas and objects of nature, depending upon their significance, may be carried out by the organization of:

(*a*) state *zapovedniki*, the territories of which are forever withdrawn from economic utilization for scientific-research and cultural-educational purposes;

(*b*) *zakazniki*, on the territory of which economic use is granted for only part of the natural objects, only in fixed seasons, for fixed periods, and only in such degree so as not to cause harm to the protected objects.

* From personal observation by author.
† From *Pravda* (October 28, 1960) p. 2.

The institutions of state *zapovedniki* and *zakazniki* are set up both for sizable territories, and for smaller areas (groves, lakes, sections of valleys and shorelines, etc.) and individual objects (waterfalls, caves, unique geological formations, rare or historically valuable trees, etc.), which are declared accordingly as being protected unique areas and natural monuments.

Prospecting on the territory of state *zapovedniki* is permitted only within the limits of the plans for their scientific-research work.

The declaration of territories as state *zapovedniki* and *zakazniki*, and also as protected unique areas and natural monuments with similar functions as state *zapovedniki*, will be carried out in a manner established by the Council of Ministers of the R.S.F.S.R.'

Appendix 10. Decree of the Council of People's Commissars (September 16, 1921) 'On the Preservation of Natural Monuments, Gardens, and Parks'*

'The Council of People's Commissars decrees:

1. Natural areas and individual objects (animals, plants, mineral deposits, etc.), which represent scientific and cultural-historical values and require protection, may be declared by the People's Commissariat for Education, with the concurrence in each individual case of the departments and institutions concerned, as inviolable natural monuments.

2. Natural areas, remarkable in their physical features, which are more significant in size may be designated as *zapovedniki* or as national parks.

Gardens and parks of historical-artistic significance, created through the efforts of skilled landscape architects, or associated with architectural structures and forming with them a single artistic whole, may be designated by the People's Commissariat for Education, with the concurrence of concerned departments, as inviolable garden/park monuments of museum-academic significance.

Note I. Auxiliary features, such as greenhouses, hot beds, nurseries, separate stations, and others, associated with gardens and parks, constitute inseparable component parts and shall be protected with them.

Note II. The designation of such gardens and parks for public use shall take place on the basis of special instructions drawn up by the People's Commissariat for Education, with the concurrence of the Guberniya Executive Committees concerned, with the exception of those designated for public use within cities.

The use of the cited auxiliary features shall take place with the consent of the corresponding organs of the People's Commissariat for Education.

3. Land in *zapovedniki* and national parks may not be reverted to cultivation or be worked for their natural resources except by decision of the People's Commissariat for Education; and in like manner in *zapovedniki* and national parks hunting or capturing animals and birds, collecting eggs and nests, and fishing is not permitted without a similar decision.

Note. Resolving questions of economic utilization and management for each of these special situations shall be carried out by the People's Commissariat for

* From V. N. Makarov, *Okhrana prirody v S.S.S.R.* (Moskva: Goskul'tprosvetizdat, 1947), pp. 57–8.

Education with the concurrence, in each individual case, of the local Guberniya Executive Committee.
4. Inviolable natural monuments and garden/park monuments of museum-academic significance shall be placed under the jurisdiction of the People's Commissariat for Education in the division for museum affairs and the preservation of monuments of art and antiquity.'

Appendix 11. Article 40 of the 1968 U.S.S.R. Land Legislation Act*

'*Preserve land.* Sections of land that have been set aside in accordance with established procedure, and within which are found natural objects of a special scientific or cultural value (representative or unusual landscapes or communities of plant and animal life; unusual geologic formations or rare types of plants and animals, etc.), are recognized as preserve land.

Any activity that infringes upon the natural complexes of preserves, or threatens the protection of natural objects having special scientific or cultural value, is forbidden both on the territory of the preserves as well as within the boundaries of protected zones established around the preserves.'

Appendix 12. Contents of 'Trudy Kavkazskogo Gosudarstvennogo Zapovednika' (Works of the Kavkaz State Preserve), Volume 8, 1965

* From *Pravda* (December 14, 1968) p. 3.

Appendix 13. Proposed natural parks in the Soviet Union (as compiled by Central Laboratory for the Conservation of Nature)*

1. Imandra natural park. In Murmansk Oblast, near Monchegorsk. About 180–200,000 hectares around Lake Imandra.

2. Sebezhskiy natural park. In southwestern Pskov Oblast. About 43,000 hectares, containing numerous small lakes.

3. Osipovich natural park. In southwestern Mogilev Oblast, Belorussia. About 60,000 hectares within the Osipovich forest.

4. Ilet′ natural park. In southeastern Mari A.S.S.R. About 15,000 hectares along Ilet′ River. Karst landscape.

5. Belaya River natural park. In northeastern Bashkir A.S.S.R., near Beloretsk. 60–90,000 hectares around Belaya River canyon.

6. Omsk natural park. In Omsk Oblast, north of Tara. About 33,000 hectares along right bank of Ob′.

7. Novosibirsk natural park. In Novosibirsk Oblast, extending about 100 km along the right bank of the Novosibirsk Reservoir.

8. Irkutsk natural park. In Irkutsk Oblast, within the triangle formed by Lake Baykal and the Irkutsk–Kultuk railroad.

9. Lake Beklemishev natural park. In Chita Oblast, about 50 km west of Chita. Several thousand hectares around the lake.

10. Lake Dzhek (Jack) London natural park. In Magadan Oblast, exact area and location unspecified. Alpine lake setting.

11. Novaya i Staraya Dzhalka natural park. In Chechen-Ingush A.S.S.R. About 10,000 hectares along the Dzhalka River.

12. Borovoye natural park. In northern Kazakhstan, near Kokchetav. Lake region in vicinity of former 83,000 hectares Borovoye *zapovednik*.

13. Dzhalal-Abad natural park. In Kirgiz S.S.R., east of Dzhalal-Abad. About 16,000 hectares in foothills of the Fergana Range.

Appendix 14. Excerpt from the 1960 Russian Republic law 'On the Conservation of Nature in the R.S.F.S.R.'†

'*Art. 11. Protection of animal life.* Useful animals, birds, fish, etc., which exist in a naturally wild state, are subject to protection and regulated use, as resources for hunting, trapping, whaling, fishing, and other industries, as destroyers of harmful animal life and as a food base for commercial and other useful animals, as objects for future domestication and fur-farms, as reserve species for developing new forms and improving breeds of domestic animals, etc.

* From L. S. Belousova, 'Ob organizatsii prirodnykh parkov v Sovetskom Soyuze' ('On the organization of natural parks in the Soviet Union'); in L. K. Shaposhnikov (ed.), *Primechatel′nyye prirodnyye landshafty S.S.S.R. i ikh okhrana* (Moskva: Izdatel′stvo 'Nauka' 1967) pp. 144–54.

† From *Pravda* (October 28, 1960) p. 2.

Appendix 15

Rare and disappearing species of wildlife are also subject to protection from destruction and extinction.

In connection with the above, it is necessary:

(*a*) to adhere strictly to established laws on hunting and fishing;
(*b*) to promote improvements in conditions for the existence and reproduction of animals by means of protecting and improving dwelling places and routes of migration;
(*c*) to regulate the use of commercial stocks, ensuring commercial densities and replenishments;
(*d*) to enrich useful species of fauna, preventing, in connection with this, the extermination of useful wild animals, fish, birds, etc.;
(*e*) to carry out measures in the fight against harmful species – agricultural and forest pests; carriers of infections; poisonous, parasitic, and other predators – which inflict damage on the economy.

It is forbidden to destroy non-commercial species of wild animals if they are not harmful to the economy or to the public health.'

Appendix 15. Composition of Soviet fish extraction

Species or type	Annual catch (1000 metric tons)		
	1964	1966	1968
Freshwater fish (carp, bream, roach, zander, pike, etc.)	378	409	397
Sturgeon	18	15	18
Pink salmon	15	21	17
Chum salmon	27	30	16
Other salmon and white fish	64	106	79
Caspian *kil'ka* (sprat)	312	364	368
Flounder, halibut, sole	187	207	270
Atlantic cod	340	357	986
Silver hake	268	225	62
Alaska pollock	214	425	567
Other cod, hake, haddock, etc.	97	687	715
Rockfish	276	248	84
Other redfish, basses, congers, etc.	281	342	413
Jacks, mullets, etc.	162	173	264
Atlantic herring	699	618	442
Pacific herring	461	323	445
Other herring, sardines, anchovies, etc.	350	425	452
Tuna	2	6	10
King crab	46	46	40
Other crustaceans	3	13	9
Aquatic plants	65	65	61

Species or type	Annual catch (1000 metric tons)		
	1964	1966	1968
All other (including mackerels, cutlassfish, sharks, squid, and unsorted and unidentified species)	211	246	369
Total catch:	4,476	5,349	6,082

From United Nations Food and Agriculture Organization, *Yearbook of Fishery Statistics*, Vol. 22 ('Catches and Landings, 1966'), 1967, pp. b-150 to b-153, and same, Vol. 26 ('Catches and Landings, 1968'), 1969, pp. b-156 to b-159.

Appendix 16. Soviet specialists associated with various commissions, and the executive board, of the International Union for Conservation of Nature and Natural Resources (I.U.C.N., 1971)*

B. N. Bogdanov	Member, Executive Board (1966–72)
V. A. Kovda	Member, Executive Board (1967–73)
L. K. Shaposhnikov	Chairman, Commission on Education
V. M. Galushin	Member, Commission on Education
V. A. Popov	Member, Commission on Education
A. K. Rustamov	Member, Commission on Education
K. P. Mirimanyan	Member, Commission on Ecology
V. A. Chichvarin	Member, Commission on Legislation
O. S. Kolbasov	Member, Commission on Legislation
A. G. Bannikov	Member, Survival Service Commission
V. G. Geptner	Member, Commission on National Parks
L. V. Motorina	Member, Landscape Planning Commission

In addition, certain of the above, together with other Soviet specialists, serve on the various specialized committees which function under these primary I.U.C.N. commissions.

Appendix 17. Endangered mammals in the U.S.S.R.†

Polar Bear (*Thalarctos maritimus*)
Ratel or Honey Bear (*Mellivora indica indica*)
Amur or Siberian Tiger (*Panthera tigris altaica*)
Turan or Caspian Tiger (*Panthera tigris virgata*)
Snow Leopard (*Panthera uncia uncia*)
East Siberian Leopard (*Panthera pardus orientalis*)
Transcaucasian Leopard (*Panthera pardus tulliana*)

* From *I.U.C.N. 1970 Yearbook* (Morges, Switzerland: I.U.C.N. 1971) pp. 90–4.

† From A. A. Nasimovich, 'Rare and endangered species of mammals and their protection in the U.S.S.R.', *Geograficheskiy sbornik no. 4* (Moscow, 1970); as translated by Field Research Projects, Coconut Grove, Flo. (1970).

Cheetah (*Acinonyx jubatus venaticus*)
Caracal or Desert Lynx (*Felis caracal michaelis*)
Sand Hill Cat (*Felis margarita thinobius*)
East Asian Red Wolf (*Cuon alpinus alpinus*)
West Asian Red Wolf (*Cuon alpinus hesperius*)
Menzbier's Marmot (*Marmota marmota menzbieri*)
Onager (*Equus hemionus onager*)
Bukhairan Deer (*Cervus elaphus bactrianus*)
Sika or Spotted Deer (*Cervus nippon hortulorum*)
Novaya Zemlya Reindeer (*Rangifer tarandus pearsoni*)
European Bison or Wisent (*Bison bonasus*)
Goral (*Nemorhedus goral raddeanus*)
Spiral Horned Goat or Markhor (*Capra falconeri heptneri*)
Bessarabian Goat (*Capra aegagrus*)
Armenian or Asia Minor Mouflon (*Ovis armeniana*)
Persian or Goitered Gazelle (*Gazella subgutturosa subgutturosa*)
Mongolian Gazelle (*Procapra gutturosa gutturosa*)
Russian Desman (*Desmana moschata moschata*)
Atlantic Walrus (*Odobenus rosmarus rosmarus*)
Kurile Seal (*Phoca kurilensis*)
Monk Seal (*Monachus monachus*)

Appendix 18. Excerpt from the 1960 Russian Republic law 'On the Conservation of Nature in the R.S.F.S.R.'*

'*Art. 3. The conservation of mineral resources.* Reserves of solid, liquid, and gaseous mineral resources found in the earth are subject to protection as sources for providing mineral raw materials and fuels for the national economy; also protected are classical and supporting geological outcroppings which serve to determine the age of rock formations and which have important scientific and industrial significance.

Ministries, departments, regional economic councils, and enterprises subordinate to them which are engaged in the extraction of mineral resources are required, under the control of republic geology and mineral conservation organs, to ensure that their work is carried out safely, and that deposits are worked in accordance with established norms and regulations, with consideration for their fullest and most integrated utilization and economic efficiency.

Art. 5. The conservation of forests. Forests are subject to protection and regulated use as sources of lumber and other technical raw materials and of food and fodder products, as habitats of useful animals and plants, and as an important part of the geographic environment having watershed, water regulating, soil and field protecting, climatic, health and cultural-aesthetic significance. The planning of forestry practices and timber cutting should be carried out not only for the purpose of fully satisfying the needs of the national economy and the populace for lumber, but also for the necessity of protecting and renewing the forests. Industrial timber harvesting should be mainly concentrated in densely forested regions.

All forest users are required to carry out a complex of forestry measures

* From *Pravda* (October 28, 1960), p. 2.

designed to quickly reforest cut-over areas with valuable tree species, and to protect forests from fires, unauthorized cutting, trampling by cattle, and harmful insects; they are also to clear cut-over areas promptly.

All enterprises, institutions, and citizens are required to strictly observe rules for preventing fires in forests.

The executive committees of local Councils of Workers' Deputies, forestry organizations, state and collective farms, and other land users are required to carry out measures towards improving and increasing forest resources, to plant forests in lightly forested areas, and to create shelter belts and other protective plantings.

In the course of planning and constructing new cities and major centers or reconstructing old ones, ministries, departments, regional economic councils, and the executive committees of local Councils of Workers' Deputies are required to provide for the protection of forests which could form green belts, as well as of green tracts within populated areas.

It is forbidden:

(*a*) to cut more timber than the annually prescribed amount established for each enterprise;
(*b*) to cut forests (except maintenance cutting) which have soil or field protective, watershed, or water regulating significance, the zones of which are established by the R.S.F.S.R. Council of Ministers; and also along the shores of lakes, rivers, and their tributaries which are spawning areas of valuable commercial fish;
(*c*) to use on sloping ground methods of cutting and hauling out logs which would lead to the destruction of forest soils or young trees;
(*d*) to cut in cedar groves in ways that will not ensure their natural reforestation;
(*e*) to cut forests or construction in the forest reserve without authorization, or to transfer forest lands to other uses without authorization;
(*f*) to pasture cattle in protected or restricted forest zones, in new growth areas or tree farms, in parks, forest parks, urban forests, forested areas around populated centers, or in gardens.'

Appendix 19. Composition of the forests of the U.S.S.R.[a]

Species	Forested area (million hectares)	Total reserves (billion cubic meters)
Coniferous:		
Larch	274.3	28.5
Pine	109.5	15.0
Spruce	72.1	10.6
Cedar	32.1	5.8
Fir	23.1	4.2
Other coniferous	20.3	0.5
Total coniferous	531.4	64.6
(Total coniferous in R.S.F.S.R.)	(521.4)	(63.5)
Broadleaf:		
Birch	91.8	6.8
Saksaul	19.8	0.03

Species	Forested area (million hectares)	Total reserves (billion cubic meters)
Aspen	14.5	1.6
Oak	8.5	0.7
Beech, ash	3.0	0.6
Alder, poplar, linden	5.3	0.5
Other broadleaf	6.6	0.3
Total broadleaf	149.5	10.5
(Total broadleaf in R.S.F.S.R.)	(118.7)	(9.6)
Total for U.S.S.R.	680.9	75.1
(Total for R.S.F.S.R.)	(640.1)	(73.1)

[a] Excluding forests belonging to collective farms and to other ministries and organizations.

From A. I. Mukhin (ed.), *Les-nashe bogatstvo* (Moskva: Goslesbumizdat, 1962) p. 54. Data based on 1956 forest inventory.

Appendix 20. Plea by Leonid Leonov for more effective conservation practices in the U.S.S.R.

'Concerning large chips of wood. (By Leonid Leonov. *Literaturnaya gazeta*, March 30, 1965. Complete text.) Actually, the occasion for this heart-to-heart talk was the documentary film 'Our Unfailing Friend,' which is currently being shown. This excellent product of cinematic journalism should be seen by everyone for whom living nature is more than just a term. It relates in picturesque fashion the story of wood, this presumably boundless wealth of ours, its charms and blemishes. What I wish to speak of is the waste – which is becoming habitual – of priceless living raw material, waste that can exhaust even the richest treasury.

There is an old proverb to the effect that you can't fell trees without scattering chips. But in the words of one border guard who has had ample opportunity to view the stands of trees on both sides of our boundary with Finland, in that country the waste 'is carried home in bags from the cutting area,' whereas among us it blazes up in little bonfires. You could grow wealthy on what we burn up: some of these chips are so big that two people together cannot lift them. This is why our little talk begins with these barbaric bonfires, which are so characteristic of our ubiquitous forest landscape. But let us feast our eyes together on these perpetual dark-crimson clouds, permeated with crackling flashes, with which they love to burn tannin-laden spirits, fodder yeast for cattle, resin, and paper of all kinds – including paper bearing the watermark of banknotes of the State Bank.

Let us state our preamble at once. The forest magnates, in defending themselves against accusations of wastefulness, complain of the intense emotional sufferings brought on by unmerited reproaches. The annoyance behind these hypocritical sighs is obvious to everyone! Unfortunately, our power is not formidable: a small bottle of ink and one square meter of desk – hardly a mighty staging area for launching an attack on a superbly outfitted bureaucratic stronghold with a powerful garrison. From time to time murderous boiling water, as well as cooling liquids, is poured from the stronghold on the nature enthusiasts and forest petitioners below. The latter, incidentally, reap no profit from their unsolicited intercession, only unpleasantness. Just the other day, for example, at a great public gathering a certain high-ranking Gavrilov pointed a menacing finger from his lofty station and said that, after all, we are trying, we chop and saw enthusiastically, indefatigably, and we would accomplish twice as much if only all those aforementioned scribblers would stop trying to prevent the ax from falling freely. This argument clearly neglects to point out that an iron object – i.e. an ax – without the application of intelligence can do a great deal of mischief in centralized state use. But now a certain republic-level Sidorov, sulking peevishly on his ministerial throne, grumbles about the above-mentioned category of patriot: 'Ha! They've been reading Leonov!' There's no need to threaten or to sulk like a Brahmin. It is not private bird-and-birch lovers but full-fledged citizens of this country, your real masters, who annoy you.

Because of the divergence of the views of the parties, we must one more explain, aloud and even somewhat more audibly – so as not to have to go over it again six months from now – what the fuss is all about and whence the smoke. These citizens are not demanding a ban on the felling of trees, since without wood products you can't take a step by yourself; these nuisances are not simply pushing for a place to go searching mushrooms and berries on their day off. They urgently hope that, in supplying the needs and feeding the vanity of our own generation, we leave something for posterity as well. Arrogant disregard for patent common sense and the public welfare has never accomplished anything worthwhile.

In brief, we are talking about a radical revision of the archaic view of the forest as an inexhaustible, gratuitous bounty, about a thriftier attitude toward all kinds of wealth, including timber. The chemistry of modern wood technology and the example of our flourishing forest neighbors makes it incumbent upon us, too, to give some thought to increasing the yield from each hectare of felled greenery, rather than permitting the present far from satisfactory state of affairs. The time has come for a nationwide inventory of how much scarce goods of all kinds, beginning with wood – which is equal in value to grain or fiber – is running through our fingers. It is scattered in ashes and stifling fumes, it is flushed down every available little stream in the form of lignin and sawdust, it stands in idle piles in down-river warehouses and sorting stations, it paves the beds of famous waterways in a dozen layers of waterlogged timber, it rots beside railroad embankments and along the shores of rafting channels, which soon become clogged up; and, finally, it floats into foreign seas in the form of unclaimed logs, and the resourceful Scandinavians, marvelling at the vagaries of our unaccountable cornucopia, use it to build comfortable and solid housing. . . .

Here are some statistics, deliberately presented in the lowest round numbers, for fruitful meditation on the misfortunes of our forests: give a moment to these melancholy figures. Each year approximately 300,000,000 cubic meters of state-plan timber and 50,000,000 cubic meters of timber cut on the collective farms are removed from the cutting areas of our country. Non-liquid assets of 100,000,000 cubic meters are disposed of on the spot. This figure includes – in addition to the stumps, boughs and crowns – deciduous varieties felled at the same time. Our fastidious industrial establishment demands only coniferous varieties, but stands are cleared to the last tree. The bonfires even claim every divine larch, the beauty of our Siberian forests, since the specific gravity of larch makes it unsuitable for floating. 'Won't float? Then burn it, by heaven!' To put it briefly, the carrion of this bloodless massacre will be swept away with a broom of fire, assisted by indifferent, smoke-besooted men clenching hand-rolled cigarettes in their teeth. Now, of the remaining 350,000,000 cubic meters, let's subtract a good 100,000,000 for firewood – also a bitter tribute to backwardness in chemistry – and another 20,000,000 for making cellulose, of which half, liquid and solid, promptly goes down the drain to poison fish and pollute rivers. The 230,000,000 cubic meters that are left are earmarked for pit props and packing, for poles and railroad ties, for all kinds of structural elements – with a useful end product of one cubic meter for each four deliberately wasted. Wood seems to dissolve all along the way to the consumer, and this additional bit of waste amounts to still another 170,000,000 cubic meters.

To our sorrow, the forestry institutes train students chiefly in the mechanical processing of wood – in the present methods of felling, splitting and breaking up logs. But to change this appalling ratio of four-fifths there must be a new people, many of them, with a more broadminded approach to forestry, with insight into the secret chemical transformations that take place in wood. Only forceful individuals capable of traversing thorny, untrodden paths will be able to overcome the resistance of the superannuated bureaucrats. First the people – and then the machines will follow. Along with the urgent, in my view, need for restoring the departments of landscape architecture, absolutely essential in envisioning the future but recently abolished with an imperious wave of the hand, it is necessary to see promptly to the training of these new specialists, industrial innovators who will begin to extract treasures beyond all expectations from the ordinary log, in keeping with the level of modern chemistry. And while the Ministry of Higher Education is giving fruitful thought to the matter, let us give the scholars a helping hand. Let's see, now: 100 + 170 + 10 – how much wasted wood does that make in all?

Dear Sidorovs and Gavrilovs, how can we not stoop to raise from underfoot this notorious little chip, which amounts to 280,000,000 cubic meters of organic raw material a year and can be used to create all the wonders of modern civilization? In addition, if we figure up the area of living greenery, basing ourselves on an average of 140 cubic meters per hectare, this means 20,000 square kilometers of splendid forest tracts, now cut down to no purpose every year, which could be rustling in the wind, full of birds, freshness, joy and other blessings not ordinarily included in the list of consumer goods. In the living world around us there are many such phenomena without which Man cannot exist on earth. Is it really possible, dear sirs, to throw them away? Posterity will never forgive us, and history itself will have no mercy!...

Every day we hear more and more clearly the cry of regret from the people for nature destroyed on all sides and suffering calamities from industry's victorious onslaught, though it would seem our expanses should allow space for both. Throughout the entire past decade one and the same malignant style runs through the various measures that in any way affect this disfranchised and mute nature – measures which, moreover, each time are not subject to discussion by all of us. Let us say nothing of what has been irretrievably lost in the past. But now the Yasnaya Polyana oak grove with Tolstoi's grave, one of the centers of world cultural pilgrimaging, has almost completely withered from the fiendish vapors of a neighboring chemical combine, and already a piece of vandalism against Lake Baikal is being contemplated, under a plausible pretext and, most importantly, with heavy state funds. Again our sorrow is not simply that of the dilettante – 'Farewell, beloved Baikal nor'wester, farewell, barrels of Baikal whitefish!' No reason is taken into account: Let's press on, they say, and as soon as it begins to be forgotten we can write it off as nonexistent. It will not be forgotten! Let us all bare our heads together on that dismal day when the first poison gushes into that purest basin. Our natural preserves are shrinking on all sides, although even without that, in percentage terms and in terms of the size of the preserve area, we are almost in last place! Just the other day there arose in the now-doomed Astrakhan′ Preserve, world famous as a sanctuary for birds of passage, a large reed-cellulose plant, operating, *horribile dictu*, on logs shipped from the North, from the Kama River; and the electric saw is penetrating the Northern Caucasus Preserve in order to provide employment for lumberjacks out of work in the neighborhood. It's as if after the war tanks were offered neighboring gardens to work in just so the machinery would not stand idle! The centuries-old Altai cedar is on the verge of extinction, giants that remember Yermak are falling, the unique Askania virgin soil rustles under the plowshare, as if to spite lovers of closed-off fields! Meanwhile, there is a profoundly conceived shifting of foresters from the forest belts that they guard into nearby settlements in order to bring them within the embrace of culture and introduce them harmoniously to progress. There is just one little thing they overlooked: What the devil does the Vologda forest need with a guard who lives, let us say, on Gendrikovy Pereulok beyond Taganka [in Moscow]?...

As soon as possible we must wean children from an arrogant, spoiled, consumer attitude toward nature. Oh, those happy hikers, with their kettles, their axes and their deafening trumpets, who reduce the frightened woodland fauna to trembling and long silence! Without preliminary humanistic preparation, these young people, armed with instruments of terrible, arrogant power, enter the wonderful and almost defenseless world where anything, even the jungle, can be trampled down without noticing it. This kind of expression of ironic superiority to everything, even to what is most inviolate and even, one might think, to the existence of one's makers, has always repelled me in the image of our civilization. Much would be different if our children were to take for their commandment the idea that 'I may be small, but I am infinitely strong and needed because the world contains many creatures smaller and weaker than I, and I can do them good, if only by passing them by without noticing, touching or perturbing them.' Without this basic feeling of friendship for the weaker in the world around us we shall never get champions of liberating the enslaved...

All these words of agitated hopes and regrets will perhaps now be heeded! The important thing is that for once they be pronounced loudly. Then there will be fewer littered forest trails, destroyed ant hills, broken public telephones, and more besides. There would be less cruelty on the part of children toward the speechless, tail-bearing and feathered fraternity – cruelty, which, like everything that is of man, has a tendency to develop as time goes on. How many times have we all kicked up a storm together under the windows of hard-headed, adamant officials, only to find that, as the folk tale has it, the untouchable and hard-of-hearing cat Vasily listens and goes right on eating. He chews at our suffering Nature and meows something to us that sounds like an electric saw running empty. He meows to us that, if you please, everything we have related here is pure imagination and that actually great peace and harmony are descending upon nature.

This is why this conversation was started – that such as these may not lodge themselves in nature's lap once and for all!

The latest big decision in agriculture permits us to hope for beneficial changes in the exploitation of nature also.'

Translation from the *Current Digest of the Soviet Press*, published weekly at Columbia University by the Joint Committee on Slavic Studies appointed by the American Council of Learned Societies and the Social Science Research Council.

Appendix 21. Excerpt from the 1960 Russian Republic law 'On the Conservation of Nature in the R.S.F.S.R.'*

'*Art. 4. Conservation of water resources.* Surface and underground waters are subject to protection from depletion, pollution, and obstruction, and also to the regulation of their regimes as sources of municipal and industrial water supplies, sources of energy, avenues of transportation, sources of useful vegetation, habitats for fish and water animals, hunting grounds, areas for recreation and tourism, medicinal resources, and as objects having scientific, educational, or cultural interest.

All organizations whose activities affect water resources are required:

(*a*) to carry out reclamation, forest-protective, agrotechnical, and sanitary control measures on territories used by them, so as to improve the water regime and to eliminate the possibility of harmful effects to or by the water (flooding, thermal pollution, waterlogging, salinization, soil erosion, gully formation, washouts, and so forth);

(*b*) to utilize water resources so as not to exceed established norms, and to employ irrigation, ground, and artesian waters carefully in order to prevent their unproductive use; to avoid the creation of unproductive shoals in the construction of reservoirs;

(*c*) to construct purification facilities employing artificial or natural processes at all enterprises which return waste waters into natural water bodies;

(*d*) to prevent the pollution or silting of spawning grounds, or the obstruction of the migration routes to them as a result of log floating;

(*e*) to include measures which will provide for the conservation and reproduction of fish resources in the plans for the construction of hydrotechnical installations.

It is forbidden to put into operation factories, shops, or components thereof which discharge waste water without carrying out measures to ensure its purification.'

Appendix 22. Text of the Water Legislation Principles

'Laws and Resolutions Adopted by the U.S.S.R. Supreme Soviet: Principles of water legislation of the U.S.S.R. and the Union Republics.† (*Pravda*, Dec. 12,

* From *Pravda* (October 28, 1960) p. 2.

† The draft Principles of Water Legislation of the U.S.S.R. and the Union Republics appeared in the *Current Digest of the Soviet Press*, **22**, No. 17, pp. 10–14, 23. Changes in and additions to the draft version are included herein. Substantive passages dropped from the draft are enclosed in brackets.

1970; *Izvestiya*, Dec. 11. Complete text.) As a result of the victory of the Great October Socialist Revolution, water, like all other natural resources in our country, was nationalized and became the property of the people.

State ownership of water [which came about as the result of nationalization] constitutes the basis of water relations in the U.S.S.R., creates favorable conditions for carrying out the [scientifically substantiated] planned and integrated utilization of water with the greatest national-economic effect and makes it possible to provide Soviet people with the best working, living, recreation and public health conditions.

[Population growth and] The development of social production and urban construction and the rise in the population's material well-being and cultural level are increasing the all-round requirements for water and heightening the importance of the rational utilization and conservation of water.

Soviet water legislation is called upon actively to facilitate the most effective scientifically substantiated utilization of water and its protection against pollution, obstruction and depletion.

[These Principles define the goals of Soviet water legislation and establish general provisions concerning the procedure for the use of water, its conservation prevention against the harmful action of water, and the organization of state record-keeping and planning of the utilization of water.]

I. GENERAL PROVISIONS

Art. 1. Goals of Soviet water legislation. The goals of Soviet water legislation are the regulation of water relations for the purpose of ensuring the rational utilization of water for the needs of the population and the national economy, protecting water from pollution, obstruction and depletion, preventing and eliminating the harmful action of water, improving the condition of bodies of water, protecting the rights of enterprises, organizations, institutions and citizens and strengthening legality in the field of water relations.

Art. 2. Water legislation of the U.S.S.R. and the Union Republics. Water relations in the U.S.S.R. are regulated by these Principles and by other acts of U.S.S.R. water legislation issued in accordance with these Principles, as well as by water codes and other acts of Union-republic water legislation.

Land, forest and mining relations are regulated by appropriate U.S.S.R. and Union-republic legislation.

Art. 3. State ownership of water in the U.S.S.R. In accordance with the U.S.S.R. Constitution, water in the Union of Soviet Socialist Republics is state property, i.e., the property of all the people.

Water in the U.S.S.R. is exclusively the property of the state, and only the right to the use of water is granted. Actions that infringe, directly or in a latent form, on the right of state ownership of water are prohibited.

Art. 4. The unified state water supply. All water (bodies of water) in the U.S.S.R. constitutes the unified state water supply.

The unified state water supply consists of:

(1) rivers, lakes, reservoirs, other surface bodies of water and water sources, and the water of canals and ponds;

(2) underground water and glaciers [the inland seas and territorial seas of the U.S.S.R.];

(3) the inland seas and other inland sea waters of the U.S.S.R. [underground water];

(4) the territorial waters (territorial seas) of the U.S.S.R. [glaciers].

Art. 5. The competence of the U.S.S.R. in the field of the regulation of water relations. The jurisdiction of the U.S.S.R. in the field of the regulation of water relations includes:

(1) the administration of the unified state water supply within the limits necessary to exercise the authority of the U.S.S.R. in accordance with the U.S.S.R. Constitution;

(2) the establishment of basic provisions in the field of the utilization of water, its protection from pollution, obstruction and depletion, and the prevention and elimination of the harmful action of water [as well as the planning of all-Union measures in this field];

(3) the establishment of all-Union normatives for water use and water quality and methods for evaluating water quality [determining these normatives];

(4) the establishment of a unified system of state record-keeping of water and its utilization, the registration of water use and a state water cadastre for the U.S.S.R. [and the procedure for its administration];

(5) the confirmation of schemes for the integrated utilization and conservation of water and also of [state] water-resource balances of all-Union importance;

(6) the planning of all-Union measures for the utilization and conservation of water and for the prevention and elimination of the harmful action of water [the organization of supervision over the utilization and conservation of water];

(7) state supervision over the utilization and conservation of water and the establishment of procedures for the implementation of this supervision [the regulation of the use of bodies of water situated on the territory of more than one Union republic];

(8) the determination of bodies of water the regulation of the use of which is exercised by U.S.S.R. agencies.

Art. 6. The competence of the Union republics in the field of the regulation of water relations. The jurisdiction of the Union republics in the field of the regulation of water relations, outside the limits of the competence of the U.S.S.R., includes: the administration of the unified state water supply on the republic's territory [within the rights of the Union republics]; the establishment of procedures for the use of water, its protection from pollution, obstruction and depletion and the prevention and elimination of the harmful action of water; the planning of [republic] measures for the utilization and conservation of water and for the prevention and elimination of the harmful effects of water; the confirmation of schemes for the integrated utilization and conservation of water-resource balances [of republic importance]; the exercise of state supervision over the utilization and conservation of water; and the regulation of water relations on other questions, providing they do not fall within the competence of the U.S.S.R.

Art. 7. State administration in the field of the utilization and conservation of water. State administration in the field of the utilization and conservation of water is exercised by the U.S.S.R. Council of Ministers, the Union-republic Councils of Ministers, the autonomous-republic Councils of Ministers and the executive committees of local Soviets, and also by specially empowered state agencies for the regulation of the utilization and conservation of water, directly or through basin (territorial) administrations and other state agencies, according to U.S.S.R. and Union-republic legislation [agencies for the regulation of the utilization and conservation of water, directly or through basin (territorial) administrations, according to the procedure established by U.S.S.R. legislation]. The executive committees of the local Soviets exercise state administration in the field of the utilization and conservation of water in accordance with U.S.S.R. and Union-republic legislation now in force.

[These same agencies are charged with state supervision over the utilization and conservation of water and over the observance of water legislation.]

Art. 8. State supervision over the utilization and conservation of water. The goal of state supervision over the utilization and conservation of water is to ensure the observance by all ministries and departments, all state, cooperative and public enterprises, organizations and institutions and all citizens of the established procedures for the use of water, the fulfillment of obligations for the conservation of water, the prevention and elimination of the harmful action of water, the regulations for the keeping of records on water, and other regulations established by water legislation.

State supervision over the utilization and conservation of water is exercised by the Soviets and their executive and administrative agencies, and also by specially empowered state agencies, according to the procedure established by U.S.S.R. legislation.

Art. 9. The participation of public organizations and citizens in the implementation of measures for the rational utilization and conservation of water. Trade unions, young people's organizations, conservation societies, scientific societies and other public organizations, as well as citizens, give assistance to state agencies in the implementation of measures for the rational utilization and conservation of water.

Public organizations take part in activity aimed at ensuring the rational utilization and conservation of water, in accordance with their statutes (charters) and U.S.S.R. and Union-republic legislation.

Art. 10. The siting, designing, construction and putting into operation of enterprises, installations and other facilities affecting the condition of water [Art. 8 in the draft Principles]. In the siting, designing, construction and putting into operation of new and reconstructed enterprises, installations and other facilities and in the introduction of new technological processes affecting the condition of water, the rational utilization of water is to be ensured, with priority for the satisfaction of the needs of the population for drinking water and water for everyday use. In this connection, measures are envisaged to ensure the keeping of records on water removed from bodies of water and returned thereto, the protection of water from pollution, obstruction and depletion, the prevention of the harmful action of water [and, where appropriate, to ensure], the restriction of the flooding of land to the necessary minimum [the minimum necessary flooding of land], the protection of land from salinization, rising ground water or desiccation, and also the preservation of favorable natural conditions and landscapes.

[Losses caused to enterprises, organizations and institutions, as well as to citizens, during the implementation of water-resources measures (hydraulic engineering operations, etc.) are subject to reimbursement according to the procedure established by the U.S.S.R. Council of Ministers.]*

In the siting, designing, construction and putting into operation of new and reconstructed enterprises, installations and other facilities on fishing-industry bodies of water, additional measures are to be carried out in good time to ensure the protection of fish [reserves] and of other [useful] aquatic fauna and flora and conditions for their reproduction.

The sites for construction of [the above-mentioned] enterprises, installations and other facilities affecting the condition of water are to be determined in agreement with the agencies for the regulation of the utilization and conservation of water, the executive committees of the local Soviets, the agencies exercising state sanitary supervision, the agencies for the protection of fish reserves and other agencies in accordance with U.S.S.R. and Union-republic legislation. Designs for the construction of the above-mentioned enterprises, installations and other facilities are subject to coordination with the agencies for the regulation of the utilization and conservation of water and other agencies in cases defined by U.S.S.R. legislation and according to the procedure stipulated therein.

It is forbidden to put into operation the following:

new and reconstructed enterprises, shops, aggregates and communal and other facilities that are not provided with devices to prevent the pollution and obstruction of water or the harmful action of water;

irrigation and watering systems, reservoirs and canals before the measures [devices] provided by the designs to prevent flooding, rising ground water, and the bogging up, salinization and erosion of soil [are ready or measures to prevent these conditions] have been taken;

drainage systems, before water intakes and other installations are ready according to confirmed designs;

water-removal installations, without fish-protecting devices in accordance with confirmed designs;

hydraulic engineering installations, before [without] devices for the [discharge] passage of flood water and fish are ready in accordance with confirmed designs;

drilled water wells that are not equipped with water-regulating devices and for which, in appropriate cases, sanitary protection zones have not been established.

The filling of reservoirs is prohibited before the measures envisaged in the designs for preparation of the reservoir bed have been carried out.

Art. 11. Procedure for the performance of work on bodies of water and in adjacent regions (zones) [Art. 9 in the draft Principles]. Construction [hydraulic engineering operations, the extraction of useful minerals and aquatic plants], dredging and explosive operations, the extraction of useful minerals and aquatic plants, the laying of cables, pipelines and other utilities lines, the cutting of timber and [construction] drilling, agricultural and other operations on bodies of water or in adjacent regions (zones) affecting the condition of water may be performed only with the agreement of the agencies for the regulation of the utilization and conser-

* See Art. 20.

vation of water, the executive committees of local Soviets and other agencies in accordance with U.S.S.R. and Union-republic legislation.

II. WATER USE

Art. 12. Water users [*Art. 10 in the draft Principles*]. State, cooperative and public enterprises, organizations and institutions and citizens can be water users in the U.S.S.R.

In cases stipulated by U.S.S.R. legislation, other organizations and persons may also be water users.

Art. 13. Objects of water use [*Art. 11 in the draft Principles*]. Use is granted to the bodies of water enumerated in Art. 4 of these Principles.

The [economic] utilization of bodies of water of special state importance or of special scientific or cultural value can be prohibited partially or completely according to the procedure established by the U.S.S.R. Council of Ministers and the Union-republic Councils of Ministers.

Art. 14. Types of water use [*Art. 12 in the draft Principles*]. The use of bodies of water is granted with the provision that the requirements and conditions stipulated by law for the satisfaction of the population's needs for drinking water and water for everyday, therapeutic, resort, health-improvement and other purposes and of agricultural, industrial, power-engineering, transportation, fishing-industry and other state and public needs [including the discharge of sewage] are to be observed. The utilization of bodies of water for the discharge of sewage can be permitted only in cases stipulated by U.S.S.R. and Union-republic legislation and with the observance of the special requirements and conditions stipulated therein.

A distinction is made between general water use, which is exercised without the employment of installations or technical devices affecting the condition of water, and special water use, which is exercised with the employment of such installations or devices. In certain cases, bodies of water can also be assigned for special water use without the employment of installations or technical devices when this use would affect the condition of water.

The list of types of special water use is established by the agencies for the regulation of the utilization and conservation of water.

Bodies of water may be in joint or single use.

Enterprises, organizations and institutions that have been granted the single use of bodies of water are the primary water users and, in cases established by U.S.S.R. and Union-republic legislation, have the right to authorize secondary water use by other enterprises, organizations and institutions, as well as citizens, with the agreement of the agencies for the regulation of the utilization and conservation of water.

Art. 15. Procedure and conditions for granting the use of bodies of water [*Art. 13 in the draft Principles*]. The use of bodies of water is granted above all for the satisfaction of the population's needs for drinking water and water for everyday purposes.

The single use of bodies of water is granted fully or partially on the basis of resolutions of the Union-republic Councils of Ministers or the autonomous-republic Councils of Ministers or decisions of the executive committee of the appropriate Soviet or of another specially empowered state agency, according to the procedure established by U.S.S.R. and Union-republic legislation.

Special water use is exercised on the basis of authorizations issued by the agencies for the regulation of the utilization and conservation of water, and, in cases stipulated by U.S.S.R. and Union-republic legislation, by the executive committees of local Soviets. Such authorizations are issued after coordination with [the executive committees of the local Soviets] the agencies exercising state sanitary supervision and the agencies for the protection of fish reserves [(with respect to bodies of water used in the fishing industry), and, when necessary], also with other interested agencies. The procedure for coordination and the issuance of authorizations for special water use is established by the U.S.S.R. Council of Ministers.

General water use is exercised without any authorization, according to the procedure established by Union-republic legislation. For bodies of water that have been granted for single use, general water use is permitted according to conditions established by the primary water user, with the agreement of the agencies for the regulation of the utilization and conservation of water and when necessary may be prohibited.

Water use is exercised free of charge. Special water may be subject to a fee, in cases and according to the procedure established by the U.S.S.R. Council of Ministers.

Art. 16. Time periods of water use [*Art. 14 in the draft Principles*]. Bodies of water are granted for indefinite or temporary use.

Water use without a time period set in advance is called indefinite (permanent).

Temporary use may be short-term – up to three years – or long-term – from three to 25 years. When necessary, the time periods for water use may be extended for a period not to exceed the upper limit of short-term or long-term temporary use, whichever is appropriate.

General water use is not limited in time.

Art. 17. Rights and duties of water users [*Art. 15 in the draft Principles*]. Water users have the right to use bodies of water only for those purposes for which use has been granted.

In cases stipulated by U.S.S.R. and Union-republic legislation, the rights of water users may be limited in the interests of the state, as well as in the interests of other water users. At the same time, the conditions of the use of bodies of water for the population's needs for drinking water and water for everyday purposes must not deteriorate. [The right to the use of water for the population's needs for drinking water for everyday purposes is not subject to limitation.]

Water users are obliged:

to utilize bodies of water rationally and to be concerned about the economical expenditure of water and about restoring and improving the quality of water;

to take steps for the complete cessation of the discharge into bodies of water of sewage-containing pollutants;

[in the process of using water] not to violate the rights granted to other water users, and also not to damage economic facilities or natural resources (land, forests, living things, useful minerals, etc.);

to maintain in good condition purification and other water-resource installations and technical devices affecting the condition of water, to improve their operational qualities, and in prescribed cases to keep a record of water use.

Art. 18. Grounds for discontinuing the right to water use [*Art. 16 in the draft Principles*]. The rights of enterprises, organizations, institutions and citizens to the use of water are subject to discontinuation in the following instances:

(1) termination of the need for or renunciation of water use;

(2) expiration of the time period for water use;

(3) liquidation of the enterprise, organization or institution;

(4) transfer of water-resource installations to other water users;

(5) emergence of the necessity to remove a body of water from single use.

The rights of enterprises, organizations, institutions and citizens to water use (besides the right to use water for drinking and everyday needs) may be discontinued also in cases of the violation of the regulations for the use of water and its conservation or in cases of the utilization of bodies of water for purposes not consonant with those for which it was granted [regulations on the use and conservation of water].

Other grounds for discontinuing the rights of enterprises, organizations, institutions and citizens to water use may also be stipulated by Union-republic legislation.

Art. 19. Procedure for discontinuing the right to water use [*Art. 17 in the draft Principles*]. The right to water use is discontinued through:

abrogation of the authorization for special or secondary water use;
withdrawal of bodies of water granted for single use.

The discontinuation of special water use is carried out at the decision of the agency that issued the authorization for this use.

Secondary water use may be discontinued at the decision of the primary water user, with the agreement of the agency for the regulation of the utilization and conservation of water.

The withdrawal of bodies of water from single use is carried out according to the procedure established by U.S.S.R. and Union-republic legislation.

The withdrawal of bodies of water from single use by enterprises, organizations and institutions of Union subordination is [may be] carried out on the basis of an agreement between the water users and the ministries and departments to which they are directly subordinate.

Art. 20. Compensation for losses caused by the implementation of water-resources measures or by the discontinuation of water use or changes in its conditions. Losses caused to enterprises, organizations, institutions and citizens during the implementation of water-resources measures (hydraulic engineering operations, etc.), and also by the discontinuation of water use or changes in its conditions, are subject to reimbursement in cases stipulated by and according to the procedure established by the U.S.S.R. Council of Ministers.

Art. 21. The use of bodies of water for drinking, everyday and other needs of the population [*Art. 18 in the draft Principles*]. Bodies of water in which the quality of the water corresponds to established sanitary requirements are granted for the drinking, everyday water-supply and other needs of the population.

The utilization of underground water of potable quality for needs not related to drinking and everyday water supply is, as a rule, not permitted. In areas where the necessary surface sources of water are lacking and where there are sufficient reserves of underground water of potable quality, the agencies for the regulation of the utilization and conservation of water may authorize the utilization of this water for purposes not related to drinking and everyday water supply [with the observance of the requirements established by Art. 36 of these Principles].

Art. 22. The use of bodies of water for therapeutic, resort and health-improvement purposes [*Art. 19 in the draft Principles*]. Bodies of water that, according to established procedure, come within the category of the therapeutic bodies of water are to be utilized primarily for the therapeutic and resort purposes. In exceptional cases, the agencies for the regulation of the utilization and conservation of water may authorize the utilization of bodies of water coming within the therapeutic category for other purposes, with the agreement of the appropriate public health and resort-administration agencies [except for the utilization of these bodies of water for discharging sewage].

The discharge of sewage into bodies of water coming within the therapeutic category is prohibited.

The procedure for the use of water for recreation and sports is established by U.S.S.R. and Union-republic legislation.

Art. 23. The use of bodies of water for the needs of agriculture [*Art. 20 in the draft Principles*]. The use of bodies of water for the needs of agriculture is exercised according to the procedure for both general and special water use.

Special water use applies to the irrigation, water-supply, drainage [systems, wells] and other water-resource installations and devices belonging to [specialized] state organizations, collective farms, state farms and other water users.

Collective farms, state farms and other enterprises, organizations, institutions and citizens using bodies of water for agricultural needs are obliged to observe the established plans, regulations, norms and regimen of water use, to take steps to curtail losses of water through seepage and evaporation in reclamation [irrigation and water-supply] systems, to prevent the irrational discharge of water into these systems, to prohibit the entry of fish into reclamation systems from fishing-industry bodies of water, and to create the most favorable regimen of soil moisture. [In implementing agrotechnical, land-reclamation and other measures to ensure that the condition of the land is favorable to agricultural production, the above-mentioned water users are obliged not to cause any deterioration in the condition of bodies of water.]

The irrigation of agricultural land with sewage is authorized by the agencies for the regulation of the utilization and conservation of water, with the agreement of the agencies exercising state sanitary and veterinary supervision.

The provisions of this article extend also to the irrigation and drainage of land occupied by forests, forest belts and forest nurseries.

Art. 24. The use of bodies of water for industrial purposes [*Art. 21 in the draft Principles*]. Water users using bodies of water for industrial purposes are obliged to observe the established plans, technological norms and regulations for water use, and also to take steps to curtail the expenditure of water and to discontinue the discharge of sewage, through improving production technology and water-supply schemes (the use of water-free technological processes, air cooling, recyclable water supply and other technical methods).

The executive committees of the local Soviets, in cases of natural disasters, accidents or other extraordinary circumstances, and also in cases in which enterprises exceed the established limits for the consumption of water from water lines, have the right to limit or prohibit the consumption for industrial purposes of drinking water from communal water lines and temporarily to restrict the consumption for industrial purposes of drinking water from departmental economic and drinking-water water lines in the interests of the priority satisfaction of the population's needs for drinking water and water for everyday purposes.

Underground water (fresh, mineral and thermal) not coming within the category of drinking or therapeutic water may be utilized, according to established procedure, for technical water supply, the extraction of chemical elements contained in this water, the generation of thermal power and other production needs, with the observance of the requirements of the rational utilization and conservation of water.

Art. 25. The use of bodies of water for the needs of hydraulic power engineering [*Art. 22 in the draft Principles*]. The use of bodies of water for the needs of hydraulic power engineering is exercised with consideration for the interests of other branches of the national economy, and also with the observance of the requirements of the integrated utilization of water, if no other provision has been made by a resolution of the U.S.S.R. Council of Ministers or a resolution of a Union-republic Council of Ministers or, in appropriate instances, by a decision of an agency for the regulation of the utilization and conservation of water.

Art. 26. The use of bodies of water for the needs of water transportation and timber floating [*Art. 23 in the draft Principles*]. The rivers, lakes, reservoirs, canals, inland seas [and territorial seas] and other inland sea waters of the U.S.S.R., and also the territorial waters (territorial seas) of the U.S.S.R., are waterways of general use, except in cases in which their utilization for these purposes is fully or partially prohibited or they have been granted for single use.

The procedure for assigning waterways to the navigation and timber-floating categories, as well as for establishing regulations for the operation of waterways, is determined by U.S.S.R. and Union-republic legislation.

The loose floating of timber, and also the floating of timber in bundles and baskets not pulled by ships, is prohibited:

(1) on navigable waterways;

(2) on bodies of water the list of which is confirmed by the U.S.S.R. Council of Ministers or by the Union-republic Councils of Ministers, in consideration of the special importance of these bodies of water for the fishing industry, water supply or other national-economic purposes.

On other bodies of water, the above-mentioned types of timber floating are permitted on the basis of authorizations issued by the agencies for the regulation of the utilization and conservation of water and after obtaining the agreement of the agencies exercising the protection of fish reserves [(3) on the bodies of water, without authorization from the agencies for the regulation of the utilization and conservation of water and the agreement of the agencies exercising the protection of fish reserves].

Timber-floating organizations are obliged to conduct the regular cleaning of sunken logs from waterways used for timber floating.

Art. 27. The use of bodies of water for the needs of air transportation [*Art. 24 in the draft Principles*]. The procedure for the use of bodies of water for the stationing, takeoff and landing of aircraft, as well as for other needs of air transportation, is established by U.S.S.R. legislation.

Art. 28. The use of bodies of water for the needs of the fishing industry [*Art. 25 in the draft Principles*]. [The rivers, lakes, reservoirs, inland seas and other bodies of water that are utilized for the catching and breeding of fish or other aquatic fauna and flora or are of importance for the reproduction of fish reserves are recognized as fishing-industry bodies of water.]

On fishing-industry bodies of water or on individual sections of these bodies of water that are of especially great importance for the conservation and reproduction of valuable types of fish or other objects of water industries, the rights of water users may be limited in the interests of the fishing industry. The lists of such bodies of water or sections of bodies of water and the types of limitations on water use are determined by the agencies for the regulation of the utilization and conservation of water on the basis of presentations by the agencies exercising the protection of fish reserves.

In the operation of hydraulic-engineering and other installations on fishing-industry bodies of water, measures ensuring the protection of fish reserves and conditions for their reproduction must be carried out in good time.

The procedure for the use of bodies of water for the needs of the fishing industry is established by U.S.S.R. and Union-republic legislation.

Art. 29. The use of bodies of water for the needs of the hunting industry [*Art. 26 in the draft Principles*]. On rivers, lakes and other bodies of water inhabited by wild waterfowl and valuable fur-bearing animals (beaver, muskrat, desmans, coypu, etc.), the agencies for the regulation of the utilization and conservation of water may grant preferential water-use rights to hunting-industry enterprises and organizations, with consideration for the requirements of the integrated utilization of water.

The procedure for the use of bodies of water for the needs of the hunting industry is established by U.S.S.R. and Union-republic legislation.

Art. 30. The use of bodies of water for the needs of preserves [*Art. 27 in the draft Principles*]. Bodies of water of particular scientific or cultural value are declared preserves, according to the procedure established by U.S.S.R. and Union-republic legislation, and are granted for indefinite single use as preserves, for purposes of conservation and the conduct of scientific research.

The procedure for the use of the water of preserves is determined by the statute on preserves.

The withdrawal of bodies of water from use as preserves is permitted only in cases of special necessity, on the basis of a resolution of a Union-republic Council of Ministers.

Art. 31. The use of bodies of water for the discharge of sewage [*Art. 28 in the draft Principles*]. The use of [rivers, lakes, seas and other] bodies of water for the discharge of industrial, communal, household, drainage and other sewage may be carried out only with the authorization of the agencies for the regulation of the utilization and conservation of water, after obtaining the agreement of the agencies exercising state sanitary supervision and the protection of fish reserves and of other interested agencies [and with the observance of the requirements stipulated in Art. 35 of these Principles].

The discharge of sewage is permitted only in cases in which it does not lead to an increase in the pollutant content of the particular body of water above established norms and on the condition that the water users purify the sewage up to the limits established by the agencies for the regulation of the utilization and conservation of water.

If the above requirements are violated, the discharge of sewage is to be limited, halted or prohibited by the agencies for the regulation of the utilization and conservation of water, up to and including discontinuation of the activity of individual industrial installations, shops, enterprises, organizations and institutions. In cases threatening the health of the population, the agencies exercising state sanitary supervision have the right to halt the discharge of sewage, up to and including discontinuation of the operation of production or other facilities, with notification to the agencies for the regulation of the utilization and conservation of water.*

The procedure and conditions for the use of bodies of water for the discharge of sewage are established by U.S.S.R. and Union-republic legislation.

Art. 32. The use of bodies of water for fire-fighting needs and other state and public requirements [*Art. 29 in the draft Principles*]. The removal of water for fire-fighting needs is permitted from any body of water.

The procedure for the use of bodies of water for fire-fighting needs, as well as for other state and public requirements, is established by U.S.S.R. and Union-republic legislation.

Art. 33. The operation of reservoirs [*Art. 30 in the draft Principles*]. Enterprises, organizations and institutions that operate water-raising, water-transmission or water-removal installations on reservoirs are obliged to observe the regimen for the filling and operation of reservoirs that has been established, with consideration for the interests of water users and land users, in the zones affected by the reservoirs.

The procedure for the operation of reservoirs is determined by regulations confirmed by the agencies for the regulation of the utilization and conservation of water for each reservoir, cascade or system of reservoirs, with the agreement of the agencies exercising state sanitary supervision and the protection of fish reserves and of other interested agencies.

The organization and coordination of measures to ensure the proper technical condition and comprehensive outfitting of reservoirs and supervision over the observance [by water users of the requirements of water legislation and] of the regulations for reservoir operation is carried out by the agencies for the regulation of the utilization and conservation of water [specially empowered agencies], according to the procedure established by the U.S.S.R. Council of Ministers or the Union-republic Councils of Ministers.

The provisions of this article extend also to the operation of lakes and other bodies of water used as reservoirs.

Art. 34. Regulation of the use of bodies of water situated on the territory of more than one Union republic [*Art. 31 in the draft Principles*]. The regulation of the use

* See bracketed paragraphs under Art. 38.

of bodies of water situated on the territory of two or more Union republics, in areas affecting the interests of these republics, is exercised by agreement between the agencies of the interested republics, with the exception of bodies of water for which the regulation of their use has been assigned to the competence of the U.S.S.R.

[In the use of bodies of water situated on the territory of two or more Union republics, the water users of each republic are obliged not to violate the rights and interests of the water users of the other republic or republics.]

Art. 35. The procedure for resolving disputes concerning water use [*Art. 32 in the draft Principles*]. Disputes concerning water use are resolved by the Union-republic Councils of Ministers, the autonomous-republic Councils of Ministers, the executive committees of local Soviets, the agencies for the regulation of the utilization and conservation of water and other specially empowered state agencies, according to [administrative] procedures established by U.S.S.R. and Union-republic legislation.

Disputes between water users of one Union republic and water users of another Union republic concerning water use are examined by a commission, set up on the basis of parity, of representatives of the interested Union republics. In cases in which the commission does not reach an agreed-upon decision, disputes concerning such questions are subject to examination according to a procedure determined by the U.S.S.R. Council of Ministers.

Property disputes involving water relations [the violation of rights to the use of water] are resolved by procedures established by U.S.S.R. and Union-republic legislation [arbitration or by legal procedure].

Art. 36. The use of the border waters of the U.S.S.R. [*Art. 33 in the draft Principles*]. The use of U.S.S.R. border waters is exercised on the basis of international agreements.

To the extent that the use of the Soviet section of border waters is not regulated by international agreements to which the U.S.S.R. is a party, it is exercised in accordance with U.S.S.R. and Union-republic legislation.

The procedure for the use of U.S.S.R. border waters is established by the competent agencies, in coordination with the border troops command.

III. THE CONSERVATION OF WATER AND THE PREVENTION OF ITS HARMFUL ACTION

Art. 37. The conservation of water [*Art. 34 in the draft Principles*]. All water (bodies of water) is subject to protection from pollution, obstruction and depletion that may harm [the results of which may entail a deterioration in] the health of the population or entail the diminution of fish reserves, the deterioration of water-supply conditions or other unfavorable phenomena as a consequence of changes in the physical, chemical or biological properties of the water, a reduction in [or] its capacity for natural purification or the violation of the hydrological or hydrogeological regimen of water.

[All water is also subject to protection from obstruction, depletion and other actions deleterious to its condition.]

Enterprises, organizations and institutions whose activity affects the condition of water are obliged to conduct – with the agreement of the agencies for the regulation of the utilization and conservation of water, of the executive committees of local Soviets, of the agencies exercising state sanitary supervision and the protection of fish reserves and of other interested state agencies or at the direction of specially empowered state agencies [and also with the agreement of the executive committees of local Soviets or at their direction] – technological, forest-amelioration, agrotechnical, hydraulic-engineering, sanitary and other measures to ensure the protection of water from pollution, obstruction and depletion, and also to improve the water's condition and regimen.

Measures for the conservation of water [that is of all-Union or republic importance] are stipulated in the state plans for the development of the national economy [of the U.S.S.R. and the Union republics].

Art. 38. The protection of water from pollution and obstruction [*Art. 35 in the draft Principles*]. The discharge of production, household and other types of waste materials into bodies of water is prohibited. The discharge of sewage is permitted only with the observance of the requirements stipulated in Art. 31 of these Principles. [The utilization of bodies of water for the discharge of production, household and other types of waste materials, except sewage, is prohibited.

[The discharge of sewage is permitted only in cases in which it does not lead to an increase in the pollutant content of the particular body of water above established norms and on the condition that the water users purify the sewage up to the limits established by the agencies for the regulation of the utilization and conservation of water.

[If the above requirements are violated, the discharge of sewage is to be limited, halted or prohibited by the agencies for the regulation of the utilization and conservation of water, up to and including discontinuation of the activity of individual industrial installations, shops, enterprises, organizations or institutions. In cases threatening the health of the population, the agencies exercising state sanitary supervision have the right to halt the discharge of sewage, up to and including discontinuation of the operation of production or other facilities, with notification to the agencies for the regulation of the utilization and conservation of water.]

The owners of means of water transportation, pipelines [installations built on piles] and floating and other installations on bodies of water, timber-floating organizations and other enterprises, organizations and institutions are obliged not to permit the pollution or obstruction of water as a consequence of losses of oils, wood, chemical, petroleum and other products [by petroleum products, oils, wood and other types of waste materials].

Enterprises, organizations and institutions are obliged not to permit the pollution or obstruction of the surface of drainage collection systems, the ice cover of bodies of water or the surface of glaciers by industrial, everyday and other waste materials and effluents or by petroleum and chemical products whose contact with water entails a deterioration in the quality of surface or underground water.

The administrations of state water-resource systems, collective farms, state farms and other enterprises, organizations and institutions are obliged to prevent the pollution of water by fertilizers and toxic chemicals.

For the purpose of protecting water that is used for drinking and everyday water supply and for the therapeutic, resort and health-improvement needs of the population, regions and zones of sanitary protection are established in accordance with U.S.S.R. and Union-republic legislation.

Appendix 22

Art. 39. The protection of water from depletion [*Art. 36 in the draft Principles*]. To maintain a favorable water regimen in rivers, lakes, reservoirs, underground water [seas] and other bodies of water, [particularly] to prevent the water erosion of soil, the silting up of bodies of water and the deterioration of habitat conditions for aquatic animals, to reduce fluctuations in runoff water, etc., forest water-conservation zones are established and forest-amelioration, anti-erosion, hydraulic-engineering and other measures are conducted, according to the procedure stipulated by U.S.S.R. and Union-republic legislation.

The agencies for the regulation of the utilization and conservation of water, in coordinating questions of the siting and construction of enterprises, installations and other [economic] facilities affecting the condition of water, and also in issuing authorizations for special water use, must be guided by the schemes for the integrated utilization and conservation of water and by the [state] water-resource balances, which take into consideration the interests of the water users and land users [within the respective river, lake or inland-sea basin].

If, during the performance of drilling or other mining operations connected with the prospecting, surveying and exploitation of deposits of gas, petroleum, coal or other useful minerals, underground water tables are discovered, the organizations conducting the mining operations are obliged immediately to report this to the agencies for the regulation of the utilization and conservation of water and to take steps to protect the underground water, according to established procedure.

[The utilization of underground water of potable quality for needs not connected with the drinking and everyday water supply of the population is authorized by the agencies for the regulation of the utilization and conservation of water, in cases in which there is an especially good reason for doing so.]

Self-discharging wells are to be equipped with regulating devices, closed down or dismantled, according to the procedure established by U.S.S.R. and Union-republic legislation.

Art. 40. The prevention and elimination of the harmful action of water [*Art. 37 in the draft Principles*]. Enterprises, organizations and institutions are obliged to conduct, in coordination with the agencies for the regulation of the utilization and conservation of water, the executive committees of local Soviets and other interested state agencies or at the direction of specially empowered state agencies [their direction], measures to prevent and eliminate the following harmful actions of water:

> floods, inundations and rising ground water; the destruction of banks, protective dikes and other installations; the bogging up and salinization of land; the erosion of soil, the formation of ravines, landslides, flash floods and other harmful phenomena.

The implementation of urgent measures to prevent and eliminate natural disasters caused by the harmful action of water is regulated by U.S.S.R. and Union-republic legislation.

Measures for the prevention and elimination of the harmful action of water [that are of all-Union or republic importance] are provided for in the state plans for the development of the national economy [of the U.S.S.R. and the Union republics].

IV. STATE RECORD-KEEPING [CONTROL] AND PLANNING OF THE UTILIZATION OF WATER

Art. 41. The goals of state record-keeping and planning of the utilization of water [organization of the keeping of records on water and of control over its utilization – Art. 38 in the draft Principles]. The state keeping of records on water and its utilization has the goals of establishing the quantity and quality of water and of providing data on the utilization of water for the needs of the population and the national economy.

The planning of the utilization of water must ensure the scientifically substantiated distribution of water among water users, taking into consideration the priority satisfaction of the population's needs for drinking water and water for everyday purposes, the conservation of water and the prevention of its harmful action. The planning of the utilization of water must take into consideration the data of the state water cadastre, the water-resources balances and the schemes for the integrated utilization and conservation of water.

[For the purpose of ensuring the rational and planned utilization and conservation of water and preventing its harmful action, a state water cadastre is kept, current and long-range state water-resource balances are compiled, schemes for the integrated utilization and conservation of water are worked out and state control is exercised.

[The goal of state control over the utilization of water is to ensure the observance by ministries and departments, by state, cooperative and public enterprises, organizations and institutions and by citizens of water legislation and the procedures for the utilization and conservation of water and for the prevention of its harmful action.

[State control over the utilization of water is carried out by specially empowered state agencies for the regulation of the utilization and conservation of water and by the executive committees of the local Soviets, according to the procedure established by U.S.S.R. legislation.]

Art. 42. The state water cadastre [Art. 39 in the draft Principles]. The state water cadastre includes record-keeping data on water according to quantitative and qualitative indices, the registration of water use, and record-keeping data on the utilization of water.

Art. 43. The [State] water-resource balances [Art. 40 in the draft Principles]. [State] Water-resource balances, which evaluate the presence of water and the degree of its utilization, are compiled by basins, economic regions, Union republics and the U.S.S.R. [and are taken into consideration in the current and long-range planning of the development of the U.S.S.R. national economy].

Art. 44. Schemes for the integrated utilization and conservation of water [Art. 41 in the draft Principles]. General and basin (territorial) schemes for the integrated utilization and conservation of water determine the basic water-resource and other measures to be carried out for the satisfaction of the long-range requirements for water of the population and the national economy and for the conservation of water and the prevention of its harmful action.

Art. 45. The procedure for the keeping of state records on water and its utilization and keeping the state water cadastre, for compiling [state] water-resource balances

and for working out schemes for the integrated utilization and conservation of water [*Art. 42 in the draft Principles*]. The keeping of state records on water and its utilization, the keeping of the state water cadastre, the compilation of [state] water-resource balances and the working out of schemes for the integrated utilization and conservation of water are carried out by the state, according to a single system for the U.S.S.R.

The procedure for keeping state records on water and its utilization, keeping the state water cadastre, compiling the [state] water-resource balances and working out and confirming the schemes for the integrated utilization and conservation of water is established by the U.S.S.R. Council of Ministers.

V. LIABILITY FOR THE VIOLATION OF WATER LEGISLATION

Art. 46. Liability for the violation of water legislation [*Art. 43 in the draft Principles*]. [The unauthorized disposition of water] The ceding of the right to water use and other transactions that violate, overtly or covertly, the right of state ownership of water are invalid.

Persons guilty of the above transactions or of:

the unauthorized seizure of bodies of water or the unauthorized use of water;

the removal of water in violation of water-use plans;

the pollution or obstruction of water;

the putting into operation of enterprises, communal and other facilities without installations and devices to prevent the [water] pollution and obstruction of water or its harmful action;

the negligent utilization of water (extracted or drained from bodies of water);

the violation of the water-protection regimen in drainage collection systems, leading to their pollution, the water erosion of the soil and other harmful phenomena; the unauthorized conduct of hydraulic-engineering operations; the damaging of water-resource installations and devices; the violation of regulations for the operation of water-resource installations and devices,

are criminally or administratively liable in accordance with U.S.S.R. and Union-republic legislation.

Union-republic legislation may establish liability for other types of violations of water legislation as well.

Bodies of water seized without authorization are to be returned to their authorized users, with no compensation for expenditures incurred during the unlawful use.

[*Art. 44. Compensation for losses caused by the violation of water legislation.* Enterprises, organizations, institutions and citizens are obliged to make reimbursement for losses caused by the violation of water legislation, in amounts and according to the procedure established by U.S.S.R. and Union-republic legislation. Officials and other employees through whose fault enterprises, organizations and institutions incur expenditures connected with reimbursement for losses are materially liable according to established procedure.]'

Appendix 23. Major rivers of the U.S.S.R.

River (with main tributaries indented)	Length (km)	Basin area (1000 km²)	Average annual flow at mouth (km³)	Main-stream hydropower potential (million kW)
Lena	4270	2425	488	18.4
Aldan	2242	702	164	5.5
Yenisey	3354[a]	2599	549	18.2
Angara	1826	1056	131	9.9
Ob′	3676[a]	2929	394	5.7
Irtysh	4422	1592	95	3.2
Amur	2846[a]	1843	347	6.4
Kolyma	2600[a]	644	120	5.2
Syr-Darya	2137[a]	462	21[b]	5.9[c]
Amu-Darya	1404[a]	227	62[b]	3.8
Pyandzh	921	114	35	5.8
Volga	3688	1380	252	6.2
Kama	2032	522	119	*ca.* 4
Dnepr	2285	503	53[b]	*ca.* 2
Don	1967	423	28	*ca.* 1
Pechora	1790	327	129	*ca.* 4
Northern Dvina	733[a]	360	111	*ca.* 3
For comparison:				
Mississippi	3780	3260	*ca.* 600	[d]
Columbia	1955	671	226	*ca.* 19[e]

[a] Not including headwater streams having a different name.
[b] Equals maximum flow, measured at a point other than at the mouth.
[c] For the Naryn River only. The Naryn is the name of the Syr-Darya in its upper course.
[d] Most of the hydropower potential of the Mississippi is not feasible due to the flooding which would result, and has therefore not been precisely calculated.
[e] In the United States only.

From Davydov, *Gidrografiya S.S.S.R.*, Vol. 2; Karaulov (1963) p. 34.

Appendix 24. Conversion table for hydrologic data

1 km = 0.62137 mile	1 meter³ (m³) = 35.29 feet³ (c.f.)
1 km² = 0.3861 square mile	1,000 m³/second = 31.536 km³/year
1 km³ = 0.2397 cubic mile	1 cubic foot = 0.0283 m³
1 mile = 1.6093 km	1 million gallons/day (mgd) = 3,785 m³/day
1 mile² = 2.59 km²	1 cubic foot/second (cfs) = 723.97 acre-feet/year
1 mile³ = 4.168 km³	1,000,000 acre-feet (a.f.) = 1.233 km³
1 meter = 3.281 feet	1 (unit)/second = 31,536,000 (units)/year

Appendix 25. Hydroelectric development of the Volga, Kama, and Dnepr rivers

Installation	Installed capacity (mW)	Average output (billion kWh)	Static head (meters)	Reservoir area (km^2)	Reservoir volume (km^3)	Annual flow (km^3)
Volga:						
Ivankovo	30	0.11	11	330	1.2	10
Uglich	110	0.23	11	220	1.3	14
Rybinsk	360	1.05	18	4550	25.4	35
Gor'kiy	520	1.40	13	1570	8.7	54
Cheboksary[a]	1680	3.37	15	2295	14.2	113
Kuybyshev	2300	10.50	25	5901	58.0	241
Saratov	1500	5.50	13	1831	12.9	248
Volgograd	2560	11.10	22	3117	31.4	251
(Lower Volga)[b]	—	—	(16)	—	—	(251)
Kama:						
Upper Kama[a]	700	*ca.* 0.8	17	2690	16.0	28
Kama	504	1.75	19	1730	12.2	52
Votkinsk	1000	2.25	21	1066	9.4	54
Lower Kama[a]	1058	2.56	15	2400	13.0	89
Total, Volga basin	12320	*ca.* 40.6	—	27700	203.7	—

Dnepr:						
Kiev	350	0.64	11.5	1420	3.7	43
Kanev[a]	420	0.82	10.5	*ca.* 1200	3.3	44
Kremenchug	625	1.51	17.0	2252	13.5	48
Dneprodzerzhinsk	352	1.25	12.6	567	2.4	51
Dneproges	650	3.58	35.4	410	3.1	51
Kakhovka	351	1.42	16.0	2155	18.2	49
Total, Dnepr basin	2748	9.22	—	*ca.* 8000	44.2	—

[a] Planned, or in early stages of construction.

[b] No longer under active consideration.

From Vendrov *et al.* (1964) p. 26; Askochenskiy (1967) pp. 156 and 171; *Gidrotekhnicheskoye stroitel'stvo* (1967) No. 5, p. 4; Avakyan and Sharapov (1962) p. 144; Grin (ed.) (1968) p. 105.

Appendix 26

Appendix 26. Excerpt from the 1960 Russian Republic law 'On the Conservation of Nature in the R.S.F.S.R.'*

'*Art. 4. Conservation of water resources* (see Appendix 21).

Art. 7. Protection of greenery in populated areas. Greenery planted in all populated areas, and also in surrounding green belts and along roads, is subject to protection as having hygienic, protective, and cultural-aesthetic significance.

Cutting such green plantings (except for maintenance) or transplanting them in other locations is allowable only by way of an exception with the permission of the executive committee of the local Soviet of Workers' Deputies in accordance with procedures established by the R.S.F.S.R. Council of Ministers.

Art. 10. Protection of health spas, forest–park shelter belts, and suburban green areas. In areas of recreation and recuperation for the workers (health spas, sanitary-protective zones around health spas, forest–park shelter belts, and suburban green areas), the totality of the natural conditions contributing to the curative and hygienic importance of the area is protected.

In health spa locales, in addition, the natural features which provide the basic specialization of the spa (mineral springs, mud, beaches, pine groves, etc.) are also subject to protection.

The Councils of Ministers of the autonomous republics, and the executive committees of kray, oblast, and city Soviets of Workers' Deputies, shall establish along prominent tourist routes and in the most frequented recreation areas of the workers protected zones with regulated regimes.

The development and regulation of the above cited objects and territories shall be carried out only in accordance with a comprehensive construction plan.

Ministries, departments, economic councils, and the executive committees of local Soviets of Workers' Deputies associated with planning, and design organizations associated with designing, the development or reorganization of the above cited objects and territories are required to provide for the preservation and improvement of the complex of these natural conditions.

To these ends it is necessary:

(*a*) to provide for locating the areas and features of new communal and transport construction so as to preserve the medicinal properties and quality landscapes of local areas;

(*b*) to provide for the systematic carrying-out of landscaping and reclamation measures, including preventative work to avert landslides, flash floods, avalanches and cave-ins, bank erosion, the destruction of beaches, and the littering and polluting of workers' recreation areas.'

'*Art. 12. The sanitary protection of natural resources.* Atmospheric resources, surface and underground water, soils and other ground covers are subject to sanitary protection.

The executive committees of local Soviets of Workers' Deputies, institutions, enterprises and organizations are required to carry out measures to prevent pollution of the atmosphere, surface and underground waters, soils and other ground covers, and also to prevent littering of the landscape.

* From *Pravda* (October 28, 1960) p. 2.

Household and industrial waste material and refuse is subject either to re-use in the national economy or to systematic removal and disposal.

Ministries, departments, and economic councils are required, when planning enterprises and installations which employ natural resources, to design and install technological processes which will ensure a maximum utilization of raw materials and fuels and prevent harmful waste materials from entering the atmosphere, surface waters, ground waters, or soils.

When it is not possible to introduce technological processes and forms of production which will exclude waste products from the atmosphere, water bodies, and soils, there must be set up effective cleaning, disposal, and recovery facilities.

The concentration of harmful substances in waste materials discharged into the atmosphere, water bodies, and soils should not exceed the maximum permissible limits established under consideration of all economic interests and public health standards.'

Appendix 27. U.S.S.R. air quality standards (from World Health Organization – 1968)

	Single exposure (approx. 20 minute averaging time)		24-hour averaging time	
Substance	mg/m³ STP	ppm	mg/m³ STP	ppm
Acetaldehyde	0.01	0.005	—	—
Acetic acid	0.2	0.08	—	—
Acetic anhydride	0.1	0.02	—	—
Acetone	0.35	0.15	0.35	0.15
Acetophenone	0.003	0.0006	0.003	0.0006
Acrolein	0.3	0.12	0.1	0.04
Ammonia	0.2	0.28	0.2	0.28
Amyl acetate	0.1	0.019	0.1	0.019
Amylene	1.5	0.5	1.5	0.5
Aniline	0.05	0.013	0.03	0.008
Arsenic (as As)	—	—	0.003	—
Benzene	1.5	0.5	0.8	0.25
Butane	200	85	—	—
Butanol	0.3	0.1	—	—
n-Butyl acetate	0.1	0.021	0.1	0.021
Butylene	3	1.3	3	1.3
Butyric acid	0.015	0.004	0.01	0.003
Caprolactum	0.06	0.013	0.06	0.013
Caprylic acid	0.01	0.002	0.005	0.001
Carbon black (soot)	0.15	—	0.05	—
Carbon disulfide	0.03	0.01	0.01	0.0033
Carbon monoxide	3	2.7	1	0.9
Carbon tetrachloride	4	0.7	—	—
Chlorine	0.1	0.033	0.03	0.01

Substance	Single exposure (approx. 20 minute averaging time) mg/m³ STP	ppm	24-hour averaging time mg/m³ STP	ppm
p-Chloroaniline	0.04	0.008	—	—
Chlorobenzene	0.1	0.02	0.1	0.02
Chloroprene	0.1	0.028	0.1	0.028
m-Chlorophenyl isocyanate	0.005	0.001	0.005	0.001
p-Chlorophenyl isocyanate	0.0015	0.0002	0.0015	0.0002
Chromium hexavalent (as CrO_3)	0.0015	—	0.0015	—
Cyclohexanol	0.06	0.015	0.06	0.015
Cyclohexanone	0.04	0.008	0.04	0.008
Dichloroethane	3	0.75	1	0.25
2-3-Dichloro-1-4-naphthaquinone	0.05	—	0.05	—
Diethylamine	0.05	0.02	0.05	0.02
Diketene	0.007	0.002	—	—
Dimethylaniline	0.0055	0.001	—	—
Dimethyl disulfide	0.7	0.18	—	—
Dimethylformanide	0.03	0.01	0.03	0.01
Dimethyl sulfide	0.08	0.03	—	—
Dinyl (diphenyl + its oxides)	0.01	0.0015	0.01	0.0015
Divinyl	3	1.2	1	0.4
Epichlorohydrin	0.2	0.05	0.2	0.05
Ethanol	5	2.5	5	2.5
Ethyl acetate	0.1	0.029	0.1	0.29
Ethylene	3	2.3	3	2.3
Ethylene oxide	0.3	0.15	0.03	0.015
Fluorides (as F)	0.03	—	0.01	—
Fluorides (insoluble)	0.2	—	0.03	—
Formaldehyde	0.035	0.029	0.012	0.01
Furfural	0.05	0.013	0.05	0.013
Gasoline (as C) (from crude oil)	5	1.25	1.5	0.38
Gasoline (as C) (from shale)	0.05	0.01	0.05	0.01
Hexamethylene-diamine	0.01	0.002	0.01	0.002
Hydrochloric acid (as H^+)	0.006	—	0.006	—
Hydrochloric acid (as HCl)	0.2	0.15	—	—
Hydrogen fluoride	0.02	0.03	0.005	0.008
Hydrogen sulfide	0.00 (?)	0.005	0.00 (?)	0.005
Isopropyl benzene	0.014	0.003	0.014	0.003
Isopropyl benzene hydroperoxide	0.007	0.001	0.007	0.001
Lead (as Pb)	—	—	0.0007	—
Lead sulfide (as Pb)	—	—	0.0017	—
Malathion	0.015	—	—	—

Substance	Single exposure (approx. 20 minute averaging time)		24-hour averaging time	
	mg/m³STP	ppm	mg/m³STP	ppm
Maleic anhydride	0.2	0.05	0.05	0.01
Manganese (as Mn)	0.03	—	0.01	—
Mercury (as Hg)	—	—	0.0003	—
Mesidine (2-amino-1,3,5-trimethyl benzene)	0.003	—	—	—
Methanol	1	0.75	0.5	0.38
Methyl acetate	0.07	0.023	0.07	0.023
Methyl acrylate	0.01	0.003	—	—
Methyl aniline	0.04	0.01	—	—
Methyl mercaptan	9×10^{-6}	—	—	—
Methyl methacrylate	0.1	0.025	0.1	0.025
Methyl parathion	0.008	—	—	—
α-Methylstyrene	0.04	0.01	0.04	0.01
α-naphthaquinone	0.005	0.001	0.005	0.001
Nitric acid (as HNO_3)	0.4	0.15	—	—
Nitric acid (as H^+)	0.006	—	0.006	—
Nitrobenzol	0.008	0.001	0.008	0.001
Nitrogen dioxide	0.085	0.045	0.085	0.045
Phenol	0.01	0.0026	0.01	0.0026
Phosphoric anhydride	0.15	0.026	0.05	0.0085
Phthalic anhydride	0.2	0.03	0.2	0.03
Propanol	0.3	0.12	—	—
Propylene	3	1.5	3	1.5
Pyridine	0.08	0.023	0.08	0.023
Styrene	0.003	0.0007	0.003	0.0007
Sulfur dioxide	0.5	0.19	0.15	0.58
Sulfuric acid (as H^+)	0.006	—	0.006	—
Sulfuric acid (as H_2SO_4)	0.3	—	0.1	—
Suspended particulate matter (dust)	0.5	—	0.15	—
Thiophene	0.6	0.17	—	—
Toluene	0.6	0.15	0.6	0.15
Toluene diisocyanate	0.05	0.0071	0.02	0.0029
Tributyl phosphate	0.01	—	—	—
Trichloroethylene	4	0.67	1	0.17
n-Valeric acid	0.03	0.008	0.01	0.003
Vanadium pentoxide	—	—	0.002	—
Vinyl acetate	0.2	0.06	0.2	0.06
Xylene	0.2	0.05	0.2	0.05

From *Profile Study of Air Pollution Control Activities in Foreign Countries* (NAPCA, 1970) pp. 162–3.

Notes

CHAPTER 1. INTRODUCTION: THE SOVIET CONCEPT OF CONSERVATION

1 P. Oldak, 'Priroda vzyvayet k shchedrosti', *Literaturnaya gazeta* (June 3, 1970) p. 11.
2 As quoted in Ye. Zhbanov, 'Zhivaya voda', *Izvestiya* (June 27, 1970) p. 4; as translated in *Current Digest of the Soviet Press*, **22**, No. 26 (July 28, 1970) pp. 39–40.
3 I. P. Gerasimov, D. L. Armand and K. M. Efron (eds.), *Prirodnyye resursy Sovetskogo Soyuza, ikh ispol'zovaniye i vosproizvodstvo* (Moskva: Izdatel'stvo Akademii nauk S.S.S.R., 1963) p. 8.
4 K. N. Blagosklonov, A. A. Inozemtsev and V. N. Tikhomirov, *Okhrana prirody* (Moskva: Izdatel'stvo 'Vysshaya shkola', 1967) p. 37.
5 Gerasimov *et al.* (eds.) (1963) pp. 6–7.
6 A. G. Bannikov, *Po zapovednikam Sovetskogo Soyuza* (Moskva: Izdatel'stvo 'Mysl'', 1966) p. 6.
7 *Ibid.* p. 7.
8 G. P. Motovilov, 'Protection of Nature in the U.S.S.R.', *Proceedings, Fifth World Forestry Congress*, Vol. 3, 1960, pp. 1851–2.
9 Blagosklonov *et al.* (1967) p. 7.
10 I. P. Gerasimov, 'Nuzhen general'nyy plan preobrazovaniya prirody nashey strany', *Kommunist* (1969) No. 2, p. 78.

CHAPTER 2. HISTORICAL AND INSTITUTIONAL FRAMEWORK

1 K. N. Blagosklonov, A. A. Inozemtsev and V. N. Tikhomirov, *Okhrana prirody* (Moskva: Izdatel'stvo 'Vysshaya shkola', 1967) p. 55.
2 *Ibid.* p. 61.
3 J. A. Baclawski, *The Soviet Conservation Program for the Steppe and Wooded Steppe Regions of the European Part of the U.S.S.R.* (unpublished Ph.D. dissertation, Ann Arbor: University of Michigan, 1951) pp. 118–44.
4 Blagosklonov *et al.* (1967) p. 64.
5 V. A. Chichvarin, *Mezhdunarodnyye soglasheniya po okhrane prirody* (Moskva: Izdatel'stvo 'Yuridicheskaya literatura', 1966) pp. 367–8.
6 A. I. Voyeykov, *Vozdeystviye cheloveka na prirodu*, ed. E. M. Murzayev, (Moskva: Izdatel'stvo Akademii nauk S.S.S.R., 1963).
7 G. A. Kozhevnikov, 'O neobkhodimosti ustroystva zapovednykh uchastkov dlya okhrany russkoy prirody', reprinted in *Okhrana prirody i zapovednoye delo v S.S.S.R.*, Byulleten' No. 4 (1960) pp. 90–7.
8 The work of this congress is summarized in A. V. Fedyushin, 'Voprosy okhrany prirody v R.S.F.S.R. na pervom vserossiyskom s'yezde (1929 g.)', *Okhrana prirody Sibiri i Dal'nego Vostoka*, **1** (1962) pp. 201–4.
9 The 'Great Plan...' is reviewed in detail in J. A. Baclawski (1951) *op. cit.*, pp. 5–34.

10 V. A. Kovda, *Great Construction Works of Communism and the Remaking of Nature* (Moscow: Foreign Languages Publishing House, 1953) pp. 6–7.
11 See, for example, Albert E. Burke, 'Influence of man upon nature – the Russian view: a case study' in *Man's Role in Changing the Face of the Earth*, ed. W. L. Thomas, Jr (Chicago: University of Chicago Press, 1956) p. 1048.
12 N. Mel′nikov, 'Prezhde, chem reki povernut′ vspyat′. . .', *Literaturnaya gazeta* (July 12, 1967) p. 11, as translated in *Current Digest of the Soviet Press*, **19**, No. 32 (August 30, 1967) pp. 12 and 39. A discussion of Soviet attitudes towards the natural environment and Stalin's 'Great Plan. . .' is presented by Ian M. Matley, 'The Marxist Approach to the Geographical Environment', *Annals of the Association of American Geographers*, **56**, No. 1 (March 1966) pp. 97–111.
13 They were even discussed in books intended for an international readership. See, for example, M. Ilin, *Men and Mountains* (Philadelphia: J. B. Lippincott, 1935) pp. 195–217.
14 Typical of recent articles by leading Soviet scientists on this point is I. P. Gerasimov, 'Nuzhen general′nyy plan preobrazovaniya prirody nashey strany', *Kommunist* (1969) No. 2, pp. 68–79.
15 See, for example, Blagosklonov *et al.* (1967) p. 410.
16 *23rd Congress of the Communist Party of the Soviet Union* (Moscow: Novosti Press Agency Publishing House, 1966) p. 320.
17 *Izvestiya* (May 24, 1969) p. 3.
18 B. N. Bogdanov, 'Conservation and economics', *Ekonomika sel′skogo khozyaystva* (1970) No. 2, pp. 7–11, as translated in *Current Digest of the Soviet Press*, **22**, No. 19 (June 9, 1970) pp. 7–9.
19 The activities of the All-Russian Society are presented in more detail in K. N. Blagosklonov *et al.* (1967) pp. 436–8.
20 'Obshchaya rezolyutsiya III s′yezda Geograficheskogo obshchestva Soyuza S.S.R.', *Izvestiya Vsesoyuznogo Geograficheskogo obshchestva* (1960) No. 3, pp. 216–26.
21 L. K. Shaposhnikov, 'Experience in conservation education and propaganda of nature conservation in the U.S.S.R. and in the countries of Eastern Europe', *Conservation Education* (I.U.C.N. Publication N.S., Supplementary Paper No. 7; Morges, Switzerland: I.U.C.N., 1965) p. 42.
22 The subject of conservation activities by school children is discussed in considerable detail in K. N. Blagosklonov *et al.* (1967) pp. 426–36, and in L. K. Shaposhnikov (1965) from which the foregoing has been summarized.
23 For example, see A. A. Inozemtsev, 'Priroda zhdet zabotlivykh', *Komsomol′skaya pravda* (January 6, 1968) p. 2.
24 L. K. Shaposhnikov (1965) p. 44.

CHAPTER 3. LAND AND SOIL RESOURCES

1 Data taken from *Narodnoye khozyaystvo S.S.S.R. v 1965 g.* (Moskva: Izdatel′stvo 'Statistika', 1966) p. 277.
2 'Resolution of the C.P.S.U. Central Committee and the U.S.S.R. Council of Ministers, dated November 28, 1969, on the Model Charter for the Collective Farm', *Pravda* (November 30, 1969) p. 1, as translated in *Current Digest of the Soviet Press*, **21**, No. 50 (January 13, 1970) p. 10.
3 A very thorough study of Soviet agriculture, though now somewhat out of

date, is Naum Jasny's *The Socialized Agriculture of the U.S.S.R.* (Stanford University Press, 1949); of similar scope is Lazar Volin, *A Century of Russian Agriculture: From Alexander II to Khrushchev* (Harvard University Press, 1970). An earlier study which includes tsarist agricultural practices is V. P. Timoshenko, *Agricultural Russia and the Wheat Problem* (Stanford University Press, 1932).

4 For example see W. A. D. Jackson, 'The Soviet approach to the good earth: myth and reality', *Soviet Agriculture and Peasant Affairs*, ed. Roy D. Laird (Lawrence, Kansas: University of Kansas Press, 1963) pp. 171–85.

5 I. P. Gerasimov, 'Umen′shit′ i svesti k minimumu zavisimost′ nashego sel′skogo khozyaystva ot prirodnoy stikhii', *Izvestiya Akademii nauk S.S.S.R., seriya geograficheskaya* (1962) No. 5, p. 46.

6 *Narodnoye khozyaystvo S.S.S.R. v 1965 g.* pp. 365–6.

7 *Zarya vostoka* (April 14, 1970) p. 1.

8 G. Lashkevich *et al.*, 'Novomu polyu-garantiyu plodorodiya', *Pravda* (February 14, 1968) p. 2, as translated in *Current Digest of the Soviet Press*, **20**, No. 7 (March 6, 1968) p. 33.

9 Gerasimov (1962) p. 45. Frequent articles dealing with all phases of the restoration of saline and waterlogged soils appear in the monthly reclamation journal, *Gidrotekhnika i melioratsiya* (published jointly by the U.S.S.R. Ministry of Agriculture and the U.S.S.R. Ministry of Reclamation and Water Resources Management).

10 According to I. A. Churayev, director of the U.S.S.R. Ministry of Agriculture's Chief Administration for Plant Protection, as interviewed in *Pravda* (June 20, 1970) p. 3.

11 *Izvestiya* (August 14, 1970) p. 3.

12 See, for example, L. Laskavaya, 'Zemlya i veter', *Novyy mir* (1960) No. 6, pp. 193ff.

13 Gerasimov (1962) p. 46.

14 W. A. D. Jackson, 'The Virgin and Idle Lands Program reappraised', *Annals of the Association of American Geographers*, **52**, No. 1 (March, 1962) p. 76.

15 S. I. Sil′vestrov, 'Protsessy erozii i deflyatsii pochv na sel′skokhozyaystvennykh zemlyakh i bor′ba s nimi', in *Prirodnyye resursy Sovetskogo Soyuza, ikh ispol′zovaniye i vosproizvodstvo*, ed. I. P. Gerasimov, D. L. Armand, and K. M. Efron (Moskva: Izdatel′stvo Akedemii nauk S.S.S.R., 1963) p. 128.

16 Resolution of the Central Committee of the Communist Party of the Soviet Union (C.P.S.U.) and the U.S.S.R. Council of Ministers, 'O neotlozhnykh merakh po zashchite pochv ot vetrovoy i vodnoy erozii', *Pravda* (April 2, 1967) p. 1, as translated in *Current Digest of the Soviet Press*, **19**, No. 13 (April 19, 1967) pp. 3–5.

17 The early estimates are from N. I. Sus, *Eroziya pochvy i bor′ba s neyu* (Moskva: Sel′khozgiz, 1949) p. 46. Later sources include Sil′vestrov *op. cit.*, p. 121; Gerasimov, *op. cit.*, p. 46 and S. I. Sil′vestrov, 'Trebovaniya bor′by s eroziyey pri intensifikatsii sel′skogo khozyaystva', *Izvestiya Akademii nauk S.S.S.R., seriya geograficheskaya* (*1962*), No. 5, p. 58.

18 J. A. Baclawski, *The Soviet Conservation Program for the Steppe and Wooded Steppe Regions of the European Part of the U.S.S.R.* (unpublished Ph.D. dissertation, Ann Arbor: University of Michigan, 1951) pp. 144–54. Baclawski discusses in some detail the nature of the pre-revolutionary erosion problem.

19 Sil′vestrov (1963) p. 123.
20 'O neotlozhnykh merakh...', *Pravda* (April 2, 1967) p. 1.
21 *Ibid.*
22 Sil′vestrov (1963) pp. 127–8.
23 N. B. Vernander *et al.*, 'Physical-geographic processes unfavorable for agriculture and control measures against them' (translation of a paper presented at the Symposium on Physical Geography of the Fourth Congress of the Geographical Society of the U.S.S.R.), *Soviet Geography: Review and Translation*, **5**, No. 7 (September 1964) p. 63.
24 V. Darmodekhin, 'Semnadtsat′ protiv buri', *Izvestiya* (April 10, 1960) p. 3. This report appears in *Current Digest of the Soviet Press*, **12**, No. 20 (June 15, 1960) p. 42.
25 A. Koptsov and O. Pavlov, 'Bespokoynoye pole', *Izvestiya* (September 20, 1970) p. 3.
26 The phenomenon of the *sukhovey* is explained in P. E. Lydolph, 'The Russian sukhovey', *Annals of the Association of American Geographers*, **54**, No. 3 (September 1964) pp. 291–309.
27 'O neotlozhnykh merakh...' (1967) p. 1.
28 N. Mel′nikov, 'Prezhde, chem reki povernut′ vspyat′...', *Literaturnaya gazeta* (July 12, 1967) p. 11.
29 I. Ikonitskaya, 'Yesli voznik spor...', *Izvestiya* (August 15, 1968) p. 3.
30 Such systems are described in K. V. Zvorykin, V. G. Konovalenko and G. V. Cheshikin, *Geografiya i zemel′nyy kadastr* ('Voprosy geografii, No. 67'; Moskva: Izdatel′stvo 'Mysl′', 1965); especially the article on pp. 100–09 by Vedenichev and Marakulin (see bibliography) which, together with others from this book, has been translated in *Soviet Geography: Review and Translation*, **9**, No. 3 (March 1968) pp. 162–93. No. 78 in the 'Voprosy geografii' series also deals with the evaluation of Soviet natural resources.
31 See, for example, I. V. Komar *et al.*, 'Problems of economic evaluation of natural conditions and resources', a paper presented at the Symposium on Economic Geography of the U.S.S.R. at the Fourth Congress of the Geographical Society of the U.S.S.R., as translated in *Soviet Geography: Review and Translation*, **5**, No. 9 (November 1964) pp. 44–50.
32 'Man and his planet'; summary of a symposium conducted by the periodical *Nedelya*, as reported in *Soviet News* (November 28, 1967) p. 104.
33 Among them, for example, are the articles by Lyashchenko and Bronshteyn cited in the bibliography.

CHAPTER 4. 'ZAPOVEDNIKI' IN THE U.S.S.R.

1 Ye. Lavrenko, V. G. Geptner, S. V. Kirikov and A. N. Formozov, 'Perspektivnyy plan geograficheskoy seti zapovednikov S.S.S.R. (proyekt)', *Okhrana prirody i zapovednoye delo v S.S.S.R.*, Byulleten′ No. 3 (1958) pp. 3–92.
2 Cited in L. K. Shaposhnikov (ed.), *Primechatel′nyye prirodnyye landshafty S.S.S.R. i ikh okhrana* (Moskva: Izdatel′stvo 'Nauka', 1967) p. 8.
3 L. K. Shaposhnikov and V. A. Borisov, 'Pervyye meropriyatiya Sovetskogo gosudarstva po okhrane prirody', *Okhrana prirody i zapovednoye delo v S.S.S.R.*, Byulleten′ No. 3 (1958) pp. 93–8.

4 'On the Registration and Protection of Monuments of Art, Antiquity and Nature', enacted January 7, 1924.
5 A. G. Bannikov, *Zapovedniki Sovetskogo Soyuza* (Moskva: 'Kolos', 1969) and L. S. Belousova, V. A. Borisov and A. A. Vinokurov, *Zapovedniki i natsional'nyye parki mira* (Moskva: 'Nauka', 1969).
6 S. M. Uspenskiy, 'Sovremennyye problemy okhrany prirody v arktike', *Problemy severa*, No. 7 (1963) p. 160; as translated in *Problems of the North*, No. 7 (1963) p. 177.
7 I. I. Khutortsov, 'Nauchnaya rabota v Kavkazskom zapovednike za 40 let (1924–1964 gg.)', *Trudy Kavkazskogo gosudarstvennogo zapovednika* (Vypusk VIII 1965) p. 3.
8 Sources for these figures are the same as those used in Table 4.1 for the same years; for 1933 the source is *Bol'shaya Sovetskaya Entsiklopediya*, 1st edn, **26**, p. 239.
9 Figures taken from A. G. Bannikov, *Po zapovednikam Sovetskogo Soyuza* (Moskva: 'Mysl', 1966) p. 130.
10 Khutortsov (1965) p. 3.
11 For example, see Bannikov (1966) p. 14.
12 The work of this conference is discussed in Yu. A. Isakov, 'O nauchno-issledovatel'skoy rabote gosudarstvennykh zapovednikov R.S.F.S.R.', *Izvestiya Akademii nauk S.S.S.R., seriya geograficheskaya* (1963) No. 5, pp. 145–6.
13 This idea is discussed in N. Ye. Kabanov, 'O nekotorykh voprosakh nauchno-issledovatel'skoy raboty v zapovednikakh i svyazi yeyo s zadachami razvitiya narodnogo khozyaystva', *Okhrana prirody i zapovednoye delo v S.S.S.R.*, Byulleten' No. 4, 1960, pp. 75–90.
14 Lavrenko *et al.* (1958).
15 As reported in *Pravda* (March 16, 1968) p. 2. As another example, serious violations in the Il'men preserve in the Urals were reported by L. Krainov in an article headed 'Sozdannyy prirody', *Pravda* (Nov. 21, 1967) p. 3.
16 For example, the widespread tourist use of *zapovedniki* is advocated by Oleg Gusev in 'Khodit' turistu v zapovednik?', *Turist* (1967) No. 2; in opposition is the prominent Soviet zoologist A. G. Bannikov, 'Ot zapovednika do prirodnogo parka', *Priroda* (1968) No. 4, pp. 89–97.
17 B. Petrov, 'Prishestviye iz goroda', *Izvestiya* (July 22, 1970) p. 6.
18 Cited in K. N. Blagosklonov, A. A. Inozemtsev and V. N. Tikhomirov, *Okhrana prirody* (Moskva: Izdatel'stvo 'Vysshaya shkola' 1967) p. 413.
19 This campaign is described in V. A. Kotov, 'Bor'ba s volkami v Kavkazskom zapovednike', *Trudy Kavkazskogo gosudarstvennogo zapovednika*, Vypusk VIII (1965) pp. 182–4.
20 Bannikov (1968) p. 96.
21 L. K. Shaposhnikov (ed.) (1967) p. 145.
22 *Ibid.* p. 146.
23 *Ibid.* p. 148.
24 I. A. Yevlakhov, 'Generous gift of Baykal': interview by L. Shikarev of *Izvestiya* which appeared in the issue of September 25, 1965, as translated in *Current Digest of the Soviet Press*, **17**, No. 39 (October, 20 1965), pp. 32–3.
25 M. M. Bochkarev, 'Budet li u nas natsional'nyy park?' *Pravda* (January 17, 1967) p. 4.
26 *Pravda* (June 19, 1970) p. 6.

27 L. K. Shaposhnikov, *Okhrana prirody v S.S.S.R.* (Moskva: Izdatel'stvo 'Znaniye', 1961) p. 21.
28 L. K. Shaposhnikov (ed.) (1967).
29 V. I. Komendar *et al.* (eds.), *Karpatskiye zapovedniki* (Uzhgorod: Izdatel'stvo 'Karpaty', 1966) pp. 233–57.
30 B. P. Kolesnikov (ed.), *Pamyatniki prirody* (Sverdlovsk: Ural'skiy filial Akademii nauk S.S.S.R., 1967) p. 217.
31 K. N. Blagosklonov *et al.* (1967) pp. 417–18.

CHAPTER 5. MANAGEMENT OF FISHERIES AND WILDLIFE

1 Background information on the wildlife of the U.S.S.R. is available in considerable detail in Soviet sources. Most detailed are two separate series on the fauna of the U.S.S.R., published by the Zoological Institute of the U.S.S.R. Academy of Sciences, each containing several dozen volumes. Some of these volumes, such as No. 62 on carnivores, have been translated into English by the Israel Program for Scientific Translations. Mammals of the Soviet Union are treated extensively in the three-volume work *Mlekopitayushchiye Sovetskogo Soyuza*, edited by V. G. Geptner, A. A. Nasimovich, and A. G. Bannikov (Moskva: Gosizdat 'Vysshaya shkola', 1961—), and also in a multiple-volume collection edited by S. I. Ognev, entitled *Mammals of the U.S.S.R. and Adjacent Countries*, seven volumes of which have also been translated by the Israel Program. A less detailed description of Soviet wildlife may be found in L. S. Berg, *Natural Regions of the U.S.S.R.*, perhaps supplemented by such a standard reference work as Burton's *Systematic Dictionary of Mammals of the World.*
2 Figures are taken from *Soviet Life* (August 1966) p. 29.
3 S. V. Kirikov and Yu. A. Isakov, 'The reserves of wild game; the dynamics of hunting and its prospects', as translated in *Soviet Geography: Review and Translation*, **1**, No. 10 (December 1960) p. 89.
4 A. Karyakin and A. Omelin, 'Profiteer with a gun', *Izvestiya* (February 27, 1963) p. 4: as translated in *Current Digest of the Soviet Press*, **15**, No. 9 (March 27, 1963) pp. 26–7. Further complaints about poaching and official disregard for wildlife conservation appeared in *Pravda* (March 16, 1968) p. 3, *Pravda* (August 14, 1969) p. 3, *Izvestiya* (June 25, 1970) p. 6, and *Pravda* (October 12, 1970) p. 4.
5 The elk of northern Europe and Siberia (*Alces alces*) bears a physical resemblance much closer to the animal Americans know as a moose than to the American elk. However, the Russian term for this animal (*los'*) is always translated as 'elk'. Pre-revolutionary efforts to preserve this animal are noted in Chapter 2.
6 As reported in M. A. Zablotskiy, 'Vosstanovleniye zubra v S.S.S.R. i za granitsey', *Okhrana prirody i zapovednoye delo v S.S.S.R.*, Byulleten' No. 4 (1962); p. 46 in translated edition.
7 Data from K. N. Blagosklonov, A. A. Inozemtsev and V. N. Tikhomirov- *Okhrana prirody* (Moskva: Izdatel'stvo 'Vysshaya shkola', 1967) p. 285.
8 Yu. A. Isakov, S. V. Kirikov and A. N. Formozov, 'Nazemnyye okhotnich'ye-promyslovyye zhivotnyye', *Prirodnyye resursy Sovetskogo Soyuza, ikh ispol'zovaniye i vosproizvodstvo*, ed. I. P. Gerasimov, D. L. Armand and

K. M. Efron (Moskva: Izdatel'stvo Akademii nauk S.S.S.R., 1963) pp. 197–202.

9 *Ibid.* p. 205.

10 *Ibid.* pp. 207–8. In an effort to preserve the decreasing numbers of migratory birds using the northern Siberia to Africa and South Asia flyway, a ban on the hunting of all wildfowl in the Kazakh Republic was declared for the spring 1968 season (as reported in *I.U.C.N. Bulletin*, **2**, No. 6, p. 44).

11 S. V. Kirikov, 'Zapovedniki', *Sovetskaya geografiya: itogi i zadachi* (Moskva: Gosizdat geograficheskoy literatury, 1960) p. 551.

12 Chapter 4 relates the efforts to eliminate the wolf in one of its normal environments (the Kavkaz Preserve), as described in V. A. Kotov, 'Bor'ba s volkami v Kavkazskom zapovednike', *Trudy Kavkazskogo gosudarstvennogo zapovednika*, Vypusk VIII (1965) pp. 182–4. A more sympathetic discussion of the man–wolf relationship, based on Douglas Pimlott's book *The World of the Wolf*, appeared in the Soviet natural history magazine *Priroda* in the September 1970 issue, and the January 1971 issue of *Soviet Life* contained an article stating that the current management goal was to contain the wolf, not to exterminate it.

13 *Komsomol'skaya pravda* (November 3, 1966) p. 3.

14 V. Peskov, 'Sluchay v stepi', *Komsomol'skaya pravda* (April 26, 1970) p. 4.

15 D. Bilenkin and V. Peskov, 'Priroda–khimiya–chelovek', *Komsomol'skaya pravda* (March 16, 1967) p. 2. This discussion is briefly summarized in *Current Digest of the Soviet Press*, **19**, No. 18 (May 24, 1967) p. 31.

16 The need for increased use of biological controls was discussed in articles in *Pravda* (August 5, 1970) p. 3, and *Izvestiya* (August 8, 1970) p. 2.

17 L. K. Shaposhnikov (ed.), *Yadokhimikaty i fauna* (Moskva: Izdatel'stvo 'Nauka', 1967).

18 For background information regarding the internal fish resources of the Soviet Union, the standard source is L. S. Berg's *Ryby presnykh vod S.S.S.R. i sopredel'nykh stran*, all three volumes of which have been translated into English under the title *Freshwater Fishes of the U.S.S.R. and Adjacent Countries* by the Israel Program for Scientific Translations (I.P.S.T.). A more concise description of both the internal and open-sea fishing activities of the Soviet Union is presented in L. G. Vinogradov *et al.*, 'Resursy rybnogo khozyaystva' ('Fisheries Resources'), *Prirodnyye resursy Sovetskogo Soyuza, ikh ispol'zovaniye i vosproizvodstvo*, ed. I. P. Gerasimov, D. L. Armand and K. M. Efron (Moskva: Izdatel'stvo Akademii nauk S.S.S.R., 1963) pp. 217–42. A translation of this work into English is now available. Also available in translation by the I.P.S.T. is V. Shparlinskiy, *The Fishing Industry of the U.S.S.R.* (Moscow, 1959). Current activities and problems of Soviet fisheries are chronicled in the monthly periodical *Rybnoye khozyaystvo*.

19 L. G. Vinogradov *et al.* (1963) p. 229.

20 See, for example, the articles by Yu. Smirnov, V. Ananyan, and B. Moskalenko, respectively, as listed in the bibliography.

21 A. I. Isayev, 'Vosproizvodstvo rybnykh zapasov – vazhnaya narodnokhozyaystvennaya zadacha', *Rybnoye khozyaystvo* (1961) No. 4, p. 3.

22 *Literaturnaya gazeta* (March 1, 1966) p. 2.

23 From A. B. Avakyan, 'O kompleksnom ispol'zovanii vodokhranilishch: gidrostroitel'stvo i problemy rybnogo khozyaystva S.S.S.R.', *Izvestiya Akademii nauk S.S.S.R., seriya geograficheskaya* (1966), No. 2, p. 33.

24 G. Tolmachev, 'Brakon′yerskaya arifmetika', *Pravda* (June 10, 1962) p. 2.

25 M. Malyarov, 'Lyudi, priroda, zakon', *Pravda* (March 16, 1968) p. 3. Other articles on the fish poaching problem appeared in *Pravda* on January 8, January 24, February 25, and April 26, 1969; and *Izvestiya* (August 22, 1971).

26 For example, see *Izvestiya* (July 24, 1970) p. 6.

27 *Resolution of the Presidium of the R.S.F.S.R. Supreme Soviet*: 'On Progress of Fulfillment of the October 27, 1960 Russian Republic Law "On Conservation in the Russian Republic" ', as translated in *Current Digest of the Soviet Press*, **17**, No. 46 (December 8, 1965) p. 3.

28 Reported in *Pravda* (June 26, 1967) p. 3. Lysva is located in the western foothills of the Urals, east of Perm′.

29 Ye. Pavlovskiy and G. Nikol′skiy, 'Dary morey, ozer i rek', *Pravda* (July 12, 1961) p. 6, as translated in *Current Digest of the Soviet Press*, **13**, No. 28 (August 9, 1961) pp. 32–3.

30 I. Biryukov, 'Ryba u shlagbauma', *Izvestiya* (August 11, 1968) p. 2.

31 N. I. Kozhin, 'Osetrovyye S.S.S.R. i vosproizvodstvo ikh zapasov', *Rybnoye khozyaystvo* (1963) No. 9, p. 13. A more detailed account of this problem is presented in L. S. Berdichevskiy, 'Ratsional′noye ispol′zovaniye rybnykh resursov Kaspiyskogo basseyna', *Izvestiya Akademii nauk S.S.S.R., seriya geograficheskaya* (1961) No. 3, pp. 28–36.

32 *Sovetskaya Rossiya* (July 14, 1970) p. 4.

33 I. Dudenkov, 'Nazrevshiye problemy Volgo–Kaspiya', *Pravda* (May 27, 1962) p. 2, as translated in *Current Digest of the Soviet Press*, **14**, No. 21 (June 20, 1962) p. 25.

34 *Literaturnaya gazeta* (June 17, 1970) p. 11.

35 Ivan Tsatsulin, 'Poka ne pozdno', *Literaturnaya gazeta* (September 6, 1966) p. 2.

36 The results of the work with these and other species are concisely summarized in L. V. Shaposhnikov, 'Acclimatization of fur-bearing animals in connection with problems of preservation and enrichment of the U.S.S.R.'s fauna', as translated in *Conservation of Natural Resources and the Establishment of Reserves in the U.S.S.R.*, Bulletin No. 4 (Jerusalem: Israel Program for Scientific Translations, 1962) pp. 31–42. A more detailed account is A. I. Yanushevich (ed.), *Acclimatization of Animals in the U.S.S.R.*, as translated by the I.P.S.T. (Jerusalem, 1966).

37 Reported in Yu. A. Isakov *et al.* (1963) p. 197.

38 V. G. Geptner, 'Otvet opponentam', *Okhota i okhotnich′ye khozyaystvo* (1964) No. 6, pp. 17–19.

39 K. N. Blagosklonov *et al.* (1967) p. 245.

40 A recent (although apparently incomplete) résumé of Soviet conservation agreements with other nations is V. A. Chichvarin, *Mezhdunarodnyye soglasheniya po okhrane prirody* (Moskva: Izdatel′stvo 'Yuridicheskaya literatura', 1966), which has a summary in English at the end. For those dealing specifically with fish and other marine life, see O. A. Mathisen and D. E. Bevan, *Some International Aspects of Soviet Fisheries* (Columbus: Ohio State University Press, 1968) pp. 28–53.

41 I.U.C.N. Bulletin, N.S. **2**, No. 15 (April–June 1970), p. 129.

CHAPTER 6. TIMBER AND MINERAL RESOURCES

1 According to the 1961 inventory of forest lands. Other inventories were taken

in 1956 and 1927. These inventories are described in K. V. Algvere, *Forest Economy in the U.S.S.R.* (Stockholm: Royal College of Forestry, 1966) pp. 62–74.

2 Background information on the forests of the U.S.S.R. is available in numerous Soviet monographs among which the following might be noted, as they are available in English: V. P. Tseplyayev, *Lesa S.S.S.R.* (Moskva: Sel′khozgiz, 1961) translated as *The Forests of the U.S.S.R.*, by the Israel Program for Scientific Translations (1965); and Algvere, (1966). Current forestry and timber industry activities are chronicled in the monthly journal, *Lesnoye khozyaystvo*, published by the State Forestry Committee. Most issues have a special section devoted to articles on forest conservation problems.

3 G. Taraskevich, 'Razumno ispol′zovat′ lesnyye bogatstva', *Kommunist*, No. 16 (November 1963) p. 86.

4 P. V. Vasil′yev 'Lesnyye resursy i lesnoye khozyaystvo', *Prirodnyye resursy Sovetskogo Soyuza, ikh ispol′zovaniye i vosproizvodstvo*, ed. I. P. Gerasimov, D. L. Armand and K. M. Efron (Moskva: Izdatel′stvo Akademii nauk S.S.S.R., 1963) p. 154. In translated edition, p. 211.

5 P. V. Vasil′yev, 'Questions of the geographic study and economic use of forests' (translation of a paper presented at the Symposium on Natural Resources of the Third Congress of the Geographical Society of the U.S.S.R.), *Soviet Geography: Review and Translation*, **1**, No. 10 (December, 1960) p. 62.

6 K. V. Algvere (1966) p. 67.

7 *Ibid.* pp. 227–8.

8 P. V. Vasil′yev (1963) p. 154.

9 V. V. Pokshishevskiy, 'Na poroge tret′yego tysyacheletiya', *Nauka i zhizn′* (1968) No. 2, pp. 71–2.

10 P. V. Vasil′yev (1963) p. 144 (Table 7).

11 P. V. Vasil′yev (1960) p. 60.

12 *Soviet Life* (August 1966) p. 34. Such overcutting is expressly prohibited in the 1960 R.S.F.S.R. conservation law (see Appendix 18, Article 5).

13 I. Vinogradov and B. Olkhovskaya, 'Lesu nuzhen odin khozyain', *Izvestiya* (September 17, 1958) p. 3.

14 S. Khlatin, 'Lesovod i drovosek: kommentariy k konfliktu', *Komsomol′skaya pravda* (June 6, 1968) p. 1; as translated in *Current Digest of the Soviet Press*, **20**, No. 23 (June 26, 1968) p. 14.

15 K. N. Blagosklonov, A. A. Inozemtsev and V. N. Tikhomirov *Okhrana prirody* (Moskva: Izdatel′stvo 'Vysshaya shkola', 1967) p. 184.

16 D. L. Mozeson, 'Okhrana prirody severa', *Problemy severa*, No. 2 (1958) p. 225.

17 P. V. Vasil′yev (1963) p. 140 (Table 4).

18 Among such articles are I. Vinogradov and B. Olkhovskaya (1958); G. Taraskevich (1963) and V. Nesterov, 'Lesam nuzhen zabotlivyy khozyain', *Pravda* (January 19, 1966) p. 3.

19 Many articles and monographs exist on the mineral resources of the U.S.S.R. An earlier study is D. B. Shimkin, *Minerals – a Key to Soviet Power* (Harvard University Press, 1953), but it is now considerably out of date. Equally good but also out of date is the *Oxford Regional Economic Atlas of the U.S.S.R. and Eastern Europe* (Oxford University Press, 1956). A somewhat more recent study of the Soviet fuel base is J. A. Hodgkins, *Soviet Power: Energy*

Resources, Production and Potentials (Englewood Cliffs, New Jersey: Prentice-Hall, Inc., 1961). Other monographs have appeared in recent years on various aspects of the Soviet iron and steel, oil and gas, aluminum, and chemical industries. A comprehensive Soviet economic geography text which has been translated into English is A. Lavrishchev, *Economic Geography of the U.S.S.R.* (Moscow, 1969). Another recent work is Shabad, *Basic Industrial Resources of the U.S.S.R.* (New York: Columbia University Press, 1969).

20 For example, the lead article in the July, 1964 issue of the popular science magazine *Priroda* was entitled 'Neissyakayemyye bogatstva zemnykh nedr' ('Unlimited mineral wealth').

21 M. Odintsov and V. Rya'benko, 'Berech' sokrovishcha', *Izvestiya* (January 27, 1968) p. 2; as translated in *Current Digest of the Soviet Press*, **20**, No. 4 (February 14, 1968) pp. 24–5. Other quotations on this subject are given in Chapters 1 and 9.

22 N. Mel'nikov, 'Prezhde, chem reki povernut' vspyat'...', *Literaturnaya gazeta* (July 12, 1967) p. 11, and *Pravda* (January 31, 1970) p. 2.

23 *Komsomol'skaya pravda* (September 24, 1970).

24 Other aspects of conserving Soviet oil and gas resources, as seen from the economist's point of view, are presented in Robert W. Campbell, *The Economics of Soviet Oil and Gas* (Baltimore: Johns Hopkins Press, 1968) pp. 42–53. This discussion centers around questions of the rates, objectives and criteria of capital investment and recovery coefficients for this industry in the U.S.S.R.

25 V. Davydchenkov 'Shumyat, bushuyut pozhary pod zemley', *Izvestiya* (June 25, 1960) p. 3.

26 N. Mel'nikov, M. Agoshkov, and B. Laskorin, 'Zapasam nedr nuzhen schet', *Pravda* (January 27, 1970) p. 2.

27 F. Shutliv, 'Poleznyye iskopayemyye-nashe bogatstvo', *Partiynaya zhizn'* (1965) No. 11, p. 19.

28 K. N. Blagosklonov *et al.* (1967) p. 78.

29 V. Chivilikhin, 'Kak vam dyshitsya, gorozhane?' *Literaturnaya gazeta* (August 9, 1967) p. 10. Other examples of such atmospheric losses are given in Chapter 8.

30 F. Shutliv (1965) p. 18. Exactly what is meant by the term 'lost' (teryayetsya) was not specified. See also Mel'nikov *et al.* (1970) p. 2.

31 Odintsov and Rya'benko (1968) p. 2.

32 These problems are discussed, for example, in A. Chemonin, 'Vtoroye otkrytiye tret'yego polyusa', *Izvestiya* (September 14, 1968) p. 2.

33 N. Mel'nikov (1967) p. 11.

34 L. V. Motorina, 'Rekul'tivatsiya zemel', narushennykh promyshlennost'yu', *Izvestiya Akademii nauk S.S.S.R., seriya geograficheskaya* (1966) No. 5, p. 40.

35 *Zarya vostoka* (April 7, 1970) p. 1, as translated in *Current Digest of the Soviet Press*, **22**, No. 14 (May 5, 1970) p. 6.

36 See, for example, N. Mel'nikov *et al.* (1970) p. 2.

CHAPTER 7. WATER RESOURCE UTILIZATION AND CONSERVATION

1 G. L. Magakyan, 'The Mingechaur multi-purpose water-management project', as translated in *Soviet Geography: Review and Translation*, **2**, No. 10 (December 1961) pp. 43–50.

2 For example, in N. T. Kuznetsov and M. I. L′vovich, 'Problemy kompleksnogo ispol′zovaniya i okhrany vodnykh resursov', *Prirodnyye resursy Sovetskogo Soyuza, ikh ispol′zovaniye i vosproizvodstvo*, ed. I. P. Gerasimov, D. L. Armand and K. M. Efron (Moskva: Izdatel′stvo Akademii nauk S.S.S.R., 1963) p. 26. In translated edition, p. 30.

3 Magakyan (1961), 44–5.

4 I. P. Gerasimov, D. L. Armand and V. S. Preobrazhenskiy, 'Natural resources of the Soviet Union, their study and utilization', translation of a paper presented at the Fourth Congress of the Geographical Society of the U.S.S.R., *Soviet Geography: Review and Translation*, **5**, No. 8 (October 1964) p. 11.

5 See, for example, O. Kolbasov, 'Nuzhen zakon, okhranyayushchiy vodu', *Izvestiya* (February 28, 1964) p. 3, and M. I. L′vovich, 'O nauchnykh osnovakh kompleksnogo ispol′zovaniya i okhrany vodnykh resursov', *Vodnyye resursy i ikh kompleksnoye ispol′zovaniye* 'Voprosy geografii', No. 73 (Moskva: Izdatel′stvo 'Mysl', 1968) p. 6.

6 K. Radchenko and F. Bokhin, 'Voda. Skol′ko ona stoit?', *Izvestiya* (June 21, 1967) p. 4. 0.7 kopecks per cubic meter equals 8.63 rubles per acre-foot, or about $9.59/a.f. at the official exchange rate. The case for pricing water is also stated in a follow-up article in *Izvestiya* (July 21, 1967) p. 3, as well as in *Literaturnaya gazeta* (June 17, 1970) p. 11.

7 The standard work on the rivers of the Soviet Union is L. K. Davydov, *Gidrografiya S.S.S.R.*, in two volumes (Leningrad: Ugletekhizdat, 1955).

8 L′vovich (1968) p. 11.

9 M. Odintsov *et al.*, 'Nuzhno li zatoplyat′ Ilimskuyu dolinu?', *Izvestiya* (February 17, 1960) p. 4.

10 M. N. Khromov, 'Izmeneniya v geografii naselennykh punktov v svyazi s sozdaniyem Tsimlyanskogo vodokhranilishcha', *Izvestiya Vsesoyuznogo Geograficheskogo Obshchestva*, 1961, No. 1, pp. 79–81.

11 A. B. Avakyan and V. A. Sharapov, *Vodokhranilishcha gidroelektrostantsiy S.S.S.R.* (Moskva: Gosenergoizdat, 1962) p. 80.

12 M. I. L′vovich, 'O kompleksnom ispol′zovanii i okhrane vodnykh resursov', *Izvestiya Akademii nauk S.S.S.R., seriya geograficheskaya* (1961) No. 2, p. 38.

13 S. L. Vendrov *et al.*, 'The problem of transformation and utilization of the water resources of the Volga River and the Caspian Sea', *Soviet Geography: Review and Translation*, **5**, No. 7 (September 1964) pp. 27–8.

14 S. Khlatin, 'Lesovod i drovosek: kommentariy k konfliktu', *Komsomol′skaya pravda* (June 6, 1968) p. 1.

15 N. Urazbayev, 'Poyma dolzhna ostat′sya' *Komsomol′skaya pravda* (June 3, 1967) p. 1; and P. Loyko, 'Polyu-ekonomicheskuyu zashchitu', *Izvestiya* (September 29, 1968) p. 2.

16 Ivan Tsatsulin, 'Poka ne pozdno', *Literaturnaya gazeta* (September 6, 1966) p. 2.

17 S. V. Klopov, 'O vliyanii proyektiruyemoy Nizhne-Obskoy GES na prirodu i narodnokhozyaystvennyy kompleks Zapadno-Sibirskoy nizmennosti', *Problemy severa*, No. 9 (1965) p. 263. This issue of *Problemy severa* also contains seven other articles on the lower Ob′ problem, all of which have been translated in the journal *Problems of the North*, No. 9 (1965).

18 These conclusions are explicitly brought out in L′vovich (1968).

19 See Grin (ed.), *Vodnyye resursy i ikh kompleksnoye ispol'zovaniye* ('Voprosy geografii', No. 73; Moskva: Izdatel'stvo 'Mysl', 1968) p. 159.
20 S. L. Vendrov and G. P. Kalinin, 'Surface water resources of the U.S.S.R., their utilization and study', *Soviet Geography: Review and Translation*, **1**, No. 6 (June 1960) p. 43.
21 V. V. Pokshishevskiy, 'Na poroge tret'yego tysyacheletiya', *Nauka i zhizn'* (1968) No. 2, p. 72.
22 Information on specific irrigation projects in the Soviet Union may be found, among many sources, in R. A. Lewis, 'The irrigation potential of Soviet Central Asia', *Annals of the Association of American Geographers*, **52**, No. 1 (March 1962) pp. 99–114. Later data is presented in L. V. Dunin-Barkovskiy, 'The water problem in the deserts of the U.S.S.R.', *Soviet Geography: Review and Translation*, **9**, No. 6 (June 1968) pp. 458–68.
23 Kuznetsov and L'vovich (1963) p. 25.
24 R. A. Lewis (1962) p. 109.
25 I. P. Gerasimov, 'Basic problems of the transformation of nature in Central Asia', *Soviet Geography: Review and Translation*, **9**, No. 6 (June 1968) p. 457.
26 N. H. Greenwood, 'Developments in the irrigation resources of the Sevan-Razdan cascade of Soviet Armenia', *Annals of the Association of American Geographers*, **55**, No. 2 (June 1965) p. 299.
27 I. Valeshniy and V. Yakhnevich, 'Voda idet...mimo poley', *Pravda* (November 3, 1965) p. 2.
28 I. P. Gerasimov, 'Umen'shit' i svesti k minimumu zavisimost' nashego sel'skogo khozyaystva ot prirodnoy stikhii', *Izvestiya Akademii nauk S.S.S.R., seriya geograficheskaya* (1962) No. 5, p. 45.
29 I. P. Gerasimov (1967) p. 448. See also I. M. Matley, 'The Golodnaya Steppe: a Russian irrigation venture in Central Asia', *The Geographical Review*, **60**, No. 3 (July 1970) pp. 344–5.
30 *Pravda* (January 16, 1971) p. 2, as translated in *Current Digest of the Soviet Press*, **23**, No. 3 (February 16, 1971) pp. 22 and 40.
31 I. P. Gerasimov (1967) p. 448.
32 S. N. Bobrov, 'Preobrazovaniye Kaspiya', *Geografiya v shkole* (1961) No. 2, p. 10.
33 S. L. Vendrov *et al.* (1964) pp. 24–5.
34 S. N. Bobrov (1961) p. 8.
35 *Ibid.* A more extensive discussion of the Caspian fisheries problem, based on data available in the mid-1950s, is presented in N. C. Field, *The Role of Irrigation in the South European U.S.S.R. in Soviet Agricultural Growth: an Appraisal of the Resource Base and Development Problem* (unpublished Ph.D. dissertation; University of Washington, 1956) pp. 137–53.
36 A more complex Manych depression plan, which would utilize fresh water from the Don and Dnepr rivers, is presented in I. I. Stas', 'Kak spasat' Kaspiy', *Priroda* (1968) No. 12, pp. 76–9.
37 S. L. Vendrov, 'Geograficheskiye aspekty...', p. 43; as translated in *Soviet Geography: Review and Translation*, **4**, No. 6 (June 1963) pp. 38–9.
38 The Apollov plan is presented in some detail in Bobrov (1961).
39 I. P. Gerasimov (1967) p. 454.
40 I. M. Chernenko, 'The Aral Sea problem and its solution', *Problemy osvoyeniya pustyn'* (1968) No. 1, pp. 31–4.

41 A discussion of these points took place in the newspaper *Komsomol'skaya pravda*; see the bibliographic entries under Bezrukov and Dadabayev (1965) and Kunin (1966).
42 L. B. Bernshteyn, 'Ob ispol'zovanii energii prilov', *Izvestiya Vsesoyuznogo Geograficheskogo Obshchestva* (1962) No. 5, pp. 405–13.
43 Rostarchuk (1970) p. 4.
44 'O kontinental'nom shel'fe Soyuza S.S.R.', *Vedomosti Verkhovnogo Soveta S.S.S.R.* (1968) No. 6 (February 7, 1968), item 40, pp. 55–6; available in translation in *Current Digest of the Soviet Press*, **20**, No. 8 (March 13, 1968) pp. 22–3. This decree was supplemented in October of 1968 by a tripartite agreement with Poland and East Germany on the development of the continental shelf of the Baltic Sea.

CHAPTER 8. ENVIRONMENTAL QUALITY

1 Soviet journals which chronicle research done in the U.S.S.R. on pollution control include *Gigiyena i sanitariya* (*Hygiene and Sanitation*), *Vodosnabzheniye i sanitarnaya tekhnika* (*Water Supply and Sanitation Technology*), and others in allied fields which are concerned with pollution questions. Recent volumes of *Gigiyena i sanitariya* are available in translation done by the Israel Program for Scientific Translations.
2 N. Litvinov, 'Water Pollution in the U.S.S.R. and in Other Eastern European Countries', Conference on Water Pollution Problems in Europe: Documents, Vol. 1 (Geneva: United Nations, 1961) p. 26.
3 K. N. Blagosklonov, A. A. Inozemtsev, and V. N. Tikhomirov (eds.), *Okhrana prirody* (Moskva: Izdatel'stvo 'Vysshaya shkola', 1967) p. 124.
4 For complete text, see Appendix 22.
5 N. T. Kuznetsov and M. I. L'vovich, 'Problemy kompleksnogo ispol'zovaniya i okhrany vodnykh resursov', *Prirodnyye resursy Sovetskogo Soyuza, ikh ispol'zovaniye i vosproizvodstvo*, ed. I. P. Gerasimov, D. L. Armand, and K. M. Efron (Moskva: Izdatel'stvo Akademii nauk S.S.S.R., 1963) p. 19.
6 K. N. Blagosklonov *et al.* (1967) p. 108.
7 P. A. Spyshnov, 'Water supply and sewerage development in U.S.S.R. cities', *Vodosnabzheniye i sanitarnaya tekhnika* (1960) No. 6, pp. 1–4.
8 L. S. Gurvich and I. A. Kibal'chich, 'Materialy po izucheniyu sanitarnogo sostoyaniya vodoyemov R.S.F.S.R.', *Gigiyena i sanitariya* (1961) No. 3, pp. 15–21.
9 I. Demin and D. Bilenkin, 'Reka zovet na vyruchku', *Komsomol'skaya pravda* (April 27, 1960) p. 2.
10 V. Rostovshchikov, 'Chego boyatsya kapitany', *Izvestiya* (July 27, 1968) p. 2.
11 N. Mironov, 'Kamen' na Kame', *Pravda* (June 26, 1967) p. 3.
12 V. Zvonkov, 'Pust' prozvuchit golos obshchestvennosti', *Komsomol'skaya pravda* (April 27, 1960) p. 2.
13 *Izvestiya* (September 25, 1970) p. 5.
14 Zvonkov, *loc. cit.*
15 'O merakh po predotvrashcheniyu zagryazneniya Kaspiyskogo morya', *Pravda* (October 3, 1968) p. 2.
16 K. Kir'yanov, 'V zashchitu vody', *Pravda* (August 23, 1970) p. 3.
17 *Pravda* (January 2, 1971) p. 3.

18 *Pravda* (March 20, 1971) p. 2.
19 *Priroda* (1970) No. 9, p. 99.
20 Ya. Grushko, 'Normal′na li eta norma?', *Izvestiya* (July 6, 1960) p. 5.
21 N. T. Kuznetsov and M. I. L′vovich (1963) pp. 29–30, and K. N. Blagosklonov *et al.* (1967) p. 113. These norms, as they existed following a regulatory decree of the Ministry of Public Health put forth on July 15, 1961, are presented in translation in B. S. Levine, *U.S.S.R. Literature on Water Supply and Pollution Control*, Vol. 4 (Washington: U.S. Public Health Service, 1962) pp. 1–29.
22 I. Demin and D. Bilenkin (1960) as translated in *Current Digest of the Soviet Press*, **12**, No. 17 (May 25, 1960) p. 21.
23 N. Mel′nikov, 'Prezhde, chem reki povernut′ vspyat′...', *Literaturnaya gazeta* (July 12, 1967) p. 11.
24 Yu. Danilov, 'Zashchitim ot zagryazneniy vodu, vozdukh, pochvu', *Pravda* (June 21, 1965) p. 2.
25 *Twenty-third Congress of the Communist Party of the Soviet Union* (C.P.S.U.) (Moscow: Novosti Press Agency, 1966) pp. 202–3.
26 Oleg Volkov, 'Poyezdka na Baykal', *Literaturnaya gazeta* (January 29, 1966) p. 2.
27 *Ibid.* (as translated in *Current Digest of the Soviet Press*, **18**, No. 5 (February 23, 1966) p. 15).
28 *Literaturnaya gazeta* (February 6, 1965; March 18, 1965; April 10, 13, and 15, 1965; January 29, 1966; and June 2, 1966). See also the reference cited in the bibliography by A. Merkulov. Follow-up articles appeared in *Literaturnaya gazeta* on October 11, 1967; November 15, 1967; and January 17, 1968.
29 This letter, together with another long article by Volkov, received widespread foreign publicity as well in the English language periodical *Soviet Life* (August 1966) pp. 5–8.
30 Oleg Volkov, 'Uroki Baykala', *Literaturnaya gazeta* (October 11, 1967) p. 12.
31 A. Ishkov, 'Okonchatel′nyye vyvody delat′ rano', *Literaturnaya gazeta* (January 17, 1968) p. 10.
32 G. I. Galaziy, *Baykal i problema chistoy vody v Sibiri* (Irkutsk: Sibirskoye otdeleniye Akademii nauk S.S.S.R., 1968) p. 51.
33 'Zabota o Baykale', *Izvestiya* (February 8, 1969) p. 2.
34 These clearance zones, apparently first worked out in 1948, are presented for 272 different types of industrial concerns in B. S. Levine (1962) Vol. 4, pp. 165–76. See also *Profile Study . . .* (1970) pp. 155ff.
35 M. Ye. Lyakhov, 'Okhrana atmosfery ot zagryazneniya', *Prirodnyye resursy Sovetskogo Soyuza, ikh ispol′zovaniye i vosproizvodstvo*, ed. I. P. Gerasimov, D. L. Armand and K. M. Efron (Moskva: Izdatel′stvo Akademii nauk S.S.S.R., 1963) pp. 64–5. In translated edition, p. 85.
36 Yu. D. Lebedev, 'Sovremennoye sostoyaniye i ocherednyye zadachi v oblasti sanitarnoy okhrany atmosfernogo vozdukha naselennykh mest v S.S.S.R.', *Gigiyena i sanitariya* (1960) No. 1, p. 6.
37 Lyakhov (1963) p. 66.
38 V. Chivilikhin, 'Kak vam dyshitsya, gorozhane?' *Literaturnaya gazeta* (August 9, 1967) p. 10.
39 R. W. Campbell, *The Economics of Soviet Oil and Gas* (Baltimore: Johns Hopkins Press, 1968) p. 126.

40 I. L. Varshavskiy, 'Avtomobil' i gorod', *Izvestiya* (February 15, 1967) p. 3.
41 *Izvestiya* (May 1, 1970) p. 3 and *Nedelya* (1970) No. 9, p. 1.
42 A discussion of electric cars appeared in *Pravda* (January 16, 1971) p. 3, and was translated in *Current Digest of the Soviet Press*, **23**, No. 3 (February 16, 1971) pp. 16–17.
43 Danilov (1965) p. 2. See also Chivilikhin (1967) p. 10.
44 *Vedemosti Verkhovnogo Soveta R.S.F.S.R.* (November 4, 1965) p. 924.
45 N. Gulin and A. Belevitskiy, 'Industriya chistogo vozdukha', *Pravda* (March 24, 1969) p. 3.
46 Sinev (1971) p. 7.
47 *Atomic Energy in the Soviet Union* (1963) p. 79.
48 'Rekam byt' chistymi', *Izvestiya* (November 14, 1968) p. 1.
49 A map showing the expansion of Moscow and the area of its outlying green belt is presented in *Soviet Geography: Review and Translation*, **1**, No. 10 (December 1960) pp. 90–3.
50 This ensued from a decree of the R.S.F.S.R. Council of Ministers on May 24, 1966, entitled 'On the Conditions of and Measures for Improving the Safeguarding of Historical and Cultural Monuments'; see V. I. Kochemasov, 'God poiskov, god nadezhd', *Istoriya S.S.S.R.* (1967) No. 5, pp. 197–203.
51 *Pravda* (December 20, 1969) as translated in *Current Digest of the Soviet Press*, **22**, No. 1 (February 3, 1970) pp. 7–13.
52 See, for example, Bordukov (1968).

CHAPTER 9. ATTITUDES, PROBLEMS AND TRENDS

1 P. Oldak, 'Priroda vzyvayet k shchedrosti', *Literaturnaya gazeta* (June 3, 1970) p. 11, as translated in *Current Abstracts of the Soviet Press*, **2**, No. 6 (June 1970) p. 1.
2 Speech by N. S. Khrushchev in Moscow to departing settlers of the Virgin Lands, *Pravda* (January 8, 1955) p. 2, as translated in *Current Digest of the Soviet Press*, **7**, No. 1 (February 16, 1955) p. 12.
3 *Izvestiya* (September 14, 1963) p. 2.
4 *Izvestiya* (January 3, 1964) p. 3.
5 Boris Urlanis, 'Sushchestvuyut li problemy narodonaseleniya?' *Literaturnaya gazeta* (November 23, 1965) p. 4.
6 The 'hard line' was put forth in articles by (among others) V. Cheprakov, P. Pod'yachikh, and S. Strumilin in *Literaturnaya gazeta* (March 13, 1965, February 22, 1966, and May 28, 1966, respectively, all p. 4).
7 V. A. Kovda, 'The problem of biological and economic productivity of the Earth's land areas' (paper given at a Leningrad conference, 1966) as translated in *Soviet Geography: Review and Translation*, **12**, No. 1 (January 1971) p. 20.
8 G. Gerasimov, 'Bez predubezhdeniy', *Literaturnaya gazeta* (March 3, 1966) p. 4.
9 M. Kolganov, 'Ugrozhayet li miru perenaseleniye?', *Literaturnaya gazeta* (December 25, 1965) p. 4.
10 Robert C. Cook, 'Soviet population theory from Marx to Kosygin', *Population Bulletin*, **23**, No. 4 (October 1967) p. 86.
11 *Literaturnaya gazeta* (December 11, 1968) p. 10.
12 See, for example, I. Adabashev, *Global Engineering* (Moscow: Progress Publishers, 1966), or N. Rusin and L. Flit, *Man Versus Climate* (Moscow:

Peace Publishers; undated, *ca.* 1967), both put out in English to publicize such schemes abroad. An article of similar nature from the magazine *Yunost'* was translated in *Current Abstracts of the Soviet Press*, **2**, No. 3 (March 1970) pp. 27–8.

13 F. Fraser Darling and John P. Milton (eds.), *Future Environments of North America* (New York: Natural History Press, 1966) as reviewed by I. P. Gerasimov and I. V. Komar in *Izvestiya Akademii nauk S.S.S.R., seriya geograficheskaya* (1968) No. 3, pp. 141–6. The quotation on p. 144 in the original is taken from the translation of the review in *Soviet Geography: Review and Translation*, **9**, No. 9 (November 1968) p. 801.

14 I. P. Gerasimov, 'Konstruktivnaya geografiya; tseli, metody, rezul'taty', *Izvestiya Vsesoyuznogo Geograficheskogo Obshchestva* (1966) No. 5, p. 392.

15 *Ibid.* p. 393.

16 Reference is made to the discussion of these efforts in Chapter 2, and to the corresponding citations from K. N. Blagosklonov, A. A. Inozemtsev and V. N. Tikhomirov, *Okhrana prirody* (Moskva: Izdatel'stvo 'Vysshaya shkola', 1967) pp. 426–36, and from the article by L. K. Shaposhnikov (1965).

17 I. P. Gerasimov, 'Nuzhen general'nyy plan preobrazovaniya prirody nashey strany', *Kommunist* (1969) No. 2, p. 75. See also the quotation cited in Chapter 6 by Odintsov and Rya'benko.

18 'Rekam byt' chistymi' (editorial), *Izvestiya* (November 14, 1968) p. 1, as translated in *Current Digest of the Soviet Press*, **20**, No. 46 (December 4, 1968) p. 19.

Bibliography

CHAPTER 1. INTRODUCTION: THE SOVIET CONCEPT OF CONSERVATION

Bannikov, A. G. *Po zapovednikam Sovetskogo Soyuza* (*In the Nature Preserves of the Soviet Union*). Moskva: Izdatel'stvo 'Mysl'', 1966, pp. 5–7.

Blagosklonov, K. N., Inozemtsev, A. A. and Tikhomirov, V. N. *Okhrana prirody* (*Conservation of Nature*). Moskva: Izdatel'stvo 'Vysshaya shkola', 1967, Introduction.

Gerasimov, I. P. 'Nuzhen general'nyy plan preobrazovaniya prirody nashey strany' ('We need a general plan for the transformation of nature in our country'), *Kommunist* (1969) No. 2, pp. 68–79. Translated in *Current Digest of the Soviet Press*, **21**, No. 8 (March 12, 1969) pp. 3–6 and 28.

Gerasimov, I. P., Armand, D. L. and Efron, K. M. (eds.) *Prirodnyye resursy Sovetskogo Soyuza, ikh ispol'zovaniye i vosproizvodstvo* (*Natural Resources of the Soviet Union, their Use and Renewal*). Moskva: Akademii nauk S.S.S.R., 1963. Translated edition by Freeman and Co., 1971.

Motovilov, G. P. 'Protection of Nature in the U.S.S.R.', *Proceedings, Fifth World Forestry Congress* (Vol. 3). Seattle: University of Washington, 1960, pp. 1850–4.

Oldak, P. 'Priroda vzyvayet k shchedrosti' ('Nature appeals to our generosity'), *Literaturnaya gazeta* (June 3, 1970) p. 11. Translated in *Current Abstracts of the Soviet Press*, **2**, No. 6 (June 1970) p. 1.

Zhbanov, Ye. 'Zhivaya voda' ('Living water'), *Izvestiya* (June 27, 1970) p. 4. Translated in *Current Digest of the Soviet Press*, **22**, No. 26 (July 28, 1970) pp. 39–40.

CHAPTER 2. HISTORICAL AND INSTITUTIONAL FRAMEWORK

Abramov, L. S., Armand D. L., Nasimovich, A. A. and Rakhilin, V. K. 'V. I. Lenin i okhrana prirody nashey strany' ('V. I. Lenin and the conservation of nature in our country'), *Izvestiya Akademii nauk S.S.S.R., seriya geograficheskaya* (1970) No. 2, pp. 62–74.

Algvere, Karl V. *Forest Economy in the U.S.S.R.* Stockholm: Royal College of Forestry, 1966.

Baclawski, J. A. *The Soviet Conservation Program for the Steppe and Wooded Steppe Regions of the European Part of the U.S.S.R.* (unpublished Ph.D. dissertation). Ann Arbor: University of Michigan, 1951.

Blagosklonov, K. N., Inozemtsev, A. A. and Tikhomirov, V. N. *Okhrana prirody* (*Conservation of Nature*). Moskva: Izdatel'stvo 'Vysshaya shkola', 1967.

Bogdanov, B. N. 'Conservation and economics', *Ekonomika sel'skogo khozyaystva*, 1970, No. 2, pp. 7–11. Translated in *Current Digest of the Soviet Press*, **22**, No. 19 (June 9, 1970) pp. 7–9.

Chapter 2

Borisov, P. G. *Fisheries Research in Russia: a Historical Survey*. Translated from the Russian and published by the Israel Program for Scientific Translations. Jerusalem: I.P.S.T., 1964.

Burke, Albert E. 'Influence of man upon nature – the Russian view: a case study', *Man's Role in Changing the Face of the Earth*, pp. 1036–51. Ed. William L. Thomas, Jr. Chicago: University of Chicago Press, 1956.

Bykhovskiy, B. and Gladkov, N. A. 'Prava prirody' ('The rights of nature'), *Izvestiya* (July 5, 1967) p. 5. Translated in *Current Digest of the Soviet Press*, **19**, No. 27 (July 26, 1967) pp. 26–7.

Chefranova, N. A. 'Okhrana prirody v epokhu Petra Pervogo' ('The conservation of nature in the era of Peter the Great'), *Okhrana prirody i zapovednoye delo v S.S.S.R.*, Byulleten′ No. 6 (1960) pp. 111–17.

Chichvarin, V. A. *Mezhdunarodnyye soglasheniya po okhrane prirody* (*International Agreements on the Conservation of Nature*). Moskva: Izdatel′stvo 'Yuridicheskaya literatura', 1966.

— 'Osnovy i kharakter mezhdunarodno-pravovoy okhrany prirody Arktiki′ ('Principles and character of international law governing conservation in the Arctic'), *Problemy severa*, No. 11 (1967) pp. 244–9. Translated in *Problems of the North*, No. 11 (1968) pp. 303–9.

Dement′yev, G. P. 'Deyatel′nost′ Komissii po okhrane prirody AN S.S.S.R. za pervyy god yeyo sushchestvovaniya' ('Activities of the Commission for the Conservation of Natural Resources of the Academy of Sciences of the U.S.S.R. in the first year of its existence'), *Okhrana prirody i zapovednoye delo v S.S.S.R.*, Byulleten′ No. 2 (1957) pp. 3–15. Translated in *Conservation of Natural Resources and the Establishment of Reserves in the U.S.S.R.*, Bulletin No. 2 (1960) pp. 1–12.

Dolgushin, I. Yu. 'Po vole cheloveka: tsepnyye reaktsii v geograficheskoy srede i preobrazovaniye prirody' ('By the will of man: chain reactions in the geographic environment and the transformation of nature'), *Priroda* (1964) No. 11, pp. 10–22.

Fedyushin, A. V. 'Voprosy okhrany prirody v R.S.F.S.R. na pervom vserossiyskom s′yezde (1929g.)' ('Questions of conservation in the R.S.F.S.R. at the First All-Russian Conference (1929)'), *Okhrana prirody Sibiri i Dal′nego Vostoka*, Vol. 1 (1962) pp. 201–4.

Gerasimov, I. P. 'Izucheniye, ratsional′noye ispol′zovaniye i okhrana prirodnykh resursov', *Sovetskaya geografiya: itogi i zadachi*, pp. 413–19. Moskva: Geografgiz, 1960. Translated as Chapter 34, 'The study, rational utilization, and preservation of natural resources', *Soviet Geography: Accomplishments and Tasks*, pp. 261–4. New York: American Geographical Society, 1962.

— 'Nuzhen general′nyy plan preobrazovaniya prirody nashey strany' ('We need a general plan for the transformation of nature in our country'), *Kommunist* (1969) No. 2, pp. 68–79. Translated in *Current Digest of the Soviet Press*, **21**, No. 8 (March 12, 1969) pp. 3–6 and 28.

— 'Preobrazovaniye prirody nashey strany' ('The transformation of nature in our country'), *Priroda* (1962) No. 3, pp. 9–16.

— 'Sovetskaya geograficheskaya nauka i problemy preobrazovaniya prirody' ('Soviet geographic science and problems of the transformation of nature'), *Izvestiya Akademii nauk S.S.S.R., seriya geograficheskaya* (1961) No. 5, pp. 6–17. Translated in *Soviet Geography: Review and Translation*, **3**, No. 1 (January 1962) pp. 27–38.

Bibliography

Gerasimov, I. P., Armand, D. L. and Preobrazhenskiy, V. S. 'Natural resources of the Soviet Union, their study and utilization', *Soviet Geography: Review and Translation*, **5**, No. 8 (October, 1964) pp. 3–15.

Gladkov, N. A. 'Bogatstva prirody: zabotlivo okhranyat′, razumno ispol′zovat′, vosstanavlivat′ i umnozhat' ('The riches of nature: to be carefully protected, wisely used, replenished, and increased') *Priroda* (1962) No. 2, pp. 3–10.

— 'Okhrana prirody i geografiya' ('The conservation of nature and geography'), *Vestnik Moskovskogo Universiteta, seriya geografiya* (1961) No. 1, pp. 23–30.

Grin, M. F. 'Kommunizm i preobrazovaniye prirody' ('Communism and the transformation of nature') *Priroda* (1962) No. 1, pp. 25–36.

Ilin, M. *Men and Mountains.* Philadelphia: J. B. Lippincott, 1935.

Inozemtsev, A. A. 'Priroda zhdet zabotlivykh' ('Nature awaits concerned people'), *Komsomol′skaya pravda* (January 6, 1968) p. 2. Translated in *Current Digest of the Soviet Press*, **20**, No. 3 (February 7, 1968) p. 29.

Kazantsev, N. D. and Kolotinskaya, Ye. N. *Pravovaya okhrana prirody v S.S.S.R.* (*Conservation Law in the U.S.S.R.*). Moskva: Gosizdat yuridicheskoy literatury, 1962.

Kirikov, S. V. *Izmeneniya zhivotnogo mira v prirodnykh zonakh S.S.S.R.*, XIII–XIX*vv.* (*Changes in the Animal Life of the Natural Zones of the U.S.S.R.*, XIII–XIX *Centuries*). 2 vols. Moskva: Izdatel′stvo Akademii nauk S.S.S.R., 1959.

Kolbasov, O. S. 'Leninskiye idei ob okhrane prirody' ('Lenin's ideas on the conservation of nature') *Priroda* (1958) No. 4, pp. 41–4.

— (ed.) *Okhrana prirody – sbornik zakonodatel′nykh aktov* (*Conservation of Nature – Collection of Legislative Acts*). Moskva: Gosizdat yuridicheskoy literatury, 1961.

Kolotinskaya, Ye. N. *Pravovaya okhrana prirody v S.S.S.R.* (*Conservation Law in the U.S.S.R.*). Moskva: Izdatel′stvo Moskovskogo Universiteta, 1962.

Kovda, V. A. *Velikiye stroyki kommunizma i preobrazovaniye prirody* (*Great Construction Projects of Communism and the Transformation of Nature*). Moskva: Izdatel′stvo voennogo Ministerstva S.S.S.R., 1951.

— *Velikiy plan preobrazovaniya prirody* (*The Great Plan for the Transformation of Nature*). Moskva: Izdatel′stvo Akademii nauk S.S.S.R., 1952.

— *Great Construction works of Communism and the Remaking of Nature.* Moscow: Foreign Languages Publishing House, 1953.

Kozhevnikov, G. A. 'O neobkhodimosti ustroystva zapovednykh uchastkov dlya okhrany russkoy prirody' ('On the need for establishing preserves for the protection of Russia's natural resources'), *Okhrana prirody i zapovednoye delo v S.S.S.R.*, Byulleten′ No. 4 (1960) pp. 90–7. Translated in *Conservation of Natural Resources and the Establishment of Reserves in the U.S.S.R.*, Bulletin No. 4 (1962) pp. 73–8.

Leonov, Leonid. 'O bol′shoy shchepe' ('Concerning large chips of wood'), *Literaturnaya gazeta* (March 30, 1965) p. 2. Translated in *Current Digest of the Soviet Press*, **17**, No. 20 (June 9, 1965) pp. 13–15.

L′vovich, M. I. 'O kompleksnom ispol′zovanii i okhrane vodnykh resursov' ('On the complex utilization and conservation of water resources'), *Izvestiya Akademii nauk S.S.S.R., seriya geograficheskaya* (1961) No. 2, pp. 37–45. Translated in *Soviet Geography: Review and Translation* **3**, No. 10 (December, 1962) pp. 3–11.

— *et al.* 'Vodnyy balans S.S.S.R. i perspektivy yego preobrazovaniya' ('The water balance of the U.S.S.R. and prospects for its transformation'), *Izvestiya Akademii nauk S.S.S.R., seriya geograficheskaya* (1961) No. 6, pp. 36–46. Translated in *Soviet Geography: Review and Translation*, **3**, No. 10 (December 1962) pp. 12–25.

Makarov, V. N. *Okhrana prirody v S.S.S.R.* (*Conservation of Nature in the U.S.S.R.*). Moskva: Goskul′tprosvetizdat, 1947.

Maksimov, A. A. 'Istoriya razvitiya sel′skokhozyaystvennogo landshafta v lesnoy zone yevropeyskoy chasti S.S.S.R.' ('History of the development of the agricultural landscape in the forest zone of the European part of the U.S.S.R.') *Okhrana prirody i zapovednoye* delo v S.S.S.R., Byulleten' No. 7 (1962) pp. 102–32.

Matley, Ian M. 'The Marxist approach to the geographical environment', *Annals of the Association of American Geographers*, **56**, No. 1 (March, 1966) pp. 97–111.

Mel′nikov, N. 'Prezhde, chem reki povernut′ vspyat′...' ('Before reversing the rivers...'), *Literaturnaya gazeta* (July 12, 1967) p. 11. Translated in *Current Digest of the Soviet Press*, **19**, No. 32 (August 30, 1967) pp. 12 and 39.

'Ob okhrane prirody v R.S.F.S.R.' ('On the conservation of nature in the R.S.F.S.R.'), *Pravda* (October 28, 1960) p. 2. Translated in *Current Digest of the Soviet Press*, **12**, No. 44 (November 30, 1960) pp. 3–5.

'Obshchaya rezolyutsiya III s′yezda Geograficheskogo obshchestva Soyuza S.S.R.' ('General Resolution of the Third Congress of the Geographical Society of the U.S.S.R.'), *Izvestiya Vsesoyuznogo Geograficheskogo obshchestva* (1960) No. 3, pp. 216–226. Translated in *Soviet Geography: Review and Translation*, **2**, No. 1 (January 1961) pp. 63–75.

Polyanskaya, G. N. (ed.) *Pravovyye voprosy okhrany prirody v S.S.S.R.* (*Legal Questions of Conservation in the U.S.S.R.*). Moskva: Gosizdat yuridicheskoy literatury, 1963.

Saushkin, Yu. G. 'Ob "envayronmentalizme" i "possibilizme"' ('On environmentalism and possibilism'), *Vestnik Moskovskogo Universiteta, seriya geografiya* (1960) No. 3, pp. 67–9. Translated in *Soviet Geography: Review and Translation*, **2**, No. 2 (February 1961) pp. 72–4.

— *Velikoye preobrazovaniye prirody Sovetskogo Soyuza* (*The Great Transformation of Nature in the Soviet Union*). Moskva: Geografgiz, 1951.

Shaposhnikov, L. K. 'Experience in conservation education and propaganda of nature conservation in the U.S.S.R. and in the countries of Eastern Europe', *Conservation Education* (I.U.C.N. Publication New Series, Supplementary Paper No. 7) pp. 41–8. Morges, Switzerland: International Union for Conservation of Nature and Natural Resources (I.U.C.N.), 1965.

Shaposhnikov, L. K. and Borisov, V. A. 'Pervyye meropriyatiya Sovetskogo gosudarstva po okhrane prirody' ('The first measures of the Soviet Government for the conservation of nature'), *Okhrana prirody i zapovednoye delo v S.S.S.R.*, Byulleten′ No. 3 (1958) pp. 93–8. Translated in *Conservation of Natural Resources and the Establishment of Reserves in the U.S.S.R.*, Bulletin No. 3 (1961) pp. 85–91.

Shaposhnikov, L. K., Pokrovskiy, V. S., and Prokof′yev, I. Yu. *Okhrana prirody i narodnoye khozyaystvo* (*Conservation and the National Economy*). Moskva: Obshchestvo 'Znaniye' R.S.F.S.R., 1966.

Shcherbakov, D. I. and Abramov, L. S. 'Berech′ prirodnyye resursy strany' ('Protect the natural resources of the country'), *Priroda* (1958) No. 3, pp. 3–10.

Tsvetkov, M. A. *Izmeneniye lesistosti Yevropeyskoy Rossii* (*Changes in the Forested Area of European Russia*). Moskva: Izdatel'stvo Akademii nauk S.S.S.R., 1957.

23rd Congress of the Communist Party of the Soviet Union. Moscow: Novosti Press Agency Publishing House, 1966.

Voronov, A. G. and Gladkov, N. A. 'Geografiya i okhrana prirody' ('Geography and the conservation of nature'), in *Sovetskaya geografiya v period stroitel'stva kommunizma*, pp. 121–8. Moskva: Geografgiz, 1963.

Voyeykov, A. I. *Vozdeystviye cheloveka na prirodu* (*The Influence of Man on Nature*) ed. E. M. Murzayev. Moskva: Izdatel'stvo Akademii nauk S.S.S.R., 1963.

Zhakov, S. I. 'Perspektivnyye preobrazovaniya prirody i rezhim atmosfernogo uvlazhneniya yevropeyskoy territorii S.S.S.R.' ('The long-term transformation of nature and the regime of atmospheric moisture in the European part of the U.S.S.R.'), *Vestnik Moskovskogo Universiteta, seriya geografiya* (1964) No. 1, pp. 37–43. Translated in *Soviet Geography: Review and Translation*, **5**, No. 3 (March 1964) pp. 52–60.

CHAPTER 3. LAND AND SOIL RESOURCES

Aksenenok, G., Kozyr', M., Pavlov, I. and Krasnov, N. 'Stoit li deneg zemlya? ...s tochki zreniya prava' ('Does land cost money?...from the point of view of the law'), *Pravda* (May 31, 1966) p. 2. Reviewed in *The New York Times* (June 1, 1966) p. 12.

Aleksankin, A. V. 'Razvitiye melioratsii zemel' v Belorussii' ('The development of drainage reclamation in Belorussia'), *Gidrotekhnika i melioratsiya*, **19**, No. 10, pp. 38–62.

Armand, D. L. 'Geograficheskiye issledovaniya po uluchsheniyu ispol'zovaniya sel'skokhozyaystvennykh zemel', *Sovetskaya geografiya: itogi i zadachi*, pp. 462–74. Moskva: Geografgiz, 1960. Translated as Chapter 38, 'Geographic research in the improvement of the utilization of agricultural lands', *Soviet Geography: Accomplishments and Tasks*, pp. 290–7. New York: American Geographical Society, 1962.

Armand, D. L., Zvorykin, K. V. and Kuznetsov, G. A. (eds.) *Kachestvennyy uchet i otsenka zemel'* (*Qualitative inventory and evaluation of land*). 'Voprosy geografii', No. 43. Moskva: Geografgiz, 1958.

Aver'yanov, S. F., Minayeva, Ye. N. and Timoshkina, V. A. 'Povysheniye produktivnosti sel'skokhozyaystvennykh zemel' putem orosheniya i osusheniya' ('Increasing the productivity of agricultural land by means of irrigation and drainage'), *Prirodnyye resursy Sovetskogo Soyuza, ikh ispol'zovaniye i vosproizvodstvo*, pp. 97–119. Ed. I. P. Gerasimov, D. L. Armand, and K. M. Efron. Moskva: A.N. S.S.S.R. 1963. Translated by Freeman and Co., 1971.

Baclawski, J. A. *The Soviet Conservation Program for the Steppe and Wooded Steppe Regions of the European Part of the U.S.S.R.* (unpublished Ph.D. dissertation) Ann Arbor: University of Michigan, 1951.

Blagosklonov, K. N., Inozemtsev, A. A. and Tikhomirov, V. N. *Okhrana prirody* (*Conservation of Nature*). Moskva: Izdatel'stvo 'Vysshaya shkola', 1967.

Bogdanov, B. N. 'Conservation and economics', *Ekonomika sel'skogo khozyaystva* (1970) No. 2, pp. 7–11. Translated in *Current Digest of the Soviet Press*, **22**, No. 19 (June 9, 1970) pp. 7–9.

Chapter 3

Braude, I. D. *Eroziya pochv, zasukha i bor'ba s nimi v TsChO* (*Soil Erosion, Drought, and Efforts to Combat them in the Central Chernozem Region*). Moskva: Izdatel'stvo 'Nauka', 1965.

Bronshteyn, M. 'Ekonomicheskaya otvetstvennost' ('Economic responsibility'), *Izvestiya* (September 5, 1968) p. 3. Translated in *Current Digest of the Soviet Press*, **20**, No. 43 (November 13, 1968) pp. 14–15.

Chernyak, I. 'Poteri vozmeshchat' spolna' ('Compensate losses in full'), *Izvestiya* (November 12, 1968) p. 2. Translated in *Current Digest of the Soviet Press*, **20**, No. 49 (December 25, 1968) pp. 12–13.

Churayev, I. A. 'Na strazhe polya' ('On guard over the fields'), as interviewed in *Pravda* (June 20, 1970) p. 3. Translated in *Current Digest of the Soviet Press*, **22**, No. 25 (July 21, 1970) pp. 27–8.

Darmodekhin, V. 'Semnadtsat' protiv buri' ('Seventeen against the storm'), *Izvestiya* (April 10, 1960) p. 3. Translated in *Current Digest of the Soviet Press*, **12**, No. 20 (June 15, 1960) p. 42.

Dolgilevich, M. I. 'Pyl'nyye buri na Ukraine' ('Dust storms in the Ukraine'), *Izvestiya Akademii nauk S.S.S.R. seriya geograficheskaya* (1966) No. 1, pp. 34–40.

Gayel', A. G. Doskach, A. G. and Trushkovskiy, A. A. 'O pyl'nykh buryakh v marte–aprele 1960 g.' ('On the dust storms of March–April 1960'), *Izvestiya Akademii nauk S.S.S.R., seriya geograficheskaya* (1961) No. 1, pp. 57–67.

Gerasimov, I. P. 'Umen'shit' i svesti k minimumu zavisimost' nashego sel'skogo khozyaystva ot prirodnoy stikhii' ('Reducing the dependence of Soviet agriculture on natural elements to a minimum'), *Izvestiya Akademii nauk S.S.S.R., seriya geograficheskaya* (1962) No. 5, pp. 43–51. Translated in *Soviet Geography: Review and Translation*, **4**, No. 2 (February 1963) pp. 3–11.

Gubar', N. S., Krivonosov, I. M., Rozin, V. A. and Seliverstov, M. N. *Sel'skokhozyaystvennyye melioratsii v nechernozemnoy polose* (*Agricultural Reclamation in the Non-Chernozem Zone*). Moskva: Izdatel'stvo 'Kolos', 1964.

Ikonitskaya, I. 'Yesli voznik spor...' ('If a dispute arises'), *Izvestiya* (August 15, 1968) p. 3. Translated in *Current Digest of the Soviet Press*, **20**, No. 43 (November 13, 1968) p. 13.

Ivanovskiy, A. I. 'Preobrazovaniye prirody i puti razvitiya zemledeliya na kraynem severe' ('The transformation of nature and ways of developing agriculture in the far north'), *Problemy severa*, Vypusk VII (1963) pp. 5–21.

Jackson, W. A. D. 'The Soviet approach to the good earth: myth and reality', *Soviet Agriculture and Peasant Affairs*, pp. 171–85. Ed. Roy D. Laird. Lawrence, Kansas: University of Kansas Press, 1963.

— 'The virgin and idle lands program reappraised', *Annals of the Association of American Geographers*, **52**, No. 1 (March 1962) pp. 69–79.

Jasny, Naum. *The Socialized Agriculture of the U.S.S.R.* Stanford University Press, 1949.

Komar, I. V., Mints, A. A., Pomus, M. I. and Freykin, Z. G. 'Problems of economic evaluation of natural conditions and resources' (translation of a paper presented at the Symposium on Economic Geography of the U.S.S.R. at the Fourth Congress of the Geographical Society of the U.S.S.R.), *Soviet Geography: Review and Translation*, **5**, No. 9 (November 1964) pp. 44–50.

Koptsov, A. and Pavlov, O. 'Bespokoynoye pole' ('Restless fields'), *Izvestiya* (September 20, 1970) p. 3. Translated in *Current Digest of the Soviet Press*, **22**, No. 38 (October 20, 1970) pp. 7 and 10.

Bibliography

Kosov, B. F. 'Bor'ba s ovragami' ('The fight against ravine erosion'), *Priroda* (1963) No. 4, pp. 55–9.

Kozmenko, A. S. *Bor'ba s eroziyey pochvy* (*The Fight against Soil Erosion*). Moskva: Sel'khozgiz, 1957.

— *Osnovy protivoerozionnoy melioratsii* (*Principles of Erosion Control Reclamation*). Moskva: Sel'khozgiz, 1954.

Kravchenko, I. V. 'Chernyye buri' ('Dust storms'), *Priroda* (1961) No. 12, pp. 67–72.

Kulikov, V. A. 'Pyl'nyye buri na yuge Ukrainy vesnoy 1960 g.' ('Dust storms in southern Ukraine in the spring of 1960'), *Pochvovedeniye* (1961) No. 6, pp. 11–18. Translated in *Soviet Soil Science* (1961) No. 6, pp. 598–603.

Kunyavskiy, M. and Tatarskiy, P. 'Poteryali oazis' ('We lost an oasis'), *Izvestiya* (August 22, 1968) p. 4. Translated in *Current Digest of the Soviet Press*, **20**, No. 43 (November 13, 1968) pp. 13–14.

Lashkevich, G., Levitskiy, P., Novikov, I. and Simurov, A. 'Novomu polyugarantiyu plodorodiya' ('Guarantee fertility to new fields'), *Pravda* (February 14, 1968) p. 2. Translated in *Current Digest of the Soviet Press*, **20**, No. 7 (March 6, 1968) pp. 33–4.

Laskavaya, L. 'Zemlya i veter' ('Land and wind'), *Novyy mir* (1960) No. 6, pp. 186–200. Translated in *Current Digest of the Soviet Press*, **12**, No. 36 (October 5, 1960) pp. 7–11.

Ledger, R. 'Urban gully erosion in the U.S.S.R.', *Soviet Studies*, **19**, No. 3 (January 1968) pp. 426–9.

Lyashchenko, S. 'Zemlya i gorod' ('Land and cities'), *Literaturnaya gazeta* (August 2, 1967) pp. 10–11. Translated in *Current Digest of the Soviet Press*, **19**, No. 31 (August 23, 1967) pp. 13–14.

Lyaskovets, E. 'Zemlya zabotu tsenit' ('Land appreciates care'), *Izvestiya* (September 22, 1961) p. 3. Translated in *Current Digest of the Soviet Press*, **13**, No. 38 (October 18, 1961) pp. 26–7.

Lydolph, Paul E. 'The Russian sukhovey', *Annals of the Association of American Geographers*, **54**, No. 3 (September 1964) pp. 291–309.

'Man and his planet', summary of a symposium conducted by the periodical *Nedelya*, as reported in *Soviet News* (publication of the Soviet Embassy in London) (November 28, 1967) pp. 102–4.

Mel'nikov, N. 'Prezhde, chem reki povernut' vspyat'...' ('Before reversing the rivers'), *Literaturnaya gazeta* (July 12, 1967) p. 11. Translated in *Current Digest of the Soviet Press*, **19**, No. 32 (August 30, 1967) pp. 12 and 39.

Mints, A. A. *Ekonomicheskaya otsenka prirodnykh resursov i usloviy proizvodstva* (*Economic evaluation of natural resources and conditions of production*), 'Geografiya S.S.S.R.', Vypusk 6. Moskva: VINITI, 1968.

Molodkin, P. F. and Sementsov, I. V. 'Pyl'nyye buri na Donu' ('Dust storm on the Don'), *Priroda* (1969) No. 7, pp. 124–5.

Motorina, L. V. 'Rekul'tivatsiya zemel', narushennykh promyshlennost'yu' ('The recultivation of land disturbed by industry'), *Izvestiya Akademii nauk S.S.S.R., seriya geograficheskaya* (1966) No. 5, pp. 40–7.

Narodnoye khozyaystvo S.S.S.R. (*The National Economy of the U.S.S.R.*). Moskva: Izdatel'stvo 'Statistika', annual.

Nefedov, V. D. and Lapidovskiy, K. M. 'Melioratsiya zemel' nechernozemnoy zony' ('Drainage reclamation in the non-Chernozem zone'), *Gidrotekhnika i melioratsiya*, **20**, No. 1 (January 1968) pp. 61–77.

'O neotlozhnykh merakh po zashchite pochv ot vetrovoy i vodnoy erozii' ('On urgent measures for protecting soils from wind and water erosion'), *Pravda* (April 2, 1967) pp. 1 and 3. Translated in *Current Digest of the Soviet Press*, **19**, No. 13 (April 19, 1967) pp. 3–5.

'Osnovy zemel'nogo zakonodatel'stva Soyuza S.S.R. i soyuznykh respublik' ('Principles of land legislation for the U.S.S.R. and the Union Republics') *Pravda* (December 14, 1968) pp. 2–3. Translated in *Current Digest of the Soviet Press*, **21**, No. 1 (January 22, 1969) pp. 14–20.

Parkhomenko, I. I. (ed.) *Zemel'nyye resursy i lesnyye resursy S.S.S.R.* (*Land and forest resources of the U.S.S.R.*), 'Geografiya S.S.S.R.', Vypusk 1. Moskva: Institut nauchnoy informatsii Akademii nauk S.S.S.R., 1965.

Pokshishevskiy, V. V. 'O khozyaystvennoy otsenke prirodnykh resursov i usloviy' ('On the economic evaluation of natural resources and conditions'), *Ekonomicheskaya geografiya: toponimika*, pp. 35–59. Moskva: Izdaniye MGPI, 1960.

'Proyekt: Osnovy zemel'nogo zakonodatel'stva Soyuza S.S.R. i soyuznykh respublik' ('Draft principles of land legislation of the U.S.S.R. and the union republics') *Izvestiya* (July 26, 1968) pp. 3–4. Translated in *Current Digest of the Soviet Press*, **20**, No. 30 (August 14, 1968) pp. 11–15.

Ratsional'noye ispol'zovaniye zemli (*Rational Use of Land*). Moskva: Sel'khozizdat, 1962.

'Resolution of the C.P.S.U. Central Committee and the U.S.S.R. Council of Ministers, dated November 28, 1969, on the model charter for the collective farm', *Pravda* (November 30, 1969) p. 1. Translated in *Current Digest of the Soviet Press*, **21**, No. 50 (January 13, 1970), pp. 9–15.

Sil'vestrov, S. I. 'Protsessy erozii i deflyatsii pochv na sel'skokhozyaystvennykh zemlyakh i bor'ba s nimi' ('The processes of erosion and soil deflation on agricultural lands and the fight against them'), *Prirodnyye resursy Sovetskogo Soyuza, ikh ispol'zovaniye i vosproizvodstvo* pp. 119–36, ed. I. P. Gerasimov, D. L. Armand, and K. M. Efron. Moskva: Izdatel'stvo Akademii nauk S.S.S.R., 1963. Translated edition by Freeman and Co., 1971.

— 'Trebovaniya bor'by s eroziyey pri intensifikatsii sel'skogo khozyaystva' ('Erosion control needs and the intensification of agriculture'), *Izvestiya Akademii nauk S.S.S.R., seriya geograficheskaya* (1962) No. 5, pp. 58–64. Translated in *Soviet Geography: Review and Translation*, **4**, No. 2 (February 1963) pp. 18–24.

Skorodumov, A. S. *Eroziya pochv i bor'ba s ney* (*Soil erosion and the fight against it*). Kiev: Izdatel'stvo AN Ukrainskoy S.S.R., 1955.

Sobolev, S. S. *Bor'ba s eroziyey pochv i yeyo preduprezhdeniye v rayonakh osvoyeniya tselinnykh i zalezhnykh zemel'* (*The Fight against Soil Erosion and its Prevention in the Regions of the Development of Virgin and Idle Lands*). Moskva: Sel'khozgiz, 1957.

— *Razvitiye erozionnykh protsessov na territorii Evropeyskoy chasti S.S.S.R. i bor'ba s nimi* (*The Development of Erosional Processes in the European Part of the U.S.S.R. and the Fight against them*), 2 vols. Moskva: Izdatel'stvo Akademii nauk S.S.S.R., 1948, 1960.

Sus, N. I. *Eroziya pochvy i bor'ba s neyu* (*Soil Erosion and Efforts to Combat it*). Moskva: Sel'khozgiz, 1949.

Timoshenko, V. P. *Agricultural Russia and the Wheat Problem*. Stanford University Press, 1932.

Vedenichev, P. F. and Marakulin, P. P. 'Osnovnyye polozheniya metodiki ekonomicheskoy otsenki sel'skokhozyaystvennykh ugodiy' ('Basic principles of economic evaluation of agricultural lands'), *Voprosy geografii*, No. 67 (1965) pp. 100–9. Translated in *Soviet Geography: Review and Translation*, **9**, No. 3 (March 1968) pp. 172–80.

Vernander, N. B. *et al.* 'Physical-geographic processes unfavorable for agriculture and control measures against them' (Translation of a paper presented at the Symposium on Physical Geography at the Fourth Congress of the Geographical Society of the U.S.S.R.), *Soviet Geography: Review and Translation* **5**, No. 7 (September 1964) pp. 61–7.

Volin, L. *A Century of Russian Agriculture: from Alexander II to Khrushchev*. Cambridge, Massachusetts: Harvard University Press, 1970.

'Voprosy geografii', No. 78, *Otsenka prirodnykh resursov* (*Evaluation of natural resources*). Moskva: Izdatel'stvo 'Mysl', 1968.

Yerofeyev, B. V. *Sovetskoye zemel'noye pravo* (*Soviet Land Law*). Moskva: Izdatel'stvo 'Vysshaya shkola', 1965.

'Zemlya – bogatstvo vsenarodnoye' ('The land is a national treasure'), *Pravda* (May 25, 1970) p. 1 (editorial). Translated in *Current Digest of the Soviet Press*, **22**, No. 21 (June 23, 1970) p. 21.

Zhirkov, K. F. 'O pyl'nykh buryakh v stepyakh Zapadnoy Sibiri i Kazakhstana' ('Dust storms in the steppes of Western Siberia and Kazakhstan'), *Izvestiya Akademii nauk S.S.S.R., seriya geograficheskaya* (1963) No. 6, pp. 50–5. Translated in *Soviet Geography: Review and Translation* **5**, No. 5 (May 1964) pp. 33–41.

Zvonkov, V. V. *Vodnaya i vetrovaya eroziya zemli* (*Water and Wind Erosion of the Land*). Moskva: Izdatel'stvo Akademii nauk S.S.S.R., 1962.

Zvorykin, K. V., Konovalenko, V. G. and Cheshikin, G. V. (eds.) *Geografiya i zemel'nyy kadastr* (*Geography and the land cadastre*). 'Voprosy geografii', No. 67. Moskva: Izdatel'stvo 'Mysl', 1965.

CHAPTER 4. 'ZAPOVEDNIKI' IN THE U.S.S.R.

Atlas S.S.S.R. Moskva: Ministerstvo geologii i okhrany nedr S.S.S.R., 1962.

Bannikov, A. G. 'Ot zapovednika do prirodnogo parka' ('From a nature preserve to a natural park'), *Priroda* (1968) No. 4, pp. 89–97.

— *Po zapovednikam Sovetskogo Soyuza* (*In the Nature Preserves of the Soviet Union*). Moskva: Izdatel'stvo 'Mysl', 1966. Translated by the Israel Program for Scientific Translations, 1969.

— *Zapovedniki Sovetskogo Soyuza* (*Natural Preserves of the Soviet Union*). Moskva: Izdatel'stvo 'Kolos', 1969.

Belousova, L. S., Borisov, V. A. and Vinokurov, A. A. *Zapovedniki i natsional'nyye parki mira* (*Natural Preserves and National Parks of the World*). Moskva: Izdatel'stvo 'Nauka', 1969.

Blagosklonov, K. N., Inozemtsev, A. A. and Tikhomirov, V. N. *Okhrana prirody* (*Conservation of Nature*). Moskva: Izdatel'stvo 'Vysshaya shkola', 1967.

Bochkarev, M. M. 'Budet li u nas natsional'nyy park?' ('Will we have a national park?'), *Pravda* (January 17, 1967) p. 4. Translated in *Current Digest of the Soviet Press*, **19**, No. 3 (February 8, 1967) p. 28.

Chapter 4

Borissoff, V. A. 'Soviet system of protected natural areas', *National Parks and Conservation Magazine*, **45**, No. 6 (June 1971) pp. 8–14.

—*Priroda* (1968) No. 10, pp. 79–85.

Brouwer, G. A. *The Organization of Nature Protection in the Various Countries.* New York: American Committee for International Wildlife Protection (Special Publication No. 9), 1938.

Gerasimov, I. P. *et al.* 'Sovremennyye geograficheskiye problemy organizatsii otdykha' ('Current geographical problems in recreational planning'), *Izvestiya Akademii nauk S.S.S.R., seriya geograficheskaya* (1969) No. 4, 41–50. Translated in *Soviet Geography: Review and Translation*, **11**, No. 3 (March 1970) pp. 189–98.

Gusev, O. 'Khodit′ turistu v zapovednik?' ('Should tourists go into nature preserves?'), *Turist* (1967) No. 2.

Isakov, Yu. A. 'O nauchno-issledovatel′skoy rabote gosudarstvennykh zapovednikov R.S.F.S.R.' ('On the scientific research work of the state nature preserves of the R.S.F.S.R.'), *Izvestiya Akademii nauk S.S.S.R., seriya geograficheskaya* (1963) No. 5, pp. 145–6.

Kabanov, N. Ye. 'O nekotorykh voprosakh nauchno-issledovatel′skoy raboty v zapovednikakh i svyazi yeyo s zadachami razvitiya narodnogo khozyaystva' ('On certain questions of scientific research work in nature preserves and its connection with the tasks of developing the national economy'), *Okhrana piridy i zapovednoye delo v S.S.S.R.*, Byulleten′ No. 4 (1960) pp. 75–90. Translated in *Conservation of Natural Resources and the Establishment of Reserves in the U.S.S.R.*, Bulletin No. 4 (1962) pp. 61–72.

'Kavkazskiy zapovednik' ('The Kavkaz preserve'), *Geografiya v shkole*, (1967) No. 2, pp. 94–6.

Khutortsov, I. I. 'Nauchnaya rabota v Kavkazskom zapovednike za 40 let (1924–1964gg.)' ('Scientific work in the Kavkaz preserve for 40 Years (1924–1964)'), *Trudy Kavkazskogo gosudarstvennogo zapovednika*, Vypusk VIII, pp. 3–6. Krasnodar: Krasnodarskoye knizhnoye izdatel′stvo, 1965.

Kirikov, S. V. 'Zapovedniki', *Sovetskaya geografiya: itogi i zadachi*, pp. 547–58. Moskva: Geografgiz, 1960. Translated as Chapter 43, 'Nature preserves', *Soviet Geography: Accomplishments and Tasks*, pp. 346–54. New York: American Geographical Society, 1962.

Kolesnikov, B. P. (ed.) *Pamyatniki prirody* (*Natural Monuments*). Sverdlovsk: Ural′skiy filial Akademii nauk S.S.S.R., 1967.

Komendar, V. I. *et al.* (eds.) *Karpatskiye zapovedniki* (*Carpathian Nature Preserves*). Uzhgorod: Izdatel′stvo 'Karpaty', 1966.

Kondratenko, A. I. *Gosudarstvennyye zapovedniki v okhrane prirody* (*State Preserves in the Conservation of Nature*). Moskva: Izdatel′stvo 'Kolos', 1967.

Kotov, V. A. 'Bor′ba s volkami v Kavkazskom zapovednike' ('The fight against wolves in the Kavkaz preserve'), *Trudy Kavkazskogo gosudarstvennogo zapovednika* Vypusk VIII, pp. 182–4. Krasnodar: Krasnodarskoye knizhnoye Izdatel′stvo, 1965.

Kozhevnikov, G. A. 'O neobkhodimosti ustroystva zapovednykh uchastkov dlya okhrany russkoy prirody' ('On the need for establishing preserves for the protection of Russia's natural resources'), *Okhrana prirody i zapovednoye delo v S.S.S.R.*, Byulleten′ No. 4 (1960) pp. 90–8. Translated in *Conservation of Natural Resources and the Establishment of Reserves in the U.S.S.R.*, Bulletin No. 4 (1962) pp. 73–8.

Krainov, L. 'Sozdannyy prirody' ('Created by nature'), *Pravda* (November 21,

1967) p. 3. Translated in *Current Digest of the Soviet Press*, **19**, No. 48 (December 20, 1967) p. 9.

Lavrenko, Ye. M., Geptner, V. G. Kirikov, S. V. and Formozov, A. N. 'Perspektivnyy plan geograficheskoy seti zapovednikov S.S.S.R. (proyekt): ('Long-term plan for a geographic network of natural reserves in the U.S.S.R. (project)'), *Okhrana prirody i zapovednoye delo v S.S.S.R.*, Byulleten' No. 3 (1958) pp. 3–92. Translated in *Conservation of Natural Resources and the Establishment of Reserves in the U.S.S.R.*, Bulletin No. 3 (1961) pp. 1–84.

Leonov, Leonid. 'Preserve means inviolable', *Sovetskaya Rossiya* (May 24, 1970) p. 3. Translated in *Current Digest of the Soviet Press*, **22**, No. 23 (July 7, 1970) pp. 5 and 7.

List of National Parks and Equivalent Reserves, Part two, pp. 42–7. Morges, Switzerland: International Commission on National Parks of the International Union for Conservation of Nature and Natural Resources (I.U.C.N.), 1962.

Motovilov, G. P. 'Protection of Nature in the U.S.S.R.', *Proceedings, Fifth World Forestry Congress*, **3**, pp. 1850–4. Seattle: University of Washington, 1960.

Pelevin, V. I. (ed.) *Okhrana prirody* (*Conservation of Nature*). Moskva: Izdatel'-stvo 'Prosveshcheniye', 1966.

Petrov, B. 'Prishestviye iz goroda' ('Arrival from the city'), *Izvestiya* (July 22, 1970) p. 6. Translated in *Current Digest of the Soviet Press*, **22**, No. 29 (August 18, 1970) pp. 27–8.

Rakhilin, V. 'Rukami ne trogat'...' ('Please do not touch...'), *Turist* (1967) No. 6, p. 20.

Semikhatova, L. I. 'Gosudarstvennyy altayskiy zapovednik' ('The Altay state preserve'), *Zemlevedeniye*, **36**, No. 2 (1934) pp. 113–58.

Shaposhnikov, L. K. 'Nature reserves in the Soviet Union', *National Parks – A World Need*, pp. 65–71. New York: American Committee for International Wild Life Protection, Special Publication No. 14, 1962.

— *Okhrana prirody v S.S.S.R.* (*Conservation of Nature in the U.S.S.R.*). Moskva: Izdatel'stvo 'Znaniye', 1961.

— (ed.) *Primechatel'nyye prirodnyye landshafty S.S.S.R. i ikh okhrana* (*Outstanding Natural Landscapes of the U.S.S.R. and their Preservation*). Moskva: Izdatel'stvo 'Nauka', 1967.

— and Borisov, V. A. 'Pervyye meropriyatiya Sovetskogo gosudarstva po okhrane prirody' ('The first measures of the Soviet Government for the conservation of nature'), *Okhrana prirody i zapovednoye delo v S.S.S.R.*, Byulleten' No. 3 (1958) pp. 93–8. Translated in *Conservation of Natural Resources and the Establishment of Reserves in the U.S.S.R.*, Bulletin No. 3 (1961) pp. 85–91.

Skokova, N. and Nikolayev, Yu. 'Labs under the trees', *Soviet Life* (August 1966) pp. 9–22.

Sokolov, G. A. 'Sozdat' zapovednik na beregu Yeniseyskogo morya' ('For the creation of a natural preserve on the shores of the Yenisey Sea'), *Priroda* (1967) No. 10, pp. 106–9.

Solov'yev, A. I. (ed.). *Zapovedniki S.S.S.R.* (*Nature Preserves of the U.S.S.R.*). 2 Vols. Moskva: Geografgiz, 1951.

United Nations List of National Parks and Equivalent Reserves, 2nd edn. (prepared by the I.U.C.N.). Brussels: Hayez Publishers, 1971.

Uspenskiy, S. M. 'Sovremennyye problemy okhrany prirody v arktike' ('Current problems of nature conservation in the Arctic'), *Problemy severa* No. 7 (1963) pp. 154–61. Translated in *Problems of the North*, No. 7. Ottawa: National Research Council, 1963.

Yevlakhov, I. A. 'Baykala shchedryy dar' ('Generous gift of Baykal'), *Izvestiya* (September 25, 1965) p. 6. Translated in *Current Digest of the Soviet Press*, **17**, No. 39 (October 20, 1965) pp. 32–3.

'Zapovednik' ('Nature preserve'), *Kratkaya geograficheskaya entsiklopediya* (Vol. 2, pp. 42–5). Moskva: Izdatel'stvo 'Sovetskaya entsiklopediya', 1961.

'Zapovednik' ('Nature preserve'), *Malaya Sovetskaya entsiklopediya*, 3rd edn (Vol. 3, pp. 1007–9). Moskva: Izdatel'stvo 'Bol'shaya Sovetskaya entsiklopediya', 1959.

'Zapovednik' ('Nature preserve'), *Sel'skokhozyaystvennaya entsiklopediya* (Vol. 2, pp. 56–64). Moskva: Sel'khozgiz, 1951.

'Zapovedniki' ('Nature preserves'), *Bol'shaya Sovetskaya entsiklopediya*, 1st edn (Vol. 26, pp. 238–46). Moskva: Izdatel'stvo 'OGIZ R.S.F.S.R.', 1933.

'Zapovedniki' ('Nature preserves'), *Bol'shaya Sovetskaya Entsiklopediya*, 2nd edn (Vol. 16, pp. 439–44). Moskva: Gosudarstvennoye nauchnoye izdatel'stvo, 1952.

Zapovedniki S.S.S.R. Moskva: Glavnogo upravleniya okhotnich'yego khozyaystva i zapovednikov pri Sovete Ministrov R.S.F.S.R., 1964. Translated as *Game and Wildlife Preserves in the U.S.S.R.* Coconut Grove, Florida: Field Research Projects ('Natural area studies no. 1'), 1965.

CHAPTER 5. MANAGEMENT OF FISHERIES AND WILDLIFE

'Akklimatizatsiya pushnykh zverey i promyslovykh ryb v S.S.S.R.' ('The acclimatization of fur-bearing animals and commercial fish in the U.S.S.R.'), *Priroda* (1958) No. 11, pp. 100–4.

Ananyan, V. 'V zashchitu foreli' ('In defense of the trout'), *Pravda* (February 25, 1969) p. 2. Translated in *Current Digest of the Soviet Press*, **21**, No. 8 (March 12, 1969) pp. 23–4.

Antonnikov, A. F. 'Gidrostroitel'stvo i rybnoye khozyaystvo' ('Hydro-engineering projects and the fishing industry'), *Priroda* (1970) No. 9, pp. 2–7.

Avakyan, A. B. 'O kompleksnom ispol'zovanii vodokhranilishch: gidrostroitel'stvo i problemy rybnogo khozyaystva S.S.S.R.' ('On the complex use of reservoirs: hydro-construction and problems of the U.S.S.R.'s commercial fisheries'), *Izvestiya Akademii nauk S.S.S.R., seriya geograficheskaya* (1966) No. 2, pp. 28–37.

Bannikov, A. G. 'Nastoyashcheye i budushcheye dikikh kopytnykh' ('The present and future of wild ungulates'), *Priroda* (1963) No. 8, pp. 67–72.

— 'Okhrana redkikh zhivotnykh v Sovetskom Soyuze' ('The preservation of rare animal life in the Soviet Union'), *Priroda* (1968) No. 8, pp. 82–8.

Baranskiy, N. N. *et al.* (eds.) *Okhrana prirody biogeografiya* (*Conservation of nature and biogeography*). ('Voprosy geografii', No. 48). Moskva: Geografgiz, 1960.

Berdichevskiy, L. S. 'Ratsional'noye ispol'zovaniye rybnykh resursov Kaspiyskogo basseyna' ('Rational use of the fish resources of the Caspian basin'), *Izvestiya Akademii nauk S.S.S.R., seriya geograficheskaya* (1961) No. 3, pp. 28–36.

Bibliography

Berg, L. S. *Priroda S.S.S.R.*, Moskva, 1937. Translated by Olga A. Titelbaum as *Natural Regions of the U.S.S.R.*, New York: MacMillan, 1950.

— *Ryby presnykh vod S.S.S.R. i sopredel'nykh stran.* 3 vols. Moskva: Izdatel'stvo Akademii nauk S.S.S.R., 1948–9. Translated into English as *Freshwater Fishes of the U.S.S.R. and Adjacent Countries* by the Israel Program for Scientific Translations, 1962–5.

Bilenkin, D. and Peskov, V. 'Priroda – khimiya – chelovek' ('Nature, chemistry, and man'), *Komsomol'skaya pravda* (March 16, 1967) p. 2. Summarized in *Current Digest of the Soviet Press*, **19**, No. 18 (May 24, 1967) p. 31.

Biryukov, I. 'Ryba u shlagbauma' ('Fish at the gates'), *Izvestiya* (August 11, 1968) p. 2. Translated in *Current Digest of the Soviet Press*, **20**, No. 32 (August 28, 1968) p. 17.

Blagosklonov, K. N., Inozemtsev, A. A. and Tikhomirov, V. N. *Okhrana prirody* (*Conservation of Nature*). Moskva: Izdatel'stvo 'Vysshaya shkola', 1967.

Borisov, P. G. *Fisheries Research in Russia: a Historical Survey*. Translated from the Russian and published by the Israel Program for Scientific Translations; Jerusalem: I.P.S.T., 1964.

Burton, Maurice. *Systematic Dictionary of Mammals of the World.* New York: Crowell, 1962.

Chichvarin, V. A. *Mezhdunarodnyye soglasheniya po okhrane prirody* (*International Agreements on the Conservation of Nature*). Moskva: Izdatel'stvo 'Yuridicheskaya literatura', 1966.

Dement'yev, G. P. (ed.). *Okhotnich'ye – promyslovyye zveri* (*Sport and Commercial Hunting Animals*). Moskva: Rossel'khozizdat, 1965.

Dudenkov, I. 'Nazrevshiye problemy volgo–kaspiya' ('Urgent problems of the Volga–Caspian'), *Pravda* (May 27, 1962) p. 2. Translated in *Current Digest of the Soviet Press*, **14**, No. 21 (June 20, 1962) p. 25.

Fauna S.S.S.R. (New Series nos. 1–94). Compiled by Zoologicheskiy institut Akademii nauk S.S.S.R. Moskva: Izdatel'stvo Akademii nauk S.S.S.R., 1935.

Geptner, V. G. 'Otvet opponentam' ('Answer to opponents'), *Okhota i okhotnich'-ye khozyaystvo* (1964) No. 6, pp. 17–19.

Geptner, V. G., Nasimovich, A. A. and Bannikov, A. G. (eds.) *Mlekopitayushchiye Sovetskogo Soyuza* (*Mammals of the Soviet Union*). 3 vols. Moskva: Gosizdat 'Vysshaya shkola', 1961.

Helin, Ronald A. 'Soviet fishing in the Barents Sea and the North Atlantic', *The Geographical Review*, **54**, No. 3 (July 1964) pp. 386–408.

Isakov, Yu. A. and Formozov, A. N. 'Zoogeografiya sushi', *Sovetskaya geografiya: itogi i zadachi*, pp. 547–58. Moskva: Geografgiz, 1960. Translated as Chapter 16, 'Zoogeography of the land', *Soviet Geography: Accomplishments and Tasks*, pp. 128–40. New York: American Geographical Society, 1962.

Isakov, Yu. A., Kirikov, S. V. and Formozov, A. N. 'Nazemnyye okhotnich'ye – promyslovyye zhivotnyye' ('Sport and commercial land animals'), *Prirodnyye resursy Sovetskogo Soyuza, ikh ispol'zovaniye i vosproizvodstvo*, pp. 180–209. Ed. I. P. Gerasimov, D. L. Armand, and K. M. Efron. Moskva: Izdatel'stvo Akademii nauk S.S.S.R., 1963. Translated edition by Freeman and Co., 1971.

Isayev, A. I. 'Vosproizvodstvo rybnykh zapasov – vazhnaya narodnokhozyaystvennaya zadacha' ('Increasing stocks of fish – an important economic task'), *Rybnoye khozyaystvo* (1961) No. 4, pp. 3–7.

Kalugin, S. G. 'Razvedeniye zubrov v Kavkazskom zapovednike' ('Breeding wisents in the Kavkaz preserve'), *Trudy Kavkazskogo gosudarstvennogo zapovednika*, Vypusk VIII, pp. 155–60. Krasnodar: Krasnodarskoye knizhnoye izdatel'stvo, 1965.

Karyakin, A. and Omelin, A. 'Baryga s ruzh'yem' ('Profiteer with a gun'), *Izvestiya* (February 27, 1963) p. 4. Translated in *Current Digest of the Soviet Press*, **15**, No. 9 (March 27, 1963) pp. 26–7.

Kipper, Z. M., Katkov, S. N. and Novikova, A. I. 'Vozmozhnoye vliyaniye proyektiruyemoy Nizhne-Obskoy GES na rybnoye khozyaystvo' ('Possible effect of the planned Lower Ob' dam on the fishing industry'), *Problemy severa*, No. 12 (1967) pp. 147–55. Translated in *Problems of the North*, No. 12 (1968).

Kirikov, S. V. *Promyslovyye zhivotnyye, prirodnaya sreda i chelovek* (*Game Animals, the Natural Environment and Man*). Moskva: Izdatel'stvo 'Nauka', 1966.

— 'Zapovedniki', *Sovetskaya geografiya: itogi i zadachi*, pp. 547–58. Moskva: Geografgiz, 1960. Translated as Chapter 43, 'Nature preserves', *Soviet Geography: Accomplishments and Tasks*, pp. 346–54. New York: American Geographical Society, 1962.

Kirikov, S. V. and Isakov, Yu. A. 'The reserves of wild game; the dynamics of hunting and its prospects', as translated in *Soviet Geography: Review and Translation*, **1**, No. 10 (December 1960) pp. 64–90.

Kiselev, V. K. and Luk'yanenko, V. I. 'Neft' i ryba' ('Oil and fish'), *Literaturnaya gazeta* (September 27, 1966) p. 2. Translated in *Current Digest of the Soviet Press*, **18**, No. 39 (October 19, 1966) p. 40.

Kotov, V. A. 'Bor'ba s volkami v Kavkazskom zapovednike' ('The fight against wolves in the Kavkaz preserve'), *Trudy Kavkazskogo gosudarstvennogo zapovednika*, Vypusk VIII, pp. 182–4. Krasnodar: Krasnodarskoye knizhnoye izdatel'stvo, 1965.

Kozhin, N. I. 'Osetrovyye S.S.S.R. i vosproizvodstvo ikh zapasov' ('Sturgeon of the U.S.S.R. and increasing their stocks'), *Rybnoye khozyaystvo* (1963) No. 9, pp. 12–15.

Kukhtin, P. 'Salmon poachers', *Soviet Life* (1968) No. 4, pp. 29–31.

Lavrov, N. P. and Pokrovsky, V. S. 'Ecological relationships between local fauna and species commercially introduced in the U.S.S.R.', *Towards a New Relationship of Man and Nature in Temperate Lands* (I.U.C.N. Publications new series, No. 9), pp. 181–93. Morges, Switzerland: International Union for Conservation of Nature and Natural Resources (I.U.C.N.), 1967.

Malyarov, M. 'Lyudi, priroda, zakon' ('People, nature, and the law'), *Pravda* (March 16, 1968) p. 3. Translated in *Current Digest of the Soviet Press*, **20**, No. 11 (April 1, 1968), pp. 23–4.

Malyutin, V. S. 'Osetr: nastoyashcheye i budushcheye' ('The present and future of the sturgeon'), *Priroda* (1968) No. 3, pp. 83–9.

Marakov, S. V. 'Kotiki' ('Fur seals'), *Priroda* (1964) No. 9, pp. 57–64.

Mathisen, O. A. and Bevan, D. E. *Some International Aspects of Soviet Fisheries*. Columbus: Ohio State University Press, 1968.

Mil'shteyn, V. V. (ed.) *Osetrovyye S.S.S.R. i ikh vosproizvodstvo* (*Sturgeon of the U.S.S.R. and their Propagation*). Moskva: Izdatel'stvo 'Pishchevaya promyshlennost', 1967.

Moskalenko, B. 'Snova o sud'be Baykal'skogo omulya' ('Once more about the fate of the Lake Baykal Omul'), *Pravda* (November 17, 1967) p. 3. Translated in *Current Digest of the Soviet Press*, **19**, No. 48 (December 20, 1967) p. 9.

Bibliography

Nikol′skiy, G. 'Dremlyushchiye bogatyri' ('Dormant giants'), *Pravda* (July 22, 1970) p. 3. Translated in *Current Digest of the Soviet Press*, **22**, No. 29 (August 18, 1970) pp. 26–7.

'Nikomu ne pozvoleno' ('No one is permitted'), *Pravda* (January 24, 1969) p. 2. Translated in *Current Digest of the Soviet Press*, **21**, No. 4 (February 12, 1969) p. 28.

Ognev, S. I. *Zveri S.S.S.R. i prilezhashchikh stran.* Vols. 1–9. Moskva: Izdatel′stvo Akademii nauk S.S.S.R., 1928–50. Vols. 1–7 have been translated by the Israel Program for Scientific Translations as *Mammals of the U.S.S.R. and Adjacent Countries.*

'On progress of fulfillment of the October 27, 1960, Russian Republic Law "On conservation in the Russian Republic" ', as translated in *Current Digest of the Soviet Press*, **17**, No. 46 (December 8, 1965) pp. 3–5.

Pavlovskiy, Ye. N. (ed.) *Rybnoye khozyaystvo vnutrennikh vodoyemov S.S.S.R.* (*Fisheries of the internal waters of the U.S.S.R.*). Moskva: Izdatel′stvo Akademii nauk S.S.S.R., 1963.

Peskov, V. 'Sluchay v stepi' ('Incident in the steppe'), *Komsomol′skaya pravda* (April 26, 1970) p. 4. Summarized by the Associated Press, same date.

Pryde, P. R. 'Soviet pesticides', *Environment*, **13**, No. 9 (November 1971) pp. 16–24.

Shaposhnikov, L. K. (ed.) *Yadokhimikaty i fauna* (*Pesticides and Fauna*). Moskva: Izdatel′stvo 'Nauka', 1967.

Shaposhnikov, L. V. 'Akklimatizatsiya pushnykh zverey v svyazi s voprosami sokhraneniya i obogashcheniya fauna S.S.S.R.' ('The acclimatization of fur-bearing animals in connection with the question of preserving and enriching the fauna of the U.S.S.R.'), *Okhrana prirody i zapovednoye delo v S.S.S.R.*, Byulleten′ No. 4 (1960) pp. 37–52. Translated in *Conservation of Natural Resources and the Establishment of Reserves in the U.S.S.R.*, Bulletin No. 4 (1962) pp. 31–42.

Shibayev, Y. V. 'Vyzhivut li eti ptitsy?' ('Will these birds survive?'), *Priroda* (1970) No. 6, pp. 94–5.

Shubnikova, O. N. 'Resursy okhotnich′ye-promyslovykh zvery Yakutii i ikh ispol′zovaniye' ('Resources of fur-bearing and game animals in Yakutia and their utilization'), *Izvestiya Akademii nauk S.S.S.R., seriya geograficheskaya* (1967) No. 3, pp. 43–7. Translated in *Soviet Geography: Review and Translation*, **8**, No. 10 (December 1967) pp. 793–9.

Smirnov, Yu. 'Kak pomoch′ lososyu' ('How can we help the salmon?'), *Pravda* (January 8, 1969) p. 3. Translated in *Current Digest of the Soviet Press*, **21**, No. 2 (January 29, 1969) pp. 21–2.

Soviet Life (August 1966) pp. 26–9.

Tolmachev, G. 'Brakon′yerskaya arifmetika' ('Poachers' arithmetic'), *Pravda* (June 10, 1962) p. 2. Translated in *Current Digest of the Soviet Press*, **14**, No. 23 (July 4, 1962) pp. 23–4.

Tsatsulin, I. 'Poka ne pozdno' ('Before it's too late'), *Literaturnaya gazeta* (September 6 and 8, 1966) p. 2. Translated in *Current Digest of the Soviet Press*, **18**, No. 38 (October 12, 1966) pp. 15–17.

United Nations, Food and Agriculture Organization. *Yearbook of Fishery Statistics*, **26** ('Catches and Landings', 1968), 1969, pp. b-156–b-159.

Uspenskiy, S. M. 'Belyy medved′ dolzhen zhit' ('The polar bear should live'), *Priroda* (1967) No. 2, pp. 81–4.

— (ed.) *Promyslovaya fauna kraynego severa i yeyo ispol'zovaniye* (*Industrial Fauna of the Far North and its Utilization*). *Problemy severa*, No. 11. Moskva: Izdatel'stvo 'Nauka', 1967. Translated as *Problems of the North*, No. 11. Ottawa: National Research Council, 1968.

Uspenskiy, S. M. and Shaposhnikov, L. K. 'Okhrana zhivotnogo mira Arktiki' ('Preservation of Arctic wildlife'), *Priroda* (1957) No. 6, pp. 29–34.

Vinogradov, L. G. *et al.* 'Resursy rybnogo khozyaystva' ('Fisheries resources'), *Prirodnyye resursy Sovetskogo Soyuza, ikh ispol'zovaniye i vosproizvodstvo*, pp. 217–42. Ed. I. P. Gerasimov, D. L. Armand, and K. M. Efron. Moskva: AN SSSR, 1963. Translated edition by Freeman and Co., 1971.

Vinokurov, A. A. 'O vliyanii khozyaystvennoy deyatel'nosti cheloveka na fauna i biologiyu ptits' ('Regarding the influence of man's economic activities on fauna and the biology of birds'), *Okhrana prirody i zapovednoye delo v S.S.S.R.*, Byulleten' No. 6, 1960, pp. 39–51.

Voronova, L. D., Torina, I. G. and Churkina, N. M. 'Vliyaniye yadokhimikatov i mineral'nykh udobreniy na poleznykh zhivotnykh' ('The impact of pesticides and mineral fertilizers on useful animals'), *Okhrana prirody i zapovednoye delo v S.S.S.R.*, Byulleten' No. 7, 1962, pp. 73–87.

Yanushevich, A. I. (ed.) *Akklimatizatsiya zhivotnykh v S.S.S.R.* Alma-Ata: Izdatel'stvo Akademii nauk Kazakhskoy S.S.R., 1963. Translated as *Acclimatization of Animals in the U.S.S.R.* Jerusalem: Israel Program for Scientific Translations, 1966.

Yeliseyev, I. 'Berezhlivyy promysel' ('An economical business'), *Izvestiya* (January 10, 1969) p. 3. Translated in *Current Digest of the Soviet Press*, **21**, No. 2 (January 29, 1969) p. 20.

Yurgenson, P. B. (ed.) *Sbornik materialov po rezul'tatam izucheniya mlekopitayushchikh v gosudarstvennykh zapovednikakh.* Moskva: Izdatel'stvo Ministerstva sel'skogo khozyaystva S.S.S.R., 1956. Translated as *Studies on Mammals in Government Preserves* by the Israel Program for Scientific Translations, 1961.

Zablotskiy, M. A. 'Vosstanovleniye zubra v S.S.S.R. i za granitsey' ('The restoration of wisents in the U.S.S.R. and other countries'), *Okhrana prirody i zapovednoye delo v S.S.S.R.*, Byulleten' No. 4, 1960, pp. 52–71. Translated in *Conservation of Natural Resources and the Establishment of Reserves in the U.S.S.R.*, Bulletin No. 4, 1962, pp. 43–58.

CHAPTER 6. TIMBER AND MINERAL RESOURCES

Algvere, K. V. *Forest Economy in the U.S.S.R.* Stockholm: Royal College of Forestry, 1966.

Blagosklonov, K. N., Inozemtsev, A. A. and Tikhomirov, V. N. *Okhrana prirody* (*Conservation of Nature*). Moskva: Izdatel'stvo 'Vysshaya shkola', 1967.

Bogdanov, B. N. 'Conservation and economics', *Ekonomika sel'skogo khozyaystva* (1970) No. 2, pp. 7–11. Translated in *Current Digest of the Soviet Press*, **22**, No. 19 (June 9, 1970) pp. 7–9.

Campbell, R. W. *The Economics of Soviet Oil and Gas.* Baltimore: Johns Hopkins Press, 1968.

Chemonin, A. 'Vtoroye otkrytiye tret'yego polyusa' ('The second discovery of the third pole'), *Izvestiya* (September 14, 1968) p. 2.

Chivilikhin, V. 'Kak vam dyshitsya, gorozhane?' ('How is your breathing, city dwellers?'), *Literaturnaya gazeta* (August 9, 1967) p. 10. Translated in *Current Digest of the Soviet Press*, **19**, No. 33 (September 6, 1967) pp. 10–12.

Davydchenkov, V. 'Shumyat, bushuyut pozhary pod zemley...' ('Underground fires roar and rage...') *Izvestiya* (June 25, 1960) p. 3. Translated in *Current Digest of the Soviet Press*, **12**, No. 26 (July 27, 1960) pp. 30–1.

Forestry and Forest Industry in the U.S.S.R. (Report of a U.S. Department of Agriculture Technical Exchange). Washington: U.S. Government Printing Office, 1961.

Grin, M. F. 'Neissyakayemyye bogatstva zemnykh nedr' ('Inexhaustible mineral wealth'), *Priroda* (1964) No. 7, pp. 2–13.

Hodgkins, J. A. *Soviet Power: Energy Resources, Production and Potentials.* Englewood Cliffs, N.J.: Prentice-Hall, 1961.

Kapitonov, Ye. I. 'Kurskaya magnitnaya anomaliya i yeyo osvoyeniye' ('The Kursk magnetic anomaly and its development'), *Geografiya v shkole* (1963) No. 1, pp. 19–23. Translated in *Soviet Geography: Review and Translation*, **4**, No. 5 (May 1963) pp. 10–15.

Khlatin, S. 'Lesovod i drovosek: kommentariy k konfliktu' ('The forester and the woodcutter: commentary on a conflict'), *Komsomol'skaya pravda* (June 6, 1968) pp. 1–2. Translated in *Current Digest of the Soviet Press*, **20**, No. 23 (June 26, 1968) pp. 13–16.

Kiselev, V. K. and Luk'yanenko, V. I. 'Neft' i ryba' ('Oil and fish'), *Literaturnaya gazeta* (September 27, 1966) p. 2. Translated in *Current Digest of the Soviet Press*, **18**, No. 39 (October 19, 1966) p. 40.

Kuzmenko, S., Sabirov, R. and Seleznev, V. 'Pogasite fakely' ('Extinguish the torches'), *Pravda* (August 2, 1963) p. 2. Translated in *Current Digest of the Soviet Press*, **15**, No. 31 (August 28, 1963) p. 27.

Leonov, Leonid. 'O bol'shoy shchepe' ('Concerning large chips of wood'), *Literaturnaya gazeta* (March 30, 1965) p. 2. Translated in *Current Digest of the Soviet Press*, **17**, No. 20 (June 9, 1965) pp. 13–15.

Lopatina, Ye. 'Lyseyushchiye Karpaty' ('The balding Carpathians'), *Literaturnaya gazeta* (August 30, 1966) p. 2. Translated in *Current Digest of the Soviet Press*, **18**, No. 36 (September 28, 1966) pp. 12–13.

Mel'nikov, N. 'Prezhde, chem reki povernut' vspyat'...' ('Before reversing the rivers'), *Literaturnaya gazeta* (July 12, 1967) p. 11. Translated in *Current Digest of the Soviet Press*, **19**, No. 32 (August 30, 1967) pp. 12 and 39.

Mel'nikov, N., Agoshkov, M., and Laskorin, B. 'Zapasam nedr nuzhen schet' ('Mineral reserves must be accounted for'), *Pravda* (January 27, 1970) p. 2. Translated in *Current Digest of the Soviet Press*, **22**, No. 4 (February 24, 1970) pp. 11 and 15.

Motorina, L. V. 'Rekul'tivatsiya zemel', narushennykh promyshlennostyu' ('The recultivation of land disturbed by industry'), *Izvestiya Akademii nauk S.S.S.R., seriya geograficheskaya* (1966) No. 5, pp. 40–7.

Motovilov, G. *et al.* 'Dvoye sporyat o lese...' ('Two argue about the forest'), *Komsomol'skaya pravda* (January 30, 1969) p. 2.

Mozeson, D. L. 'Okhrana prirody severa' ('Conservation of Nature in the north'), *Problemy severa*, No. 2 (1958) pp. 222–5. Translated in *Problems of the North*, No. 2 (1961) pp. 235–8.

Mukhin, A. I. (ed.) *Les-nashe bogatstvo* (*Forests – Our Wealth*). Moskva: Goslesbumizdat, 1962.

Nesterov, V. 'Lesam nuzhen zabotlivyy khozyain' ('Forests need a solicitous master'), *Pravda* (January 19, 1966) p. 3. Translated in *Current Digest of the Soviet Press*, **18**, No. 3 (February 9, 1966) p. 30.

Odintsov, M. and Rya′benko, V. 'Berech′ sokrovishcha' ('Watch over the treasure house'), *Izvestiya* (January 27, 1968) p. 2. Translated in *Current Digest of the Soviet Press*, **20**, No. 4 (February 14, 1968) pp. 24–5.

Oxford Regional Economic Atlas of the U.S.S.R. and Eastern Europe. Oxford University Press, 1956.

Parkhomenko, I. I. (ed.) *Zemel′nyye resursy i lesnyye resursy S.S.S.R.* (*Land and forest resources of the U.S.S.R.*). 'Geografiya S.S.S.R.' Vypusk 1. Moskva: Institut nauchnoy informatsii Akademii nauk S.S.S.R., 1965.

Plaksin, I. 'Kompleksno ispol′zovat′ rudy' ('Make full use of ores'), *Pravda* (June 3, 1966) p. 5. Translated in *Current Digest of the Soviet Press*, **18**, No. 22 (June 22, 1966) pp. 30–2.

'Po-khozyayski ispol′zovat′ bogatstva nedr' ('Make full use of mineral resources'), *Izvestiya* (July 30, 1958) p. 1. Translated in *Current Digest of the Soviet Press*, **10**, No. 30 (September 3, 1958) p. 26.

Pokshishevskiy, V. V. 'Na poroge tret′yego tysyacheletiya' ('On the threshold of the third millenium'), *Nauka i zhizn′* (1968) No. 2, pp. 70–3. Translated as 'The economic geography of the U.S.S.R. by the year 2000', *Soviet Geography: Review and Translation*, **9**, No. 9 (November 1968), pp. 770–6.

Pyatnitskiy, S. S. 'Berech′ lesa Karpat ot vetrovala' ('Protect the forests of Carpathia from blowdown'), *Priroda* (1959) No. 12, pp. 51–4.

Shabad, Theodore. *Basic Industrial Resources of the U.S.S.R.* New York: Columbia University Press, 1969.

Shutliv, F. 'Poleznyye iskopayemyye – nashe bogatstvo' ('Mineral resources – our wealth'), *Partiynaya zhizn′* (1965) No. 11, pp. 15–20. Translated in Joint Publications Research Service (J.P.R.S.) document No. 31,592, dated August 18, 1965.

Soviet Life (August 1966).

Taraskevich, G. 'Razumno ispol′zovat′ lesnyye bogatstva' ('Use forest resources wisely'), *Kommunist* (1963) No. 16, pp. 85–9. Translated in *Current Digest of the Soviet Press*, **16**, No. 5 (February 26, 1964) pp. 15, 16 and 28.

Timofeyev, N. V. (ed.) *Les-natsional′noye bogatstva sovetskogo naroda* (*Forests – National Treasure of the Soviet People*). Moskva: Izdatel′stvo 'Lesnaya promyshlennost', 1967.

Tseplyayev, V. P. *Lesa S.S.S.R.* Moskva: Sel′khozgiz, 1961. Translated as *The Forests of the U.S.S.R.* Jerusalem: Israel Program for Scientific Translations, 1965.

Tsvetkov, M. A. *Izmeneniye lesistosti Yevropeyskoy Rossii* (*Changes in the Forested Area of European Russia*). Moskva: Izdatel′stvo Akademii nauk S.S.S.R., 1957.

Vasil′yev, P. V. *Ekonomika ispol′zovaniya i vosproizvodstva lesnykh resursov* (*The Economics of the Use and Reproduction of Forest Resources*). Moskva: Izdatel′stvo Akademii nauk S.S.S.R., 1963.

— 'Lesnyye resursy i lesnoye khozyaystvo' ('Forest resources and forest management'), *Prirodnyye resursy Sovetskogo Soyuza, ikh ispol′zovaniye i vosproizvodstvo*, pp. 137–56. Ed. I. P. Gerasimov, D. L. Armand, and K. M. Efron. Moskva: A.N. S.S.S.R. 1963. Translated edition by Freeman and Co., 1971.

— 'Questions of the geographical study and the economic use of forests' (Translation of a paper presented at the Symposium on Natural Resources of the

Third Congress of the Geographical Society of the U.S.S.R.), *Soviet Geography: Review and Translation*, **1**, No. 10 (December 1960) pp. 50–63.

Vinogradov, I. and Olkhovskaya, B. 'Lesu nuzhen odin khozyain' ('The forests need one master'), *Izvestiya* (September 17, 1958) pp. 3–4. Translated in *Current Digest of the Soviet Press*, **10**, No. 37 (October 22, 1958) pp. 17–19.

Vorontsov, A. I. 'Razmnozheniye vrediteley lesa v svyazi s lesokhozyaystvennoy deyatel′nost′yu cheloveka' ('The multiplication of forests pests due to man's afforestation activities'), *Okhrana prirody i zapovednoye delo v S.S.S.R.*, Byulleten′ No. 4, 1960, pp. 16–25. Translated in *Conservation of Natural Resources and the Establishment of Reserves in the U.S.S.R.*, Bulletin No. 4 (1962) pp. 12–18.

Voroshilov, Yu. I and Nedotko, P. A. 'Ispol′zovaniye mineral′nogo topliva i izmeneniye prirodnoy sredy' ('The use of mineral fuels and changes in the natural environment'), *Okhrana prirody i zapovednoye delo v S.S.S.R.*, Byulleten′ No. 6, 1960, pp. 5–14.

Zhukov, A. B. *Lesa S.S.S.R.* (*Forests of the U.S.S.R.*). 5 vols. Moskva: Izdatel′stvo 'Nauka', 1966.

Zon, Raphael. 'The Union of Soviet Socialist Republics', *A World Geography of Forest Resources* (American Geographical Society Special Publication No. 33), pp. 393–419. Ed. S. Haden-Guest, J. K. Wright, and E. M. Teclaff. New York: The Ronald Press Company, 1956.

CHAPTER 7. WATER RESOURCE UTILIZATION AND CONSERVATION

Apollov, B. A., Gyul′, K. K., and Zavriyev, V. G. (eds.) *Problemy Kaspiyskogo morya* (*Problems of the Caspian Sea*). Baku: Izdatel′stvo Akademii nauk Azerbaydzhanskoy S.S.R., 1963.

Askochenskiy, A. N. *Orosheniye i obvodneniye v S.S.S.R.* (*Irrigation and Water Supply Systems in the U.S.S.R.*). Moskva: Izdatel′stvo 'Kolos', 1967.

Avakyan, A. B. and Sharapov, V. A. *Vodokhranilishcha gidroelektrostantsiy S.S.S.R.* (*Reservoirs of the Hydroelectric Stations of the U.S.S.R.*). Moskva: Gosenergoizdat, 1962.

Bernshteyn, L. B. 'Ob ispol′zovanii energii prilov' ('On the utilization of tidal power'), *Izvestiya Vsesoyuznogo Geograficheskogo Obshchestva* (1962) No. 5, pp. 405–13. Translated in *Soviet Geography: Review and Translation*, **4**, No. 5 (May 1963) pp. 16–25.

— 'Perspektivy ispol′zovaniya prilivnoy energii Belogo Morya' ('The future prospects of using the tidal energy of the White Sea'), *Problemy severa*, No. 9 (1965) pp. 167–81. Translated in *Problems of the North*, No. 9 (1965) pp. 181–96.

Bezrukov, B. and Dadabayev, N. 'Sumerki nad Aralom' ('Twilight over the Aral Sea'), *Komsomol′skaya pravda* (September 16, 1965) p. 2. Translated in *Current Digest of the Soviet Press*, **17**, No. 38 (October 13, 1965) pp. 21–2.

Bobrov, S. N. 'Preobrazovaniye Kaspiya' ('The transformation of the Caspian Sea'), *Geografiya v shkole* (1961) No. 2, pp. 5–15. Translated in *Soviet Geography: Review and Translation*, **2**, No. 7 (September 1961) pp. 47–59.

Brattsev, L. A., Vityazeva, V. A. and Podoplelov, V. P. (eds.) *O vliyanii perebroski stoka severnykh rek v basseyn Kaspiya na narodnoye khozyaystvo Komi A.S.S.R.* (*Effects of Diverting the Flow of Northern Rivers into the Caspian Basin on the Economy of the Komi A.S.S.R.*). Leningrad: Izdatel′stvo 'Nauka', 1967.

Butler, W. E. 'The Soviet Union and the continental shelf', *American Journal of International Law*, **63**, No. 1 (January 1969) pp. 103–7.
Chernenko, I. M. 'The Aral Sea problem and its solution', *Problemy osvoyeniya pustun'* (1968) No. 1, pp. 31–4. Translated in *Soviet Geography: Review and Translation*, **9**, No. 6 (June 1968) pp. 489–92.
Chupakhin, V. 'Yeshche raz o sud'be Arala' ('Once more on the fate of the Aral Sea'), *Kazakhstanskaya pravda* (February 6, 1969) p. 2.
Dadabayev, N. 'Beda prishla k Aralu' ('Misfortune has come to the Aral Sea'), *Komsomol'skaya pravda* (August 11, 1968) p. 2.
Davydov, L. K. *Gidrografiya S.S.S.R.* (*Hydrography of the U.S.S.R.*). 2 vols. Leningrad: Ugletekhizdat, 1955.
Davydov, M. M. *Perspektivy kompleksnogo ispol'zovaniya stoka sibirskikh rek* (*Prospects for the Complex Utilization of the Flow of the Siberian Rivers*). Moskva: Izdatel'stvo 'Znaniye', 1957.
Dunin-Barkovskiy, L. V. 'Razvitiye irrigatsii i sud'ba Aral'skogo morya' ('The development of irrigation and the fate of the Aral Sea'), *Problemy preobrazovaniya prirody Sredney Asii*, pp. 75–84. Ed. I. P. Gerasimov, A. S. Kes' and V. N. Kunin. Moskva: Izdatel'stvo 'Nauka', 1967.
— 'The water problem in the deserts of the U.S.S.R.', *Problemy osvoyeniya pustyn'* (1967) No. 1, pp. 13–22. Translated in *Soviet Geography: Review and Translation*, **9**, No. 6 (June 1968) pp. 458–68.
Field, N. C. *The Role of Irrigation in the South European U.S.S.R. in Soviet Agricultural Growth: An Appraisal of the Resource Base and Development Problem* (unpublished Ph.D. dissertation). Seattle: University of Washington, 1956.
Gerasimov, I. P. 'Basic problems of the transformation of nature in Central Asia', *Problemy osvoyeniya pustyn'* (1967) No. 5, pp. 3–17. Translated in *Soviet Geography: Review and Translation*, **9**, No. 6 (June 1968) pp. 444–58.
— 'Nuzhen general'nyy plan preobrazovaniya prirody nashey strany' ('We need a general plan for the transformation of nature in our country'), *Kommunist* (1969) No. 2, pp. 68–79. Translated in *Current Digest of the Soviet Press*, **21**, No. 8 (March 12, 1969) pp. 3–6 and 28.
— 'Umen'shit i svesti k minimumu zavisimost' nashego sel'skogo khozyaystva ot prirodnoy stikhii' ('Reducing the dependence of Soviet agriculture on natural elements to a minimum'), *Izvestiya Akademii nauk S.S.S.R., seriya geograficheskaya* (1962) No. 5, pp. 43–51. Translated in *Soviet Geography: Review and Translation*, **4**, No. 2 (February 1963), pp. 3–11.
— Armand, D. L. and Preobrazhenskiy, V. S. 'Natural resources of the Soviet Union, their study and utilization', *Soviet Geography: Review and Translation*, **5**, No. 8 (October 1964) pp. 3–15.
Grebenkin, V. G. 'O zatoplenii zemel' pri obrazovanii vodokhranilishch GES' ('On land flooded by the formation of hydroelectric station reservoirs'), *Gidrotekhnika i melioratsiya* (1966) No. 3, pp. 16–21.
Greenwood, N. H. 'Developments in the irrigation resources of the Sevan-Razdan cascade of Soviet Armenia', *Annals of the Association of American Geographers*, **55**, No. 2 (June 1965) pp. 291–307.
Grin, A. M. (ed.) *Vodnyye resursy i ikh kompleksnoye ispol'zovaniye* (*Water resources and their multiple-purpose utilization*). 'Voprosy geografii', No. 73. Moskva: Izdatel'stvo 'Mysl', 1968.
Karaulov, N. A. 'Gidroenergeticheskiye resursy i voprosy, svyazannyye s ikh

ispol′zovaniyem' ('Hydroelectric resources and problems regarding their use'). *Prirodnyye resursy Sovetskogo Soyuza, ikh ispol′zovaniye i vosproizvodstvo*, pp. 32–42. Ed. I. P. Gerasimov, D. L. Armand, and K. M. Efron. Moskva: Izdatel′stvo A.N. S.S.S.R., 1963. Translated edition by Freeman and Co., 1971.

Khlatin, S. 'Lesovod i drovosek: kommentariy k konfliktu' ('The forester and the woodcutter: commentary on a conflict'), *Komsomol′skaya pravda* (June 6, 1968) pp. 1–2. Translated in *Current Digest of the Soviet Press*, **20**, No. 23 (June 26, 1968) pp. 13–16.

Khromov, M. N. 'Izmeneniya v geografii naselennykh punktov v svyazi s sozdaniyem Tsimlyanskogo vodokhranilishcha' ('Changes in the geography of population centers in connection with the creation of the Tsimlyansk reservoir'), *Izvestiya Vsesoyuznogo Geograficheskogo Obshchestva* (1961) No. 1, pp. 79–81. Translated in *Soviet Geography: Review and Translation*, **2**, No. 10 (December 1961) pp. 57–63.

Klopov, S. V. 'O vliyanii proyektiruyemoy Nizhne-Obskoy GES na prirodu i narodnokhozyaystvennyy kompleks Zapadno-Sibirskoy nizmennosti′ ('On the influences of the proposed Lower Ob′GES on nature and on the economic complex of the West Siberian lowland'), *Problemy severa*, No. 9 (1965) pp. 261–67. Translated in *Problems of the North*, No. 9 (1965) pp. 291–8.

Kolbasov, O. 'Nuzhen zakon, okhranyayushchiy vodu' ('A water conservation law is needed'), *Izvestiya* (February 28, 1964) p. 3. Translated in *Current Digest of the Soviet Press*, **16**, No. 9 (March 25, 1964) pp. 31–2.

— *Zakonodatel′stvo o vodopol′zovanii v S.S.S.R.* (*Laws on Water Use in the U.S.S.R.*). Moskva: Izdatel′stvo 'Yuridicheskaya literatura', 1965.

Kolodin, M. V. 'The results of Soviet research on desalting water', *Problemy osvoyeniya pustyn′* (1967) No. 6, pp. 33–41. Translated in *Soviet Geography: Review and Translation*, **9**, No. 6 (June 1968) pp. 493–503.

Kunin, V. N. 'Chto proizoydet s nashimi vnutrennimi moryami?' ('What will be the future of our inland seas?'), *Priroda* (1967) No. 1, pp. 36–46.

— 'Sumerki li eto?' ('Is it twilight?'), *Komsomol′skaya pravda* (February 10, 1966) p. 2. Translated in *Current Digest of the Soviet Press*, **18**, No. 9 (March 23, 1966) pp. 23–4.

Kuznetsov, N. T. and L′vovich, M. I. 'Problemy kompleksnogo ispol′zovaniya i okhrany vodnykh resursov' ('Problems of the multi-purpose use and conservation of water resources'), *Prirodnyye resursy Sovetskogo Soyuza, ikh ispol′zovaniye i vosproizvodstvo*, pp. 11–32. Ed. I. P. Gerasimov, D. L. Armand, and K. M. Efron. Moskva: A.N. S.S.S.R., 1963. Translated by Freeman and Co., 1971.

Lebed, A. and Yakovlev, B. *Soviet Waterways* (Series 1, No. 36). Munich: Institute for the Study of the U.S.S.R., 1956.

Lewis, R. A. 'The irrigation potential of Soviet Central Asia', *Annals of the Association of American Geographers*, **52**, No. 1 (March 1962) pp. 99–114.

Loyko, P. 'Polyu – ekonomicheskuyu zashchitu' ('Give the field economic protection'), *Izvestiya* (September 29, 1968) p. 2. Translated in *Current Digest of the Soviet Press*, **20**, No. 44 (November 20, 1968) p. 10.

L′vovich, M. I. 'O kompleksnom ispol′zovanii i okhrane vodnykh resursov' ('On the complex utilization and conservation of water resources'), *Izvestiya Akademii nauk S.S.S.R., seriya geograficheskaya* (1961) No. 2, pp. 37–45. Translated in *Soviet Geography: Review and Translation*, **3**, No. 10 (December 1962), pp. 3–11.

— 'O nauchnykh osnovakh kompleksnogo ispol′zovaniya i okhrany vodnykh resursov' ('Scientific principles of the complex utilization and conservation of water resources'), *Vodnyye resursy i ikh kompleksnoye ispol'zovaniye* ('Voprosy geografii', No. 73), pp. 3–32. Ed. A. M. Grin. Moskva: Izdatel'stvo 'Mysl', 1968. Translated in *Soviet Geography: Review and Translation*, **10**, No. 3 (March 1969) pp. 95–118.

Magakyan, G. L. 'The Mingechaur multi-purpose water management project', *Geografiya i khozyaystvo* (1961) No. 9, pp. 16–20. Translated in *Soviet Geography: Review and Translation*, **2**, No. 10 (December 1961) pp. 43–50.

Matley, I. M. 'The Golodnaya steppe: a Russian irrigation venture in Central Asia', *The Geographical Review*, **60**, No. 3 (July 1970) pp. 328–46.

Micklin, P. P. 'Soviet plans to reverse the flow of rivers: the Kama–Vychegda–Pechora project', *Canadian Geographer*, **13**, No. 3, pp. 199–215.

Mikirtitchian, L. 'Lake Seven development projects in Soviet Armenia', *Studies on the Soviet Union* (New Series, Vol. 1), 1962, No. 3, pp. 92–106.

Mirkin, S. *Vodnyye melioratsyy v S.S.S.R. i puti ikh razvitiya* (*Reclamation in the U.S.S.R. and Ways of its Development*). Moskva: Izdatel'stvo Akademii nauk S.S.S.R., 1960.

Nefedov, V. D. 'General'naya skhema kompleksnogo ispol'zovaniya i okhrany vodnykh resursov S.S.S.R.' ('A general plan for the multi-purpose use and conservation of the U.S.S.R.'s water resources'), *Gidrotekhnika i melioratsiya* (1963) No. 3, pp. 61–4.

Nesteruk, F. Ya. *Razvitiye gidroenergetiki S.S.S.R.* (*Development of the Hydroelectric Energy of the U.S.S.R.*). Moskva: Izdatel'stvo Akademii nauk S.S.S.R., 1963.

Odintsov, M. *et al.* 'Nuzhno li zatoplyat' Ilimskuyu dolinu?' ('Must the Ilim valley be flooded?'), *Izvestiya* (February 17, 1960) p. 4. Translated in *Soviet Geography: Review and Translation*, **1**, No. 6 (June 1960) pp. 68–70.

'O kontinental'nom shel'fe Soyuza S.S.R.' ('On the continental shelf of the U.S.S.R.'), *Vedomosti Verkhovnogo Soveta S.S.S.R.*, 1968, No. 6 (February 7, 1968) Item 40, pp. 55–6. Translated in *Current Digest of the Soviet Press*, **20**, No. 8 (March 13, 1968) pp. 22–3.

'Osnovy vodnogo zakonodatel'stva Soyuza S.S.R. i soyuznykh respublik' ('Principles of water legislation of the U.S.S.R. and the Union Republics'), *Pravda* (December 12, 1970) pp. 2–3. Translated in *Current Digest of the Soviet Press*, **22**, No. 52 (January 26, 1971) pp. 7–12.

Ovsyannikov, N. G. *Vodnyye resursy – nashe bogatstvo* (*Our Water Resource Wealth*). Moskva: Izdatel'stvo 'Sovetskaya Rossiya', 1968.

Pokshishevskiy, V. V. 'Na poroge tret'yego tysyacheletiya' ('On the threshold of the third millenium'), *Nauka i zhizn'* (1968) No. 2, pp. 70–3. Translated as 'The economic geography of the U.S.S.R. by the year 2000', *Soviet Geography: Review and Translation*, **9**, No. 9 (November 1968) pp. 770–6.

Radchenko, K. and Bokhin, F. 'Voda Skol'ko ona stoit?' ('How much does water cost?'), *Izvestiya* (June 21, 1967) p. 4. Translated in *Current Digest of the Soviet Press*, **19**, No. 25 (July 12, 1967), pp. 28–29.

Rostarchuk, M. 'Gornyak ukhodit v more' ('The miner goes to sea'), *Izvestiya* (September 2, 1970) p. 4. Translated in *Current Digest of the Soviet Press*, **22**, No. 35 (September 29, 1970) pp. 23–4.

Russo, G. A. 'Problema ratsional'nogo ispol'zovaniya stoka severnykh rek' ('The problem of rationally using the flow of the Northern rivers'), *Gidrotekhnicheskoye stroitel'stvo* (1961) No. 7, pp. 11–16.

Shishkin, N. I. 'O perebroske stoka Vychegdy i Pechory v basseyn Volgi' ('On the diversion of the Vychegda and Pechora rivers to the basin of the Volga'), *Izvestiya Akademii nauk S.S.S.R., seriya geograficheskaya* (1961) No. 5, pp. 86–94. Translated in *Soviet Geography: Review and Translation*, **3**, No. 5 (May 1962) pp. 46–57.

Shul′ts, V. L. 'The Aral Sea problem', *Soviet Hydrology: Selected Papers* (1968) No. 5, pp. 489–92.

Stas′, I. I. 'Kak spasat′ Kaspiy' ('How to save the Caspian'), *Priroda* (1968) No. 12, pp. 76–9.

Taskin, G. A. 'The falling level of the Caspian Sea in relation to the Soviet economy', *The Geographical Review*, **44**, No. 4 (October 1954) pp. 508–27.

Tsatsulin, I. 'Poka ne pozdno' ('Before it is too late'), *Literaturnaya gazeta* (September 6 and 8, 1966) p. 2. Translated in *Current Digest of the Soviet Press*, **18**, No. 38 (October 12, 1966) pp. 15–17.

Tsinzerling, V. V. *Orosheniye na Amu-Dar′ye* (*Irrigation on the Amu-Darya*). Moskva: Upravleniya Vodnogo Khozyaystva Sredney Asii, 1927.

Urazbayev, N. 'Poyma dolzhna ostat′sya' ('The flood-plain must remain'), *Komsomol′skaya pravda* (June 3, 1967) p. 1. Translated in *Current Digest of the Soviet Press*, **19**, No. 24 (June 5, 1967) p. 16.

Valeshniy, I. and Yakhnevich, V. 'Voda idet...mimo poley' ('Water flows past the fields'), *Pravda* (November 3, 1965) p. 2. Translated in *Current Digest of the Soviet Press*, **17**, No. 44 (November 24, 1965) pp. 31–2.

Valesyan, L. A. 'Nekotoryye voprosy Sevanskoy problemy' ('Some questions on the Sevan problem'), *Izvestiya Vsesoyuznogo Geograficheskogo Obshchestva* (1962) No. 2, pp. 115–24.

Vasil′chikov, N. V. 'Morskoye gornoye delo' ('Mineral extraction from the sea'), *Geografiya v shkole* (1966) No. 3, pp. 12–16.

Vendrov, S. L. 'Geograficheskiye aspekty problemy perebroski chasti stoka Pechory i Vychegdy v basseyn r. Volgi' ('Geographical aspects of the problem of diverting part of the flow of the Pechora and Vychegda rivers to the Volga basin'), *Izvestiya Akademii nauk S.S.S.R., seriya geograficheskaya* (1963) No. 2, pp. 35–45. Translated in *Soviet Geography: Review and Translation*, **4**, No. 6 (June 1963) pp. 29–41.

— 'K prognozy izmeneniy prirodnykh usloviy Severnogo Priob′ya v sluchaye sooruzheniya Nizhne-Obskoy GES' ('A forecast of changes in natural conditions in the Northern Ob′ basin in case of construction of the Lower Ob′ hydro-project'), *Izvestiya Akademii nauk S.S.S.R., seriya geograficheskaya* (1965) No. 5, pp. 37–49. Translated in *Soviet Geography: Review and Translation* **6**, No. 10 (December 1965) pp. 3–18.

Vendrov, S. L. and Kalinin, G. P. 'Surface water resources of the U.S.S.R., their utilization and study', *Soviet Geography: Review and Translation*, **1**, No. 6 (June 1960) pp. 35–49.

Vendrov, S. L., *et al.* 'The problem of transformation and utilization of the water resources of the Volga River and the Caspian Sea', *Soviet Geography: Review and Translation*, **5**, No. 7 (September 1964) pp. 23–34.

Vermishev, K. Kh. 'Sevenskaya problema' ('The problem of Lake Sevan'), *Priroda* (1958) No. 11, pp. 39–45.

'Work on the Arpa–Sevan tunnel in Armenia proceeds', *Soviet Geography: Review and Translation*, **11**, No. 3 (March 1970), pp. 218–21.

Zvonkov, V. V. (ed.) *Issledovaniye i kompleksnoye ispol′zovaniye vodnykh resursov* (*The Study and Complex Use of Water Resources*). Moskva: Izdatel'stvo Akademii nauk S.S.S.R., 1960.

— *Upravleniye poverkhnostnymi i podzemnymi vodnymi resursami i ikh ispol'zovaniye* (*The Management of Surface and Underground Water Resources and their Utilization*). Moskva: Izdatel'stvo Akademii nauk S.S.S.R., 1961.

CHAPTER 8. ENVIRONMENTAL QUALITY

Akimovich, N. N. and Ramenskiy, L. A. 'Usloviya pogody i zagryaznennost' v Odesse' ('Weather conditions and air pollution in Odessa'), *Izvestiya Akademii nauk S.S.S.R., seriya geograficheskaya* (1966) No. 5, pp. 48–51.

Atomic Energy in the Soviet Union (trip report of the U.S. atomic energy delegation to the U.S.S.R.). Oak Ridge: AEC, 1963.

Bayderin, V. 'Nauka chistoy vody' ('The science of pure water'), *Izvestiya* (June 10, 1966) p. 2. Translated in *Current Digest of the Soviet Press*, **18**, No. 23 (June 29, 1966) pp. 48–9.

Blagosklonov, K. N., Inozemtsev, A. A. and Tikhomirov, V. N. *Okhrana prirody* (*Conservation of Nature*). Moskva: Izdatel'stvo 'Vysshaya shkola', 1967.

Bogatenkov, P. 'Kak pomoch' Dnestru' ('How to help the Dniester'), *Pravda* (August 27, 1970) p. 6. Translated in *Environment*, **12**, No. 9 (November 1970) pp. 36–7.

Bordukov, I. V. 'Avtomobil' i gorod' ('The automobile and the city'), *Pravda* (September 5, 1968) p. 3. Translated in *Current Digest of the Soviet Press*, **20**, No. 36 (September 25, 1968) pp. 27–8.

Burdiyan, B. G. 'Zagryazneniye Ob'-Irtyshskogo basseyna' ('Pollution of the Ob'-Irtysh basin'), *Priroda* (1968) No. 2, pp. 88–9.

Chivilikhin, V. 'Kak vam dyshitsya, gorozhane?' ('How is your breathing, city dwellers?'), *Literaturnaya gazeta* (August 9, 1967) p. 10. Translated in *Current Digest of the Soviet Press*, **19**, No. 33 (September 6, 1967) pp. 10–12.

Danilov, Yu. 'Zashchitim ot zagryazneniy vodu, vozdukh, pochvu' ('Let us protect the water, air, and soil from pollution'), *Pravda* (June 21, 1965) p. 2. Translated in *Current Digest of the Soviet Press*, **17**, No. 24 (July 7, 1965) pp. 13–14.

Demin, I. and Bilenkin, D. 'Reka zovet na vyruchku' ('A river calls for help'), *Komsomol'skaya pravda* (April 27, 1960) p. 2. Translated in *Current Digest of the Soviet Press*, **12**, No. 17 (May 25, 1960) pp. 19–22.

Drachev, S. M. and Sinel'nikov, V. Ye. 'Okhrana intensivno ispol'zuyemykh malykh rek na primere Moskvy-reki' ('Protection of intensively used small rivers: the example of the Moscow River'), *Vestnik Moskovskogo Universiteta* (1968) No. 4, pp. 23–30.

Galaziy, G. I. *Baykal i problema chistoy vody v Sibiri* (*Lake Baykal and the Problem of Clean Water in Siberia*). Irkutsk: Akademii nauk S.S.S.R., 1968.

— 'Stoit li riskovat'?' ('Is it worth the risk?'), *Komsomol'skaya pravda* (September 19, 1968) p. 2.

Gaydar, T. and Bogatenkov, P. 'Reka i lyudi' ('A river and people), *Pravda* (November 12, 1965) p. 3. Translated in *Current Digest of the Soviet Press*, **17**, No. 45 (December 1, 1965) pp. 13 and 30.

Gorbunov, Yu. V. 'Promyvka rusla Moskvy-reki' ('Flushing the channel of the Moscow River'), *Priroda* (1970) No. 9, pp. 98–9.

Bibliography

Gribanov, L. 'Vremya pokazhet' ('Time will tell'), *Literaturnaya gazeta* (November 15, 1967) p. 11. Translated in *Current Digest of the Soviet Press*, **19**, No. 48 (December 20, 1967) p. 8.

Grushko, Ya. 'Normal'na li eta norma?' ('Is this norm normal?'), *Izvestiya* (July 6, 1960) p. 5. Translated in *Current Digest of the Soviet Press*, **12**, No. 27 (August 3, 1960) p. 25.

Gulin, N. and Belevitskiy, A. 'Industriya chistogo vozdukha' ('The industry of clean air'), *Pravda* (March 24, 1969) p. 3. Translated in *Current Digest of the Soviet Press*, **21**, No. 12 (April 9, 1969) p. 21.

Gurvich, L. S. and Kibal'chich, I. A. 'Materialy po izucheniyu sanitarnogo sostoyaniya vodoyemov R.S.F.S.R.', *Gigiyena i sanitariya* (1961) No. 3, pp. 15–21. Translated as 'The sanitary conditions of R.S.F.S.R. water basins' in B. S. Levine, *U.S.S.R. Literature on Water Supply and Pollution Control*, **3**, pp. 226–35.

Iosifov, K. 'Techet reka Vokhna' ('The Vokhna River flows'), *Izvestiya* (April 25, 1965) p. 2. Translated in *Current Digest of the Soviet Press*, **17**, No. 17 (May 19, 1965) p. 32.

Ishkov, A. 'Okonchatel'nyye vyvody delat' rano' ('Too early for final conclusions'), *Literaturnaya gazeta* (January 17, 1968) p. 10. Translated in *Current Digest of the Soviet Press*, **20**, No. 5 (February 21, 1968) p. 24.

Karmanov, V. 'Razumno ispol'zovat' vodnyye bogatstva' ('Use water resources wisely'), *Izvestiya* (April 14, 1959) p. 2. Translated in *Current Digest of the Soviet Press*, **11**, No. 15 (May 13, 1959) p. 37.

Kir'yanov, K. 'V zashchitu vody' ('In defense of water'), *Pravda* (August 23, 1970) p. 3. Translated in *Current Digest of the Soviet Press*, **22**, No. 35 (September 29, 1970) pp. 17–18.

Kiselev, V. K. and Luk'yanenko, V. I. 'Neft' i ryba' ('Oil and fish'), *Literaturnaya gazeta* (September 27, 1966) p. 2. Translated in *Current Digest of the Soviet Press*, **18**, No. 39 (October 19, 1966) p. 40.

Kochemasov, V. I. 'God poiskov, god nadezhd' ('Year of quests, year of hopes'), *Istoriya S.S.S.R.* (1967) No. 5, pp. 197–203. Translated in *Current Digest of the Soviet Press*, **19**, No. 49 (December 27, 1967) pp. 13–15.

Konstantinov, B. P. *et al.* 'Baykal zhdet' ('Baykal waits'), *Komsomol'skaya pravda* (May 11, 1966) p. 2. Translated in *Soviet Life* (August 1966) pp. 6–7.

Kostrin, K. 'Trevozhnyy vzglyad na vodu' ('An alarming look at the water problem'), *Komsomol'skaya pravda* (May 25, 1967) p. 2. Translated in *Current Digest of the Soviet Press*, **19**, No. 24 (July 5, 1967) pp. 16–17.

Kuznetsov, N. T. and L'vovich, M. I. 'Problemy kompleksnogo ispol'zovaniya i okhrany vodnykh resursov' ('Problems of the integrated use and conservation of water resources'), *Prirodnyye resursy Sovetskogo Soyuza, ikh ispol'zovaniye i vosproizvodstvo*, pp. 11–32. Ed. I. P. Gerasimov, D. L. Armand, and K. M. Efron. Moskva: A.N. S.S.S.R., 1963. Translated edition by Freeman & Co., 1971.

Lebedev, Yu. D. 'Sovremennoye sostoyaniye i ocherednyye zadachi v oblasti sanitarnoy okhrany atmosfernogo vozdukha naselennykh mest v S.S.S.R.' ('Contemporary status and future problems in the sanitary protection of community atmospheric air in the U.S.S.R.'), *Gigiyena i sanitariya* (1960) No. 1, pp. 5–11. Translated in B. S. Levine, *U.S.S.R. Literature on Air Pollution and Related Occupational Diseases*, Vol. 7, pp. 309–18.

Levine, B. S. *U.S.S.R. Literature on Air Pollution and Related Occupational*

Diseases. 11 Vols. Washington, D.C.: U.S. Public Health Service, 1960–
— *U.S.S.R. Literature on Water Supply and Pollution Control*. 8 Vols. Washington, D.C.: U.S. Public Health Service, 1961–

Litvinov, N. 'Water pollution in the U.S.S.R. and in other Eastern European countries', *Conference on Water Pollution Problems in Europe*, Vol. 1, pp. 17–63. Geneva: United Nations, 1961.

L'vovich, A. I. 'Problema okhrany rek i vodoyemov ot zagryazneniya stochnymi vodami' ('The problem of protecting rivers and water bodies from pollution by waste waters'), *Izvestiya Akademii nauk S.S.S.R., seriya geograficheskaya* (1963) No. 3, pp. 35–44.

L'vovich, M. I. 'O nauchnykh osnovakh kompleksnogo ispol'zovaniya i okhrany vodnykh resursov' ('Scientific principles of the complex utilization and conservation of water resources'), *Vodnyye resursy i ikh kompleksnoye ispol'zovaniye* ('Voprosy geografii', No. 73), pp. 3–32. Ed. A. M. Grin. Moskva: Izdatel'stvo 'Mysl', 1968. Translated in *Soviet Geography: Review and Translation*, **10**, No. 3 (March 1969) pp. 95–118.

Lyakhov, M. Ye. 'Okhrana atmosfery ot zagryazneniya' ('Protection of the atmosphere from pollution'), *Prirodnyye resursy Sovetskogo Soyuza, ikh ispol'zovaniye i vosproizvodstvo*, pp. 60–6. Ed. I. P. Gerasimov, D. L. Armand and K. M. Efron. Moskva: A.N. S.S.S.R., 1963. Translation by Freeman and Co., 1971.

Mel'nikov, N. 'Prezhde, chem reki povernut' vspyat'. . .' ('Before reversing the rivers'), *Literaturnaya gazeta* (July 12, 1967) p. 11. Translated in *Current Digest of the Soviet Press*, **19**, No. 32 (August 30, 1967) pp. 12 and 39.

Merkulov, A. 'Trevoga o Baykale' ('Alarm from Baykal'), *Pravda* (February 28, 1965) p. 4. Translated in *Current Digest of the Soviet Press*, **17**, No. 9 (March 24, 1965) pp. 25–6.

Micklin, P. P. 'The Baykal controversy: a resource use conflict in the U.S.S.R.', *Natural Resources Journal*, **7**, No. 4 (October 1967) pp. 485–98.

Mironov, N. 'Kamen' na Kame' ('Stone on the Kama'), *Pravda* (June 26, 1967) p. 3. Translated in *Current Digest of the Soviet Press* **19**, No. 26 (July 19, 1967) p. 36.

Mote, V. L. *Geography of Air Pollution in the U.S.S.R.* (unpublished Ph.D. dissertation). Seattle: University of Washington, 1971.

Nagibina, T. 'Organization of water pollution control measures in the U.S.S.R. and Eastern European countries', *Conference on Water Pollution Problems in Europe*, **2**, pp. 293–314. Geneva: United Nations, 1961.

Nasimovich, A. 'O neobkhodimosti kompleksnogo podkhoda k ispol'zovaniyu prirodnykh resursov' ('On the necessity of an integrated approach to the utilization of natural resources'), *Geograficheskiy sbornik* **2**, pp. 133–43. Ed. Yu. Medvedkov. Moskva: Institut nauchnoy informatsii Akademii nauk S.S.S.R., 1966.

Nuttonson, M. Y. (ed.) *A.I.C.E. Survey of U.S.S.R. Air Pollution Literature* (2 vols.). Silver Springs, Maryland: American Institute of Crop Ecology, 1969.

'O merakh po predotvrashcheniyu zagryazneniya Kaspiyskogo morya' ('On measures for preventing the pollution of the Caspian Sea'), *Pravda* (October 3, 1968) p. 2. Translated in *Current Digest of the Soviet Press*, **20**, No. 40 (October 23, 1968) pp. 21–2.

'Osnovy vodnogo zakonodatel'stva Soyuza S.S.R. i soyuznykh respublik' ('Principles of water legislation of the U.S.S.R. and the Union Republics'),

Bibliography

Pravda (December 12, 1970) pp. 2–3. Translated in *Current Digest of the Soviet Press*, **22**, No. 52 (January 26, 1971) pp. 7–12.

'Osnovy zakonodatel'stva Soyuza S.S.R. i soyuznykh respublik o zdravookhranenii' ('Principles of legislation of the U.S.S.R. and the Union Republics on public health'), *Pravda* (December 20, 1969) pp. 2–3. Translated in *Current Digest of the Soviet Press*, **22**, No. 1 (February 3, 1970) pp. 7–13.

Profile Study of Air Pollution Control Activities in Foreign Countries. National Air Pollution Control Administration, Publication No. APTD–0601. Washington, D.C. N.A.P.C.A., November 1970.

Razin, N. V. and Gangardt, G. G. 'Ispol'zovaniye i okhrana vodnykh resursov S.S.S.R.' ('Utilization and conservation of U.S.S.R. water resources'), *Gidrotekhnicheskoye stroitel'stvo* (1967) No. 6, pp. 1–8. Translated in *Hydrotechnical Construction* (1967) No. 6, pp. 497–505.

'Rekam byt' chistymi' ('Rivers must be clean'), *Izvestiya* (November 14, 1968) p. 1. Translated in *Current Digest of the Soviet Press*, **20**, No. 46 (December 4, 1968) pp. 19–20.

Rostovshchikov, V. 'Chego boyatsya kapitany' ('What captains are afraid of'), *Izvestiya* (July 27, 1968) p. 2. Translated in *Current Digest of the Soviet Press*, **20**, No. 30 (August 14, 1968) p. 24.

Semenov, B. 'Netoroplivyye proyektirovshchiki' ('Unhurried designers'), *Pravda* (December 9, 1968) p. 3. Translated in *Current Digest of the Soviet Press*, **20**, No. 49 (December 25, 1968) p. 24.

Sinev, N. M. 'Shirokiy shag dobrogo atoma' ('Advances of the peaceful atom'), *Ogonyok*, No. 51 (December 1970) pp. 6–7. Translated in *Current Digest of the Soviet Press*, **23**, No. 5 (March 2, 1971) p. 25.

Spyshnov, P. A. 'Water supply and sewerage development in U.S.S.R. cities', *Vodosnabzheniye i sanitarnaya tekhnika* (1960) No. 6, pp. 1–4, as translated in B. S. Levine, *U.S.S.R. Literature on Water Supply and Pollution Control*, Vol. 4, pp. 30–7.

'Trevoga o reke' ('Alarm about the river'), *Pravda* (June 20, 1968) p. 3. Translated in *Current Digest of the Soviet Press*, **20**, No. 25 (July 10, 1968) p. 20.

Twenty-Third Congress of the Communist Party of the Soviet Union. Moscow: Novosti Press Agency Publishing House, 1966.

Varshavskiy, I. L. 'Avtomobil' i gorod' ('The automobile and the city'), *Izvestiya* (February 15, 1967) p. 3. Translated in *Current Digest of the Soviet Press*, **19**, No. 7 (March 8, 1967) pp. 28–9.

Vendrov, S. L. *et al*. 'The problem of transformation and utilization of the water resources of the Volga River and the Caspian Sea', *Soviet Geography: Review and Translation*, **5**, No. 7 (September 1964) pp. 23–34.

'Volge byt' chistoy' ('The Volga must be clean'), *Izvestiya* (May 15, 1968) p. 1. Translated in *Current Digest of the Soviet Press*, **20**, No. 20 (June 5, 1968) p. 20.

Volkov, O. 'A dni-to idut' ('The days slip by'), *Literaturnaya gazeta* (June 2, 1966) pp. 2–3. Translated in *Current Digest of the Soviet Press*, **18**, No. 23 (June 29, 1966) pp. 46–7.

— 'Poyezdka na Baykal' ('A trip to Baykal'), *Literaturnaya gazeta* (January 29, 1966) pp. 1–2. Translated in *Current Digest of the Soviet Press*, **18**, No. 5 (February 23, 1966) pp. 14–15.

— 'Uroki Baykala' ('Lessons of Lake Baykal'), *Literaturnaya gazeta* (October 11, 1967) p. 12. Translated in *Current Digest of the Soviet Press*, **19**, No. 48 (December 20, 1967) pp. 6–8.

'Vozdukh nashikh gorodov' ('The air in our cities'), *Pravda* (February 12, 1969) p. 1. Translated in *Current Digest of the Soviet Press*, **21**, No. 7 (March 5, 1969) pp. 27–8.
Yakimov, V. 'Druzhno, soobshcha' ('Together, as a team'), *Izvestiya* (August 20, 1970) p. 3. Translated in *Current Digest of the Soviet Press*, **22**, No. 33 (September 15, 1970) pp. 14–15.
'Zabota o Baykale' ('Concern for Baykal'), *Izvestiya* (February 8, 1969) p. 2. Translated in *Current Digest of the Soviet Press*, **21**, No. 6 (February 26, 1969) p. 36. (See also C.D.S.P. **23**, No. 38 (October 19, 1971) pp. 16–17.)
'Zagryaznyayushchiye Volgu – nakazany' ('Polluters of the Volga are punished'), *Izvestiya* (September 3, 1968) p. 4. Translated in *Current Digest of the Soviet Press*, **20**, No. 36 (September 25, 1968) pp. 26–7.
Zakharov, Yu. 'Nevzgody Ladogi' ('Adversities of Ladoga'), *Pravda* (November 15, 1966) p. 2. Translated in *Current Digest of the Soviet Press*, **18**, No. 46 (December 7, 1966) p. 27.
Zhbanov, Ye. 'Zhivaya voda' ('Living water'), *Izvestiya* (June 27, 1970) p. 4. Translated in *Current Digest of the Soviet Press*, **22**, No. 26 (July 28, 1970) pp. 39–40.
Zhunko, V. 'Toplivo i gigiyena' ('Fuel and hygiene'), *Izvestiya* (August 11, 1966) p. 5. Translated in *Current Digest of the Soviet Press*, **18**, No. 32 (August 31, 1966) pp. 25–6.
Zvonkov, V. 'Pust′ prozvuchit golos obshchestvennosti' ('Let public opinion speak up'), *Komsomol′skaya pravda* (April 27, 1960) p. 2. Translated in *Current Digest of the Soviet Press*, **12**, No. 17 (May 25, 1960), p. 20.

CHAPTER 9. ATTITUDES, PROBLEMS AND TRENDS

1. *General works on Soviet conservation philosophies and goals*

Adabashev, I. *Global Engineering*. Moscow: Progress Publishers, 1966.
Blagosklonov, K. N., Inozemtsev, A. A., and Tikhomirov, V. N. *Okhrana prirody* (*Conservation of Nature*). Moskva: Izdatel′stvo 'Vysshaya shkola', 1967.
Bykhovskiy, B. and Gladkov, N. 'Prava prirody' ('The Rights of Nature'), *Izvestiya* (July 5, 1967) p. 5. Translated in *Current Digest of the Soviet Press*, **19**, No. 27 (July 26, 1967) pp. 26–7.
Dolgushin, I. Yu. 'Po vole cheloveka: tsepnyye reaktsii v geograficheskoy srede i preobrazovaniye prirody' ('By the will of man: chain reactions in the geographic environment and the transformation of nature'), *Priroda* (1964) No. 11, pp. 10–22.
Gerasimov, I. P. 'Konstruktivnaya geografiya; tseli, metody, rezul′taty' ('Constructive geography; aims, methods, and results'), *Izvestiya Vsesoyuznogo Geograficheskogo Obshchestva* (1966) No. 5, pp. 389–403. Translated in *Soviet Geography: Review and Translation*, **9**, No. 9 (November 1968) pp. 739–55.
— 'Nuzhen general′nyy plan preobrazovaniya prirody nashey strany' ('We need a general plan for the transformation of nature in our country'), *Kommunist* (1969) No. 2, pp. 68–79. Translated in *Current Digest of the Soviet Press*, **21**, No. 8 (March 12, 1969) pp. 3–6 and 28. Translated in condensed form in *Natural History* (December 1969) pp. 24–34.
— 'Sovetskaya geograficheskaya nauka i problemy preobrazovaniya prirody' ('Soviet geographic science and problems of the transformation of nature'),

Izvestiya Akademii nauk S.S.S.R., seriya geograficheskaya (1961) No. 5, pp. 6–17. Translated in *Soviet Geography: Review and Translation*, **3**, No. 1 (January 1962) pp. 27–38.

Gerasimov, I. P. and Komar, I. V. 'Sovremennyye aspekty problemy preobrazovaniya prirody v amerikanskoy nauke' ('Current aspects of the problem of transforming nature in American science'), *Izvestiya Akademii nauk S.S.S.R., seriya geograficheskaya* (1968) No. 3, pp. 141–6. Translated in *Soviet Geography: Review and Translation*, **9**, No. 9 (November 1968) pp. 979–805.

Gerasimov, I. P. and Mints, A. A. 'Ekonomicheskiy prognoz' ('Economic forecast'), *Pravda* (April 3, 1967) p. 3. Translated in *Current Digest of the Soviet Press*, **19**, No. 14 (April 26, 1967) pp. 23–4.

Gerasimov, I. P., Armand, D. L. and Efron, K. M. (eds.) *Prirodnyye resursy Sovetskogo Soyuza, ikh ispol'zovaniye i vosproizvodstvo* (*Natural Resources of the Soviet Union, their Use and Renewal*). Moskva: Akademii nauk S.S.S.R., 1963. Translated edition by Freeman and Co., 1971.

Gerasimov, I. P., Armand, D. L. and Preobrazhenskiy, V. S. 'Natural resources of the Soviet Union, their study and utilization', *Soviet Geography: Review and Translation*, **5**, No. 8 (October 1964) pp. 3–15.

Gladkov, N. A. 'Bogatstva prirody: zabotlivo okhranyat', razumno ispol'zovat', vosstanavlivat' i umnozhat' ('The riches of nature: to be carefully protected, wisely used, replenished, and increased'), *Priroda* (1962) No. 2, pp. 3–10.

Grin, M. F. 'Kommunizm i preobrazovaniye prirody' ('Communism and the transformation of nature'), *Priroda* (1962) No. 1, pp. 25–36.

Inozemtsev, A. A. 'Priroda zhdet zabotlivykh' ('Nature awaits concerned people'), *Komsomol'skaya pravda* (January 6, 1968) p. 2. Translated in *Current Digest of the Soviet Press*, **20**, No. 3 (February 7, 1968) p. 29.

Kazantsev, N. D. and Kolotinskaya, Ye. N. *Pravovaya okhrana prirody v S.S.S.R.* (*Conservation Law in the U.S.S.R.*). Moskva: Gosizdat yuridicheskoy literatury, 1962.

Kolbasov, O. S. (ed.) *Okhrana prirody – sbornik zakonodatel'nykh aktov* (*Conservation of Nature – Collection of Legislative Acts*). Moskva: Gosizdat yuridicheskoy literatury, 1961.

Leonov, Leonid. 'O bol'shoy shchepe' ('Concerning large chips of wood'), *Literaturnaya gazeta* (March 30, 1965) p. 2. Translated in *Current Digest of the Soviet Press*, **17**, No. 20 (June 9, 1965) pp. 13–15.

'Man and his planet'. Summary of a symposium conducted by the periodical *Nedelya*, as reported in *Soviet News* (publication of the Soviet Embassy in London) (November 28, 1967) pp. 102–4.

Matley, Ian M. 'The Marxist approach to the geographical environment', *Annals of the Association of American Geographers*, **56**, No. 1 (March 1966) pp. 97–111.

Mel'nikov, N. 'Prezhde, chem reki povernut' vspyat'...' ('Before reversing the rivers'), *Literaturnaya gazeta* (July 12, 1967) p. 11. Translated in *Current Digest of the Soviet Press*, **19**, No. 32 (August 30, 1967) pp. 12 and 39.

'Ob okhrane prirody v R.S.F.S.R.' ('On the conservation of nature in the R.S.F.S.R.'), *Pravda* (October 28, 1960) p. 2. Translated in *Current Digest of the Soviet Press*, **12**, No. 44 (November 30, 1960) pp. 3–5.

Oldak, P. 'Priroda vzyvayet k shchedrosti' ('Nature appeals to our generosity'), *Literaturnaya gazeta* (June 3, 1970) p. 11. Translated (condensed) in *Current Abstracts of the Soviet Press*, **2**, No. 6 (June 1970) p. 1.

Chapter 9

'On progress of fulfillment of the October 27, 1960, Russian Republic Law "On conservation in the Russian Republic" ', *Vedomosti Verkhovnogo Soveta R.S.F.S.R.*, No. 44 (370), pp. 923–7. Translated in *Current Digest of the Soviet Press*, **17**, No. 46 (December 8, 1965) pp. 3–5.

Pokshishevskiy, V. V. 'Na poroge tret'yego tysyacheletiya' ('On the threshold of the third millenium'), *Nauka i zhizn'* (1968) No. 2, pp. 70–3. Translated as 'The economic geography of the U.S.S.R. by the year 2000', *Soviet Geography: Review and Translation*, **9**, No. 9 (November 1968), pp. 770–6.

Powell D. E. 'The social costs of modernization: ecological problems in the Soviet Union', *World Politics*, **23**, No. 4 (July 1971) pp. 618–34.

Polyanskaya, G. N. (ed.) *Pravovyye voprosy okhrany prirody v S.S.S.R.* (*Legal Questions of Conservation in the U.S.S.R.*). Moskva: Gosizdat yuridicheskoy literatury, 1963.

'Rekam byt' chistymi' ('Rivers must be clean'), *Izvestiya* (November 14, 1968) p. 1. Translated in *Current Digest of the Soviet Press*, **20**, No. 46 (December 4, 1968) pp. 19–20.

Rusin, N. and Flit, L. *Man Versus Climate*. Moscow: Peace Publishers, undated (*ca.* 1967).

Sakharov, A. D. Untitled essay on world rapprochement as published in the *New York Times* (July 22, 1968) pp. 14–16.

Shaposhnikov, L. K. 'Experience in conservation education and propaganda of nature conservation in the U.S.S.R. and in the countries of Eastern Europe', *Conservation Education* (I.U.C.N. Publication N.S., Supplementary paper No. 7), pp. 41–8. Morges, Switzerland: International Union for Conservation of Nature and Natural Resources (I.U.C.N.), 1965.

Voronov, A. G. and Gladkov, N. A. 'Geografiya i okhrana prirody' ('Geography and the conservation of nature'), in *Sovetskaya geografiya v period stroitel'stva kommunizma*, pp. 121–8. Ed. V. D. Bykov *et al.* Moskva: Geografgiz, 1963.

Zhakov, S. I. 'Perspektivnyye preobrazovaniya prirody i rezhim atmosfernogo uvlazhneniya yevropeyskoy territorii S.S.S.R.' ('The long-term transformation of nature and the atmospheric moisture regime in the European part of the U.S.S.R.'), *Vestnik Moskovskogo Universiteta, seriya geografiya* (1964) No. 1, pp. 37–43. Translated in *Soviet Geography: Review and Translation*, **5**, No. 3 (March 1964) pp. 52–60.

2. *Articles relating to Soviet attitudes on birth control*

Arab-ogly, E. 'Nauchnyy raschet ili raschet na stikhiyu?' ('Scientific calculation or reliance on spontaneity?'), *Literaturnaya gazeta* (June 11, 1966) p. 4. Translated in *Current Digest of the Soviet Press*, **18**, No. 26 (July 20, 1966) pp. 11–13.

Cheprakov, V. 'Ugroza! No komu?' ('A threat! But to whom?'), *Literaturnaya gazeta* (March 13, 1965) p. 4. Translated in *Current Digest of the Soviet Press*, **17**, No. 52 (January 19, 1966) p. 11.

Cook, Robert C. 'Soviet population theory from Marx to Kosygin: a demographic turning point?', *Population Bulletin* **23**, No. 4 (October 1967) pp. 85–115.

Gerasimov, G. 'Bez predubezhdeniy' ('Without prejudice'), *Literaturnaya gazeta* (March 3, 1966) p. 4. Translated in *Current Digest of the Soviet Press*, **18**, No. 9 (March 23, 1966), pp. 11–12.

— 'Radi zdorov'ya zhenshchiny' ('For the sake of women's health'), *Literaturnaya gazeta* (January 11, 1967) p. 12. Translated in *Current Digest of the Soviet Press*, **19**, No. 3 (February 8, 1967) p. 27.

Guzevatyy, Ya. 'Chto takoye "demograficheskiy vzryv"?' ('What is a "demographic explosion"?'), *Literaturnaya gazeta* (November 30, 1965) p. 4. Translated in *Current Digest of the Soviet Press*, **17**, No. 52 (January 19, 1966) p. 13.

Knyazhinskaya, L. A. 'Problema rosta naseleniya i ispol′zovaniya prodovol′stvennykh resursov v razvivayushchikhsya stranakh' ('The problem of population growth and the use of food resources in the developing countries'), *Izvestiya Vsesoyuznogo Geograficheskogo Obshchestva* (1968) No. 5, pp. 428–35.

Kolganov, M. 'Ugrozhayet li miru perenaseleniye?' ('Does overpopulation threaten the world?'), *Literaturnaya gazeta* (December 25, 1965) p. 4. Translated in *Current Digest of the Soviet Press*, **17**, No. 52 (January 19, 1966) pp. 14–15.

Kovda, V. A. 'The problem of biological and economic productivity of the Earth's land areas' (paper given at a conference on biological productivity, Leningrad, 1966). Translated in *Soviet Geography: Review and Translation*, **12**, No. 1 (January 1971) pp. 6–23.

Pod′yachikh, P. 'Narodonaseleniye i progress' ('Population and progress') *Literaturnaya gazeta* (February 22, 1966) p. 4. Translated in *Current Digest of the Soviet Press*, **18**, No. 9 (March 23, 1966) pp. 10–11.

Pokshishevskiy, V. V. 'Bezgranichen li rost naseleniya?' ('Is population growth unlimited?'), *Priroda* (1967) No. 1, pp. 11–23.

Strumilin, S. 'Ugrozhayet li perenaseleniye nashey planete?' ('Is our planet threatened with overpopulation?'), *Literaturnaya gazeta* (May 28, 1966) p. 4. Translated in *Current Digest of the Soviet Press*, **18**, No. 26 (July 20, 1966) p. 11.

Urlanis, B. 'Sushchestvuyut li problemy narodonaseleniya?' ('Is there a population problem?') *Literaturnaya gazeta* (November, 23, 1965) p. 4. Translated in *Current Digest of the Soviet Press*, **17**, No. 52 (January 19, 1966) pp. 11–13.

Index

Index

Index